MOBILIZATION AND REASSEMBLY OF GENETIC INFORMATION

MIAMI WINTER SYMPOSIA—VOLUME 17

1. *W. J. Whelan and J. Schultz, editors:* HOMOLOGIES IN ENZYMES AND METABOLIC PATHWAYS *and* METABOLIC ALTERATIONS IN CANCER, 1970
2. *D. W. Ribbons, J. F. Woessner, Jr., and J. Schulz, editors:* NUCLEIC ACID-PROTEIN INTERACTIONS *and* NUCLEIC ACID SYNTHESIS IN VIRAL INFECTION, 1971
3. *J. F. Woessner, Jr., and F. Huijing, editors:* THE MOLECULAR BASIS OF BIOLOGICAL TRANSPORT, 1972
4. *J. Schultz and B. F. Cameron, editors:* THE MOLECULAR BASIS OF ELECTRON TRANSPORT, 1972
5. *F. Huijing and E. Y. C. Lee, editors:* PROTEIN PHOSPHORYLATION IN CONTROL MECHANISMS, 1973
6. *J. Schultz and H. G. Gratzner, editors:* THE ROLE OF CYCLIC NUCLEOTIDES IN CARCINOGENESIS, 1973
7. *E. Y. C. Lee and E. E. Smith, editors:* BIOLOGY AND CHEMISTRY OF EUCARYOTIC CELL SURFACES, 1974
8. *J. Schultz and R. Block, editors:* MEMBRANE TRANSFORMATION IN NEOPLASIA, 1974
9. *E. E. Smith and D. W. Ribbons, editors:* MOLECULAR APPROACHES TO IMMUNOLOGY, 1975
10. *J. Schultz and R. C. Leif, editors:* CRITICAL FACTORS IN CANCER IMMUNOLOGY, 1975
11. *D. W. Ribbons and K. Brew, editors:* PROTEOLYSIS AND PHYSIOLOGICAL REGULATION, 1976
12. *J. Schultz and F. Ahmad, editors:* CANCER ENZYMOLOGY, 1976
13. *W. A. Scott and R. Werner, editors:* MOLECULAR CLONING OF RECOMBINANT DNA,,1977
14. *J. Schultz and Z. Brada, editors:* GENETIC MANIPULATION AS IT AFFECTS THE CANCER PROBLEM, 1977
15. *F. Ahmad, T. R. Russell, J. Schultz, and R. Werner, editors:* DIFFERENTIATION AND DEVELOPMENT, 1978
16. *T. R. Russell, K. Brew, H. Faber, and J. Schultz, editors:* FROM GENE TO PROTEIN: INFORMATION TRANSFER IN NORMAL AND ABNORMAL CELLS, 1979
17. *W. A. Scott, R. Werner, D. R. Joseph, J. Schultz, editors:* MOBILIZATION AND REASSEMBLY OF GENETIC INFORMATION, 1980

MIAMI WINTER SYMPOSIA—VOLUME 17

MOBILIZATION AND REASSEMBLY OF GENETIC INFORMATION

edited by

Walter A. Scott **Rudolf Werner**

University of Miami School of Medicine

David R. Joseph **Julius Schultz**

The Papanicolaou Cancer Research Institute

Proceedings of the Miami Winter Symposium, January 1980
Sponsored by the Department of Biochemistry
University of Miami School of Medicine, Miami, Florida
Symposium Director: W. J. Whelan

and by

The Papanicolaou Cancer Research Institute, Miami, Florida
Symposium Director: J. Schultz

ACADEMIC PRESS 1980
A Subsidiary of Harcourt Brace Jovanovich, Publishers
New York London Toronto Sydney San Francisco

Academic Press Rapid Manuscript Reproduction

ACADEMIC PRESS, INC.
111 Fifth Avenue, New York, New York 10003

United Kingdom Edition published by
ACADEMIC PRESS, INC. (LONDON) LTD.
24/28 Oval Road, London NW1 7DX

Library of Congress Cataloging in Publication Data

Main entry under title:

Mobilization and reassembly of genetic information.

(Miami winter symposia; v. 17)
1. Genetic regulation—Congresses. 2. Gene expression—Congresses. I. Scott, Walter A. II. Miami, University of, Coral Gables, Fla. Dept. of Biochemistry. III. Papanicolaou Cancer Research Institute. IV. Series.]DNLM: 1. Genetics, Biochemical—Congresses. W33 MI202 v. 17 1980 / QU4 M687 1980]
QH450.M62 574.87′322 80-18845
ISBN 0-12-633360-2

PRINTED IN THE UNITED STATES OF AMERICA

80 81 82 83 9 8 7 6 5 4 3 2 1

CONTENTS

FREE COMMUNICATIONS

SPEAKERS, CHAIRMEN, AND DISCUSSANTS

J. Abelson, University of California at San Diego, La Jolla, California
R. Athwal, New Jersey Medical College, Newark, New Jersey
H. Aviv, Weizmann Institute of Science, Rehovot, Israel
R. Axel, College of Physicians and Surgeons of Columbia University, New York, New York
P. Berg, Stanford University Medical Center, Stanford, California
G. Benardi, Institut de Recherche en Biologie Moleculaire, Université Paris VII, Paris, France
M. Bessman, Johns Hopkins University, Baltimore, Maryland
R. E. Block, Papanicolaou Cancer Research Institute, Miami, Florida
A. Bollon, Wadley Institutes of Molecular Medicine, Dallas, Texas
M. Botchan, University of California, Berkeley, California
A. I. Bukhari, Cold Spring Harbor Laboratory, Cold Spring Harbor, New York
M. Buraczynska, Brown University, Providence, Rhode Island
R. Butow, University of Texas Health Science Center, Dallas, Texas
R. W. Chambers, New York University Medical Center, New York, New York
J. Chirgwin, Washington University School of Medicine, St. Louis, Missouri
G. M. Church, Harvard University, Cambridge, Massachusetts
S. N. Cohen, Stanford University Medical School, Stanford, California
I. B. Dawid, National Institutes of Health, Bethesda, Maryland
D. Dressler, Harvard University, Cambridge, Massachusetts
H. Faber, Papanicolaou Cancer Research Institute, Miami, Florida
G. R. Fink, Cornell University, Ithaca, New York
J. B. Flanegan, University of Florida College of Medicine, Gainesville, Florida
L. Fox, University of Miami School of Medicine, Miami, Florida
N. C. Franklin, University of Utah, Salt Lake City, Utah
D. Gillespie, Hahnemann Medical College and Hospital, Philadelphia, Pennsylvania

Names in bold indicate speakers at the conference.

H. M. Goodman, University of California Medical School, San Francisco, California

M. Guyer, National Institutes of Health, Bethesda, Maryland

A. Halbreich, Centre de Genetique Moleculaire, Gif-Sur-Yvette, France

F. Haurowitz, Indiana University, Bloomington, Indiana

F. Heffron, University of California, San Francisco, California

C. Hershberger, Eli Lilly & Company, Indianapolis, Indiana

I. Herskowitz, University of Oregon, Eugene, Oregon

S. H. Hughes, Cold Spring Harbor Laboratory, Cold Spring Harbor, New York

D. R. Joseph, Papanicolaou Cancer Research Institute, Miami, Florida

R. Kavenoff, University of California at San Diego, La Jolla, California

G. Khoury, National Institutes of Health, Bethesda, Maryland

J. S. Knepper, Johns Hopkins University, Baltimore, Maryland

R. M. Krug, Sloan-Kettering Institute for Cancer Research, New York, New York

A. Landy, Brown University, Providence, Rhode Island

S. Langberg, National Institutes of Health, Bethesda, Maryland

R. A. Lansman, University of Georgia, Athens, Georgia

D. Levy, State University of New York, Stony Brook, New York

J. W. Little, University of Arizona College of Medicine, Tucson, Arizona

K. Lueders, National Cancer Institute, Bethesda, Maryland

J. Manley, Massachusetts Institute of Technology, Cambridge, Massachusetts

K. S. McCarty, Duke University, Durham, North Carolina

B. McClintock, Carnegie Institution of Washington, Cold Spring Harbor, New York

J. C. McPherson, University of Washington, Seattle, Washington

D. Merlo, University of Missouri, Columbia, Missouri

L. W. Mozes, Papanicolaou Cancer Research Institute, Miami, Florida

R. Nutter, University of Washington, Seattle, Washington

H. Potter, Harvard University, Cambridge, Massachusetts

C. Prives, Columbia University, New York, New York

F. W. Putnam, Indiana University, Bloomington, Indiana

J. H. Rogers, University of California, Los Angeles, California

J. Ross, McArdle Laboratories, Madison, Wisconsin

G. M. Rubin, Harvard Medical School, Boston, Massachusetts

T. R. Russell, University of Miami School of Medicine, Miami, Florida

H. Saedler, Albert Ludwigs University, Freiburg, West Germany

F. Sanger, MRC Laboratory of Molecular Biology, Cambridge, England

J. Schell, Max-Planck-Institut fur Zuchtungsforschung, Koln, West Germany

T. J. M. Schopf, University of Chicago, Chicago, Illinois

J. Schultz, Papanicolaou Cancer Research Institute, Miami, Florida

W. A. Scott, University of Miami School of Medicine, Miami, Florida

P. A. Sharp, Massachusetts Institute of Technology, Cambridge, Massachusetts

M. F. Singer, National Institutes of Health, Bethesda, Maryland

A. Skalka, Roche Institute of Molecular Biology, Nutley, New Jersey

H. K. Stanford, President, University of Miami, Coral Gables, Florida
P. Starlinger, Universität zu Koln, Koln, West Germany
A. O. Summers, University of Georgia, Athens, Georgia
P. Szoka, Massachusetts General Hospital, Boston, Massachusetts
W. Szybalski, University of Wisconsin, Madison, Wisconsin
S. Tonegawa, Basel Institut fur Immunologie, Basel, Switzerland
R. A. Weinberg, Massachusetts Institute of Technology, Cambridge, Massachusetts
R. Werner, University of Miami School of Medicine, Miami, Florida
W. J. Whelan, University of Miami School of Medicine, Miami, Florida
G. Wu, Emory University School of Medicine, Atlanta, Georgia
P. Zamecnik, Worcester Foundation of Experimental Biology, Shrewsbury, Massachusetts

PREFACE

This volume, "Mobilization and Reassembly of Genetic Information," is the seventeenth in the "Miami Winter Symposia" series. In January 1969, the Department of Biochemistry of the University of Miami and the University-affiliated Papanicolaou Cancer Research Institute organized the first of these symposia. This is the twelfth year in which the symposia have been held.

As topics, we select areas of biochemistry in which recent progress offers new insights into the molecular basis of biological phenomena. Until 1977, we organized two symposia each year. The first, sponsored by the Department of Biochemistry, emphasized the basic science aspects of the chosen topic; the second, sponsored by the Papanicolaou Cancer Research Institute, dealt with the application of this research to the cancer problem. The proceedings of each were published in separate volumes. With cancer research becoming increasingly concerned with basic cellular mechanisms, the division of the symposia into basic and cancer-related research became rather academic. For this reason the 1978 meeting was organized as a single symposium and the proceedings were published in one volume, a practice we continued this year.

It has recently become evident that nucleic acid rearrangement is involved in many biological phenomena. The molecular details of these processes are beginning to emerge in prokaryotic systems; however, as is amply demonstrated by the papers in this volume, rearrangement phenomena are important throughout eukaryotic species as well.

Associated with the symposia is the Feodor Lynen Lecture, named in honor of the Department of Biochemistry's distinguished visiting professor whose untimely death in August 1979 robbed us and the world of biochemistry of one of its most distinguished practitioners. Past speakers have been George Wald, Arthur Kornberg, Harland G. Wood, Earl W. Sutherland, Jr., Luis F. Leloir, Gerald M. Edelman, A. H. T. Theorell, Paul Berg, James D. Watson, and Francis H. C. Crick. This year the Lynen lecture was given by Fred Sanger. These lectures have provided insight into the history of discovery, and have included the personal and scientific philosophies of our distinguished speakers. The Lynen lecturer for 1981 will be H. Ghobind Khorana. The theme of the symposium will be cellular responses to molecular modulators.

Our aim is to achieve rapid publication of the proceedings, and we thank the speakers for their prompt submission of manuscripts. Our thanks also go to the participants whose interest and discussions provided the interactions that bring a symposium to life, and to the many local helpers, faculty, and administrative staff who have contributed to the success of the present symposium. Special gratitude is accorded the organizers and coordinators of the program: W. J. Whelan (joint director with J. Schultz), Sandra Black, Olga Sanchez, and Audrey Lassiter, and to Virginia Salisbury who assembled many of the typescripts.

The financial assistance by the Department of Pediatrics, University of Miami School of Medicine, the Graduate School, University of Miami, The Juvenile Diabetes Research Foundation, and the long-time generous support of the Howard Hughes Medical Institute, as well as that of Eli Lilly and Company, Hoffmann La Roche, Inc., Merck and Co., Inc., and Smith Kline Corporation are gratefully acknowledged.

David R. Joseph
Julius Schultz
Walter A. Scott
Rudolf Werner

A Review of the Eleventh Feodor Lynen Lecture Given by Fred Sanger

Kenneth Ness
Department of Biochemistry
University of Miami School of Medicine
Miami, Florida 33101

"I think some of you might have got the impression that I am really obsessed with this idea of sequencing," said Dr. Frederick Sanger, anxious to set the record straight, "Well, I think that is probably true. And I must admit that I always do get a great deal of pleasure out of sequences, both in trying to determine them and in trying to find out what they mean in biological terms. I don't think anyone really needs to make any excuses for this obsession when one considers the function of sequences in living matter."

The Eleventh Feodor Lynen Lecture was - inevitably - about sequences. It was Sanger who determined the amino acid sequence of insulin, for which he received the Nobel prize in Chemistry for 1958. In more recent years, he has turned his hand to nucleic acids and developed a means of sequencing DNA. The pivotal nature of this contribution cannot be overlooked.

Dr. Sanger presented his listeners with an intriguing synopsis of some recent work done by himself and his associates at the MRC Laboratory of Molecular Biology, Cambridge, as well as some glimpses of work in progress that has already yielded provocative results. Sanger is of course the originator of the plus-and-minus method of sequencing (1) that was used to derive the entire sequence of ØX174, a major milestone in the history of molecular biology (2). The development of this technique can be traced to earlier DNA-sequencing efforts by Sanger and his coworkers, in which they used DNA polymerase I to synthesize a highly-radioactive DNA copy of a single-stranded template. The product of the labelling reaction was heterogeneous and required

ISBN 0-12-633360-2

fractionation. As the polymerase began running out of a given nucleotide triphosphate it generated a family of DNA fragments, each one beginning with the same 5' end and terminating at some point just before the position of that nucleotide in the sequence. Eventually, this became the basis of the plus-and-minus sequencing method. By depriving the polymerase of dATP, for example, a family of fragments can be synthesized each of which ends just before deoxyadenosine, and whose size, as determined by polyacrylamide gel electrophoresis, precisely indicates the location of each A in the sequence.

There are other ways to achieve the same result. They are formally similar. Maxam and Gilbert (3) have devised a procedure that uses chemical treatment to break DNA strands at a desired nucleotide. Sanger et al. (4) have published a third method in which newly-synthesized strands are terminated by insertion of a dideoxy-nucleotide triphosphate derivative. These compounds are chain-terminators because they have no 3'-hydroxyl upon which the polymerase can build.

Both of the methods developed in Sanger's laboratory use DNA polymerase I. DNA polymerase I requires a primer, and it requires a single-stranded template. Restriction enzyme fragments are normally used as primers; this determines the 5' end exactly. A number of strategies have been devised to conveniently obtain single-stranded templates, the most direct being to clone the desired sequence in a single-stranded phage. Sanger and Coulson have used such an approach to sequence most of bacteriophage lambda. They fragmented the phage DNA with a variety of restriction enzymes and then cloned these pieces with appropriate linker segments in M13, a single-stranded bacteriophage vector described by Messing and Gronenborn (5). Like some of the Charon phages, this vector carries a portion of the **E. coli lac** operon, containing a gene for β-galactosidase that is bisected by an **Eco**RI site. This gene produces blue plaques when the phage is plated on the proper host and indicator medium. Insertion of exogenous material at the **Eco**RI site abolishes gene function, so that recombinants can be selected simply by picking white plaques. A restriction fragment of M13 that borders the **Eco**RI site can then be used to prime the polymerase reaction, and the inserted material is sequenced. The attractive feature of this system is that only one primer molecule needs to be isolated. Sequences are obtained at random throughout the lambda genome. Eventually redundancy becomes a problem, but so far the technique has rapidly provided the sequence of 35,000 base pairs of lambda, as well as the entire sequence of an 1,800 nucleotide portion of the bovine mitochondrial DNA.

A powerful alternative is the exonuclease III method of Andrew Smith (6). Exonuclease III recognizes double-stranded DNA as a substrate and chews back the 3' ends from both directions. This leaves long 5' tails which are single-stranded, with a short overlapping region in the middle of the molecule. Different restriction fragments can be used as primers to sequence selected regions on these 5' tails.

It is this latter approach which has been applied to the analysis of the human mitochondrial DNA. The native molecule contains about 17,000 bases which code for two ribosomal RNAs, upwards of twenty tRNAs, and a number of proteins. Both strands are transcribed, but, except for a few of the tRNAs, the mature transcripts all hybridize to the H strand. Most of this molecule has been sequenced, and a preliminary report has been published (7). Computer analysis of the sequence allows visualization of unobstructed reading frames initiated by AUG, which presumably code for protein, and also recognition of tRNAs by their distinctive structure. Several of the genes have been at least tentatively identified: ATPase subunit 6 and cytochrome oxidase subunit 3 by comparison to the corresponding genes in the yeast mitochondrial DNA, cytochrome oxidase subunit 1 by comparison to the amino-terminal protein sequence as determined by John Walker in Sanger's laboratory, and cytochrome oxidase subunit 2 (COII) by comparison with the total sequence of the bovine protein published by Steffens and Buse (8). This last is perhaps the most solid assignment, and the DNA sequence shows over 70% homology to the bovine protein sequence.

The organization of the genome is of interest. The outstanding feature is extreme economy. The genes are tightly packed together, and there is nothing like the extensive AT-rich spacer regions found in the yeast mitochondrional DNA by Bernardi and coworkers (this volume). Closer examination reveals an intriguing picture: in some cases there is no spacer between the genes at all. A $tRNA^{Asp}$ is joined flush to the initiation codon of the COII gene; at the other end, there is a $tRNA^{Lys}$ only 25 bases downstream from the termination codon. It is possible that the tRNAs are removed from a larger precursor message by splicing, although RNA splicing has not yet been demonstrated in human mitochondria; alternatively, they might be translated into protein: but there are UAA termination codons within the $tRNA^{Asp}$ that share the same reading frame as COII, which seems to rule this out. Another possible solution is that the ribosomal binding site, if there is one, is located somewhere in the $tRNA^{Asp}$. A similar situation exists with respect to possible polyadenylation sites and the $tRNA^{Lys}$. Likewise, the 12S and 16S rRNAs are butt-jointed to tRNA genes. The maturation of the ribosomal RNA probably occurs through excision of the tRNA by an enzyme which recognizes its secondary structure. Such economy of space and sequence is almost uncomfortably viral.

The most surprising result of all is obtained when the amino acid sequence of the bovine protein is compared to the gene of its human counterpart. The homology between them is very convincing, but there are three UGA (opal) termination codons in the same reading frame. In every case, they correspond to tryptophane in the bovine sequence. UGA differs from the normal Trp codon UGG in only one base, and that is in the wobble position. A UCA anti-codon might read both triplets. The startling conclusion is that the opal termination codon is being read as tryptophane by the human mitochondrial ribosome complex.

Naturally, we would like more evidence for this. In the absence of sequence data for the human protein, the most convincing proof would be the discovery of a tRNA with the anti-codon UCA. In fact, the DNA sequence does contain a gene that appears to code for that very tRNA. Furthermore, Alex Tzagoloff and his colleagues have found UGA being read as tryptophane in their studies of the yeast mitochondrial DNA (9).

This is the first known deviation from the standard genetic code. We immediately wonder if there might be other anomalies of coding, and the evidence so far suggests that there are. The traditional code contains just one other example of an amino acid that is coded by a single triplet, so that the tRNA must discriminate between two similar bases (in this case, as for tryptophane, they are purines) in the third position. This is methionine, which is coded by AUG. Wobble theory would allow a CAU anti-codon to read both AUG and AUA, which normally stands for isoleucine. Though this does not happen in other systems, it seems to occur in human mitochondria. There are fifteen methionines in the bovine COII protein, and five of them correspond to AUA in the human DNA sequence. Only one matches up with AUG. There are three AUAs in the gene that do not match up with methionine, but neither do they correspond to isoleucine. Two coincide with leucine; one to lysine.

If we make these adjustments in the code, so that there are two codons each for tryptophane and for methionine, a beautifully symmetric picture emerges. In eight of sixteen cases, the first two bases suffice to identify an amino acid. In the other eight cases, the tRNA must only discriminate between a pyrimidine and a purine in the third position to read the triplet correctly. Wobble is maximum.

It does not seem as if the yeast mitochondria operate this way; indeed, the two systems appear to have taken very different approaches to solving a similar problem. The problem is that, since the mitochondrion has found it desirable to carry as little genetic material as possible, it consequently would like to reduce its supply of tRNAs to the minimum. The minimum of course is twenty, but how then to read all 64 triplets? The yeast mitochondrion has apparently decided not to use all 64.

This was strikingly illustrated by Macino and Tzagoloff (10), and the same result was obtained by Hensgens et al. (11), when they sequenced the gene for ATPase subunit 9 of the yeast mitochondrion and compared it to the protein sequence determined by Sebald and Wachter (12). For example, leucine (which has six possible codons) appears eleven times coded exclusively by UUA. Isoleucine appears nine times in the protein, coinciding with AUU seven times and with AUC twice. The AUA codon, read as methionine by human mitochondria, does not appear. There is an anomaly, however; the protein sequence shows a threonine where Hensgens et al. find a CUA codon in the gene. Li and Tzagoloff (13) have published the sequence of a yeast mt-tRNA^{Thr} which has eight bases in the anticodon loop instead of the usual seven, and they have shown how this extra base could cause misreading of the CUA codon.

In summary, both mitochondrial systems read the opal termination codon UGA as tryptophane. The yeast mitochondrion uses CUA for threonine instead of leucine, which the human does not do; human mitochondria use AUA for methionine, while the yeast may not use AUA at all.

Whereas the yeast mitochondrion has chosen to limit its coding options, Sanger's results suggest that the human mitochondrion prefers to wobble. All the various codons seem to be in use, though some are preferred, but only twenty-odd tRNAs are available to get the job done. One or at most a few tRNAs must recognize all of the codons that specify a given amino acid.

One cannot help but wonder how this is accomplished, and the answer may lie in the structure of the tRNA itself. It is possible to infer the structure of a tRNA molecule from the sequence of its gene. Sanger and his colleagues have done this, and the human mitochondrial tRNAs are quite unique. Many of the features which are invariant in cytoplasmic tRNA's are modified or missing. Two examples are given in Barrell et al. : a mt-RNA^{Asp} and a mt-RNA^{Lys} (Fig. 1).

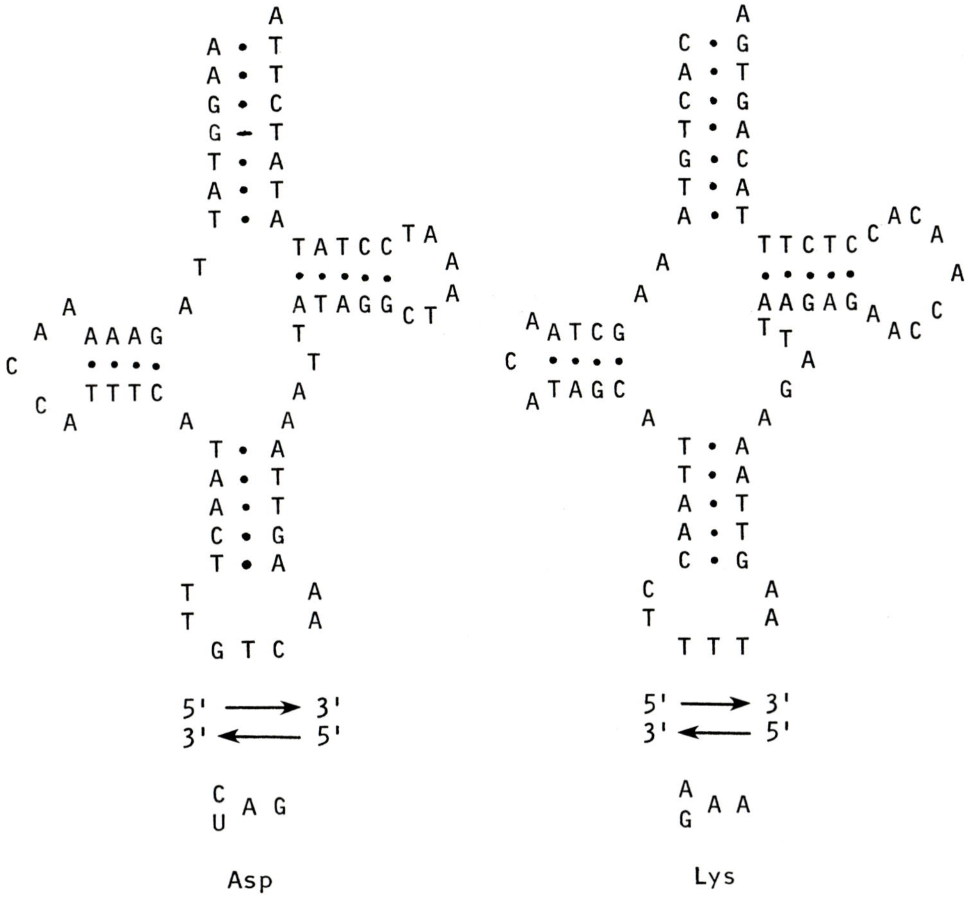

Figure 1: The DNA sequence and inferred structure of two human mitochondrial tRNA's, as presented in Barrell et al. (7). Reprinted by permission of **Nature**.

The most prominent differences concern the "D loop" and the "pseudo-uridine loop". The latter normally has seven unpaired bases. In human mt-RNA^{Asp} it has six, and in mt-RNA^{Lys} it has nine. The expected 5'-TTCG-3' sequence is missing. The D loop usually contains eight to twelve bases. In mt-RNA^{Asp} it has only five; in mt-RNA^{Lys} it is almost nonexistent. Only three bases remain unpaired.

Of course, both of the above tRNAs would be expected to wobble in the usual way. One has the anticodon GUC, which will recognize both Asp codons, GAC and GAU. The other has the anticodon UUU, which will recognize either Lys codon, AAA or GAA. There is no need to invoke a novel mechanism to explain their behavior. Presumably, the unusual usage of the UGA and the AUA codons, as tryptophane and methionine respectively, can also be adequately explained by this model. But even with maximum wobble of this kind, 31 tRNAs would be needed to recognize all the possible coding triplets, and the human mitochondrion only codes for about twenty. There is no evidence that tRNA is imported from the cytoplasm; consequently, at least some of the mitochondrial tRNAs must recognize more than two codons. In fact, wherever the first two bases are sufficient to identify an amino acid, which happens eight times in the code, Sanger's group has so far found only one tRNA for that amino acid. These are the wobbliest tRNAs ever found.

If these structural differences also apply to the other mitochondrial tRNAs - and the initial indication is that they do - then it will be interesting to see how these alterations in secondary structure affect the interactions of the D loop and the pseudo-uridine loop with each other and with the mitochondrial ribosome complex, and whether this has any implications for the specificity of codon-anticodon recognition.

The strangeness of mitochondria is challenging, and it is sure to add new fuel to arguments that have smoldered since the genetic code was broken. It has often been proposed that the present triplet code evolved from a more primitive doublet code, which perhaps the mitochondria have preserved in a less altered form. Considering the differences between human and yeast mitochondria, however, it seems more likely that mitochondrial systems have been evolving since they took up the symbiotic way of life. This is not a fossil code, but a novel item. The reason that mitochondria are different from other forms of life, as well as from each other, may explain at last why the code is universal among most organisms. Perhaps this is the key to it: that mitochondria are genetically isolated, reproduce asexually and do not exchange genetic information, leaving them free to experiment and evolve in a way that is not available to bacteria, yeast or men.

The answers to these and other questions should be forthcoming as Sanger and other workers continue to expand our understanding of the field. We anxiously await the conclusion of an important piece of work: the complete sequence of human mitochondrial DNA should soon be available, along with detailed knowledge of the structure of its tRNAs and the arrangement of its genes.

At the same time, we wonder what this remarkable man is likely to do next. Sanger himself insists that he does not know what his next project will be, but "at least there are still plenty of DNA sequences left to be determined. . ."

REFERENCES

1. Sanger, F. and Coulson, A.R. (1975) J. Mol. Biol. 94, 441.
2. Sanger, F., Air, G.M., Barrell, B.G., Brown, N.L., Coulson, A.R., Fiddes, J.C., Hutchison, C.A., Slocombe, P.M. and Smith, M. (1977) Nature 265, 687.
3. Maxam, A.M. and Gilbert, W. (1977) Proc. Natn. Acad. Sci. USA 74, 560.

4. Sanger, F., Nicklen, S. and Coulson, A.R. (1977) Proc. Natn. Acad. Sci. USA 74, 5463.
5. Messing, J. and Gronenborn, B. (1978) Nature 272, 375.
6. Smith, A.J.H. (1979) Nucl. Acids Res. 6, 831.
7. Barrell, B.G., Bankier, A.T. and Drouin, J. (1979) Nature 282, 189.
8. Steffens, G.J. and Buse, G. (1979) Hoppe-Seyler's Z. Physiol. Chem. 360, 613.
9. Macino, G., Coruzzi, G., Nobrega, F.G., Li, M., and Tzagoloff, A. (1979) Proc. Natn. Acad. Sci. USA 76, 3784.
10. Macino, G. and Tzagoloff, A. (1979) Proc. Natn. Acad. Sci. USA 76, 131.
11. Hensgens, L.A.M., Grivell, L.A., Borst, P. and Bos, J.L. (1979) Proc. Natn. Acad. Sci. USA 76, 1663.
12. Sebald, W. and Wachter, E. (1978) in 29th Mosbacher Colloquium on Energy Conservation in Biological Membranes, eds. Schafer, G. and Klingenberg, M. (Springer, Berlin), 228.
13. Li, M. and Tzagoloff, A. (1979) Cell 18, 47.

MODIFIED GENE EXPRESSIONS INDUCED BY TRANSPOSABLE ELEMENTS

Barbara McClintock

Carnegie Institution of Washington
Cold Spring Harbor Laboratory
Cold Spring Harbor, New York

I. INTRODUCTION

The selection of topics and speakers for this Symposium reflects the attention recently redirected to the broad range of genome manipulations that can occur in both prokaryotes and eukaryotes. The long-held dogma of genome stability is under attack, demanding readjustment of concepts. I use the word "redirected" advisedly because of the large body of little-known literature dealing with genome manipulation in eukaryotic organisms. Cytologists and cytogeneticists, using only the light microscope, have been responsible for most such reports. The first appeared in the latter part of the past century, with subsequent reports following continuously. The types of directed genome modification occurring in specific cells at special times in development relate to minute parts of chromosomes (as in amplification of genes, or of bands in polytene chromosomes), to larger segments of chromosomes (e.g., underreplication of heterochromatic segments of chromosomes, or their excision), to whole chromosomes (e.g., X-chromosome gains and losses by nondisjunctions, losses at anaphase, or passage through a nuclear membrane), or to whole sets of chromosomes (e.g., elimination of the germ-line chromosome set in Cecidomyidae, or condensation of the paternal set in male coccids, or the discarding by *Sciara* of the paternal set from secondary spermatocytes). The extraordinary nature of the genome fragmentation and DNA discard that accompany macronuclear development in the ciliate protozoan *Oxytricha* (*Stylonychia*), recently investi-

ISBN 0-12-633360-2

gated at the molecular level (Prescott and Murti, 1974; Lauth et al., 1976; Herrick and Wesley, 1978), was detected initially through light-microscope observations of a sequence of unanticipated kinds of behavior of its chromosomes (Ammermann, 1964, 1968).

Knowledge of genome manipulations, sources of which lie buried in the literature, could offer the molecular biologist some exceptional opportunities to characterize the responsible events and interpret their significance. This knowledge is so revealing and conceptually so important that I was tempted to review it for today's discussion. I decided, however, to direct attention to just one aspect of genome manipulation. The responsible events are well enough characterized at the phenotypic and light-microscope levels to suggest modes of exploring their nature at the molecular level. It is anticipated that the information so gained will tell much about constitutions of gene-control regions in eukaryotic organisms, and will also reveal some of the diverse mechanisms that regulate gene action during development. I shall pay particular attention to maize, the common corn plant, as it has provided a large body of information on this topic. I shall not dwell on those insertion elements that effect inactivation of a gene or of its product. Such instances are well documented in studies of IS elements, of transposons on plasmids, of Mu phage, and of suspected cases in *Drosophila* (Green, 1978; Lim, 1979). Rather, I should like to emphasize that among the many classes of transposable elements, each specific in function, one class serves to regulate types and patterns of gene expression at the locus where it is inserted and where it, or one of its derivatives, may become permanently fixed in position.

Attention was initially focused on gene-control mechanisms in maize when it was noted that many different genes suddenly began to express altered times, frequencies, and intensities of action during either plant or kernel development, or both. It was clear that these modified expressions were induced in some then unknown manner as a consequence of sequential changes in content and organization of one arm of one chromosome of the maize complement (McClintock, 1951, 1978). And the affected genes could be located in chromosomes other than the one that had undergone content change. In many instances, altered controls of gene action were readily recognized, phenotypically, because they produced conspicuous variegated patterns. Patterns of red or purple pigment in plant and kernel, or altered chlorophyll content in leaves and sheaths, were brilliantly expressed, simplifying their selection for further study. Similar patterns have been noted in many organisms, especially plants. Early genetic studies conducted with some of them generated the term "unstable genes." These studies were reviewed in 1933 by Stubbe and in 1935 by Demerec. It was recognized, however, that not all suddenly arising variegated patterns could be placed in the unstable gene category. Other mechanisms were involved, and conspicuously so with regard to "position effect" variegation in *Drosophila melanogaster*.

II. ORGANIZATION AND OPERATION OF TRANSPOSABLE GENE-CONTROL ELEMENTS IN MAIZE

The origin and expression of a number of "unstable genes" probably reflect the insertion and subsequent activity of a "foreign" element. This inference rests on instances in maize of such origins, and of the subsequent activities of the inserted element. Usually, such insertions were revealed initially by a conspicuously altered gene expression. Often the initial alteration did not inactivate the gene, but rather modified gene expression in a revealing manner. Altered expressions that proved to be heritable were selected for further investigation. Heritability indicated that the responsible initiating event had restructured a locus in some specific manner, whose phenotypic reflections were observed as new times of gene action and/or altered intensities of action at any one time. In some instances the initial event, or a subsequent restructuring of the locus, provided for "gene on-gene off" activity patterns in particular tissues (McClintock, 1965, 1971; Peterson, 1966). The linking of gene product with phenotypic expressions that were modulated by the presence at a locus of a transposable controlling element (CE) was reported recently by Dooner and Nelson (1977, 1979). They noted that an inserted element could control the time of gene action, the amount of its protein product appearing at any one time, and even the nature of the final product.

To appreciate the modes of behavior of transposable CEs in maize, some frame of reference to other systems of transposing elements would be helpful. This is not difficult to find. The actions of these components resemble those of transposons carried on plasmids of bacteria. Before pressing analogies between the systems, one should recall that transposable elements fall into recognizable classes. They have in common their ability to be transposed. In other respects, however, they differ from one another, basically according to functions they may perform. Some functions are obvious, such as those of resistance-conferring factors carried by transposons of bacterial plasmids, or the potentials for genome reassortment initiated by an integrated F plasmid in E. coli. A device for survival may account for phage and virus insertions. Other transposing elements such as Copia, 412, and 297 in Drosophila (Rubin, this Symposium), or TY1 and delta sequences in yeast (Cameron et al., 1979) have not yet revealed their individual significance although some aspects of that of TY1 may soon be available, as Gerald Fink will describe in this Symposium. The cassette style of mating-type locus exchange in yeast (Herskowitz, this Symposium) reflects yet another accomplishment of the transposition process that has much to teach us, as has the transposition mechanism associated with origins of immunoglobulins (Tonegawa, this Symposium).

The maize transposable controlling elements represent another distinct class. The distinction lies in their ability to program times and types of gene action during development, and to integrate the activity of one gene with that of another. The various derivatives of the different maize CEs may prove to be of special importance, however, by bringing about stabilization of a genome that was unstabilized, initially, by release of transposable CEs from their silent phase. It is important to stress that the initiating CEs lie silent in the maize genome until some challenge to genome stability arises. They then come into play, producing copies that can initiate numerous modifications of the genome. These are registered by structural reorderings of the genome at both gross and fine levels, and by modulation of the action of individual genes in distinctive manners. The modifying events may then cease, leaving behind a decidedly altered genome. Thus, the total effect of these modifications may result in quite new genotypes and phenotypes.

Various types of derivatives originate from an intact CE. Their origins may reflect abortive products of the transposition mechanism, for they appear with frequency only when transpositions are occurring. The response of one class of derivatives to an intact parent CE resembles that of some defective derivatives of plasmid transposons to their unmodified parents (Heffron, this Symposium), and also that of mini Mu derivatives of phage Mu to its unmodified parent (Bukhari, this Symposium). A member of this class is able to respond to trans-acting signals from an intact parent CE, either by transposition away from the gene locus, by remaining in situ after having suffered an alteration in structure, or by inducing additional alterations of the locus where it resides. In the absence of an active parent CE, this class of derivatives remains completely stable, in that a member will sustain one type of genic expression at the locus where it resides throughout generations of plants. The potential ability of the inserted element to instigate new expressions is revealed only when an active parent CE is introduced by an appropriate cross. This introduction initiates a renewed sequence of modifications. The element at the gene locus may then lose its capacity to respond to the trans-acting transposition signals from the parent CE. The element, now stably located, will continue to control action of the gene, in a defined or programmed manner. A series of new alleles of a locus may be initiated in this way. Such newly produced alleles, showing distinctive patterns of expression in various parts of the plant or kernel, have been isolated and characterized (Rhoades, 1941; Laughnan, 1948, 1955; McClintock, 1968, 1971). A knowledge of the molecular events that produced these new alleles would greatly enhance our understanding of gene-control mechanisms in higher plants and possibly in eukaryotes in general.

Intact parental CEs are readily distinguished from one another by several criteria. The transposition-induced products that are both cis- and trans-acting are unique to each. The product of one does not interfere with that of another. Again, each intact CE provides a distinctive mode of regulation of gene action that allows ready recognition and testing.

III. SIMILARITIES BETWEEN MAIZE CEs AND TRANSPOSONS ON PLASMIDS

Similarities between maize CEs and transposons on bacterial plasmids are evident at several levels. Similarities with Tn3 (Heffron, this Symposium) are appropriate for comparison. Control of transposition of Tn3 resides in its structural organization. Reverse repeats, 38 base pairs in length, define the ends of this transposon, and they must be intact if transpositions are to occur. At one end, adjacent to a repeat, is a relatively long DNA segment whose integrity also is required for the occurrence of transpositions. A deficiency within this segment will block transposition. The block may be overcome, however, if an intact Tn3 is present in addition. It is evident therefore that the product of this segment, a transposase, acts both in cis and in trans to effect transpositions. Adjacent to the repeat at the other end of Tn3 is the gene for β-lactamase. Its functioning gives significance to this transposon. Next to it is a short DNA segment that controls the frequency of occurrence of transpositions by means of its protein product, which acts both in cis and in trans to repress transposition. Deficiencies within this DNA segment give rise to altered proteins that are less effective in repressing transposition.

Each intact maize CE has components that are responsible for its transposition. These resemble those of Tn3 in that the activity of one component is required for transpositions to occur. The product of this component (a transposase?) is effective in cis to induce its own transposition, and in trans to instigate transpositions of a derivative that has lost the ability to effect its own transposition. A second component acts to regulate the time and frequency of occurrence of transpositions. Each of the maize CEs has its own mode of accomplishing this regulation, and the product of the regulatory component acts both in cis and in trans. Other components of intact CEs in maize are responsible for controlling the action of genes, and in this regard, each parent CE is distinctive in one or more ways. Considering the breadth of expression of control that these CEs and their derivatives can induce after their initial insertion at gene loci, their accomplishments must be judged impressive.

Although the molecular organization of transposable CEs in maize is unknown, it would be surprising if terminal repeats of some type were not found. They have appeared in other examined instances of cut-outs, insertions, and transpositions, involving IS elements, transposons carrying resistance factors, Mu phage, TY1, animal viruses, etc. They also define the ends of DNA fragments in the Oxytricha macronucleus.

IV. CONCLUDING STATEMENTS

It is not the purpose of this discussion to review all the subtle ways by which the different maize CEs can accomplish their results. Many of these are examined in the literature, reviews of which are given in McClintock (1965), Peterson (1970), Fincham and Sastry (1974), and Nevers and Saedler (1977). The purpose, rather, is to stress the potentials of these elements for redirecting gene activities, in manners that allow detection of the types of DNA manipulation responsible for their accomplishments. The extraordinary range in types and times of control of gene expression during development that CEs can elicit suggests to me that the basic mechanism may not be unduly diverse or complex. Indeed, variations of some common mechanism may be responsible for providing the many observed possibilities for control and integration of gene expression. For example, the many distinctive alleles of the maize *R* locus that control expression of anthocyanin pigment in parts of the plant and kernel (Styles et al., 1973), the modifying influence that some of these alleles have on others (Brink, 1974), the "presettings" and "erasures" expressed by still others (Kermicle, 1978), together with the known structural complexity of this locus (Gavazzi, 1977), suggest influences exerted by a succession of elements inserted at the locus. Because maize cannot propagate itself but must be propagated by man, and has been so propagated within the Americas for hundreds of years, opportunities for selection of desirable alleles must have been numerous. The consequences of such selection, of both stable and unstable alleles, are evident among present-day races of maize (Brown and Goodman, 1977). The *P-vv* locus (variegated pericarp and cob) with the inserted controlling element *Ac*, is an example of such selection, as are the variegated expressions on ears of "Indian corn" sold for decorative purposes.

The mechanism that activates silent CEs in maize can also initiate conditions that induce chromosome reorganization. Such conditions arise from some initial disturbance in genome equilibrium. This is true for maize and is suspected in other organisms. Strain crosses in *Drosophila* instigate genome restructuring, types of which are sometimes predictable (Kidwell et al., 1977; Thompson and Woodruff, 1978; Sved, 1979; Green, 1978; Green and Shepard, 1979). It has long been recognized that species crosses may give rise to chromosomal restructurings; and again, the type and extent may sometimes be predicted once the effects have been observed. This has been shown by Gerstel and Burns (1966, 1976) with *Nicotiana* species crosses. In another *Nicotiana* species cross, one particular gene locus becomes unstable, giving expressions resembling some of those produced by maize CEs (Smith and Sand, 1957; Sand, 1976).

There is little doubt that genomes of some if not all organisms are fragile and that drastic changes may occur at rapid rates. These can lead to new genomic organizations and modified controls of type and time of gene expression. It is reasonable to believe that such genome shocks are responsible for the release of otherwise silent elements, which can then initiate changes to overcome disruptive challenges. I have emphasized that stabilizations do occur after release and action of maize CEs. Since the types of genome restructuring induced by such elements know few limits, their extensive release, followed by stabilization, could give rise to new species or even new genera.

REFERENCES

Ammermann, D. (1964). Naturwissenschaften 51, 249.
Ammermann, D. (1968). Chromosoma (Berl.) 25, 107.
Brink, R.A. (1974). Ann. Rev. Genet. 7, 129.
Brown, W.L., and Goodman, M.M. (1977). In "Corn and Corn Improvement" (G.F. Sprague, ed.), p. 49. American Society of Agronomy.
Cameron, J.R., Loh, E.Y., and Davis, R.W. (1979). Cell 16, 739.
Demerec, M. (1935). Bot. Rev. 1, 233.
Dooner, H.K., and Nelson, O.E. (1977). Proc. Nat. Acad. Sci. (USA) 74, 5623.
Dooner, H.K., and Nelson, O.E. (1979). Proc. Nat. Acad. Sci. (USA) 76, 2369.
Fincham, J.R.S., and Sastry, G.R.K. (1974). Ann. Rev. Genet. 8, 15.
Gavazzi, G. (1977). Stadler Symp. 9, 37. Univ. of Missouri.
Green, M.M. (1978). In "The Clonal Basis of Development" (S. Subtelny and I.M. Sussex, eds.), p. 239. Academic Press, New York.
Green, M.M., and Shepard, S.H.Y. (1979). Genetics 92, 823.
Gerstel, D.U., and Burns, J.A. (1966). Genetics 56, 483.
Gerstel, D.U., and Burns, J.A. (1976). Genetica 46, 139.
Herrick, G., and Wesley, R.D. (1978). Proc. Nat. Acad. Sci. (USA) 75, 2626.
Kermicle, J.L. (1978). In "Maize Breeding and Genetics" (D.B. Walden, ed.), p. 357. Wiley Interscience, New York.
Kidwell, M.G., Kidwell, J.F., and Sved, J.A. (1977). Genetics 86, 813.
Laughnan, J.R. (1948). Genetics 33, 488.
Laughnan, J.R. (1955). Am. Nat. 89, 91.
Lauth, M.R., Spear, B.B., Heumann, J., and Prescott, D.M. (1976). Cell 7, 67.
Lim, J.K. (1979). Genetics (in press).
McClintock, B. (1951). Cold Spring Harbor Symp. Quant. Biol. 16, 13.
McClintock, B. (1965). Brookhaven Symp. Biol. 18, 162.
McClintock, B. (1968). Carnegie Inst. Wash. Year Book 66, 20.
McClintock, B. (1971). Carnegie Inst. Wash. Year Book 70, 5.
McClintock, B. (1978). Stadler Symp. 10, 25. Univ. of Missouri.
Nevers, P., and Saedler, H. (1977). Nature 268, 109.
Peterson, P.A. (1966). Genetics 54, 249.
Peterson, P.A. (1970). Genetica 41, 33.
Prescott, D.M., and Murti, K.G. (1974). Cold Spring Harbor Symp. Quant. Biol. 38, 609.

Rhoades, M.M. (1941). Cold Spring Harbor Symp. Quant. Biol. 9, 138.
Sand, S.A. (1976). Genetics 83, 719.
Smith, H.H., and Sand, S.A. (1957). Genetics 42, 560.
Stubbe, H. (1933). Bibliographia Genetica 10, 299.
Styles, E.D., Ceska, O., and Seah, K.-T. (1973). Can. J. Genet. Cytol. 15, 59.
Sved, J.A. (1979). BioScience 29, 659.
Thompson, J.N., and Woodruff, R.C. (1978). Nature 274, 317.

STUDIES WITH IS4

R.Klaer, S.Kühn, E.Tillmann, H.-J. Fritz,
L.St. Girons, P.Habermann, D.Pfeifer, P.Starlinger

Institute of Genetics
University of Cologne
FRG

INTRODUCTION

Transposition mechanisms have recently attracted widespread interest among geneticists. One of the reasons for this interest is the possibility that mechanisms of transposition may have played an important role in evolution.
Evolution has been a fascinating subject for biologists for a long time. The mechanisms of evolution are not completely clear, however. Mutation and recombination have long been thought to be the driving forces that create genetic variability. With the discoveries of molecular biology, it was learned that the gene consists of many nucleotides and that the point mutation changes one of these nucleotides at a time. It became a real question, whether the time of evolution was sufficient for the accumulation of so large a number of these to allow the creation of proteins with completely new functions from the existing ones. If these new functions have no relation to those of the progenitor proteins, the first amino acid replacements most probably do not have a selective value toward the new function. If the different deviating sequence permutations are to be tried by random genetic drift, the number of possibilities for a gene of one thousand nucleotides or more becomes exceedingly large. A way out of this difficulty would be the evolution of smaller peptides via permutation of the sequence of the corresponding nucleic acid sequences. These would then have to be combined in various combinations, and a mechanism capable to join unrelated DNA sequences would be required for this. Are the transposition mechanisms which presently become known in many organisms responsible for this type of evolution?

ISBN 0-12-633360-2

Transposition mechanisms had been known for a long time from B.McClintock´s work on maize. When we began studying similar mechanisms in bacteria, many similarities to the phenomena studied in corn became apparent. It turned out that in both organisms - and in all others that were studied since - the mechanisms involved did not join random segments of DNA in new arrangements. Specific DNA sequences are the primary objects of transposition, and rearrangements occur as a consequence of the primary transposition of the pecific sequences. In E. coli, these elements are known as IS elements and transposons (1-5) The latter carry genes that have identifiable gene products. Often these gene products confer resistance to an antibiotic to these bacteria. Thus they are easily identified. The known transposons do not occur in E.coli K12. It is understandable that the genes necessary for transposition are brought along with the transposon DNA, as has been shown for Tn3 (6, 7).

IS elements are smaller, and have not yet been shown to encode their own transposition machinery. If they do carry it, the proteins made cannot be very large, as can be seen from the published sequences of IS1 and IS2 (8, 9, 1o). Are genes carried elsewhere on the E.coli chromosome involved in transposition? If so, are these genes exclusively functioning in transposition, or is transposition but one among several functions of these genes? Why is transposition a rare event? Are the enzymes presumably functioning in transposition very inactive, or are they not made at all times during the cell cycle?

In an attempt to answer these questions, it would be desirable to alter the DNA sequence of the IS element in order to inactivate any function encoded here. If this function is the production of a protein, the mutation could not be analysed easily, because most IS elements are carried on the E.coli chromosome in several copies and the alteration of all of them simultaneously is not easy. We have therefore chosen to work with IS4 which is present in most E.coli K12 strains in one copy only (11). IS4 is interesting also because it seems to have a certain specificity in integration. Among many mutants caused by the insertion of IS elements in the gal operon of E.coli, IS4 has been found in one location only (12, 13). It is hoped therefore that a study of IS4 transposition may lead to a clearer understanding of transposition specificity.

I will describe here the present status of work on IS4.

A. The Structure Of IS4

The sequence of IS4 has been determined, using as starting material plasmid pKS57 to be described below as source of DNA and the rapid sequencing method of Maxam and Gilbert. IS4 is 1426 bp long. It has unique sites for PstI, ClaI, BalI, AvaI, and HindII, and is not cleaved by the other enzymes which recognize hexanucleotides.

The 18 terminal nucleotides of IS4 form an inverted repeat with two mismatches only. In this respect, IS4 is similar to IS1 and IS2, which also show short inverted repeats with mismatches. Transposons have much larger inverted repeats, which in some cases are known IS-elements. Two questions can be asked:

a) Are the inverted repeats essential for transposition?

b) Why are the termini of transposons matched better than the termini of IS-elements? Are the latter older on an evolutionary time-scale?

The most conspicious structures that are detected on the sequence of IS4 are two palindromes of 11 base pairs found near the termini of the IS-element. These palindromes can potentially form a stem and loop-structure. This stem and loop-structure is followed by several T-residues (on one site interrupted by a C) at the interior 3′-side. These structures are rather similar to known rho-independent transcription stop signals (14). IS4 insertions in galT are very polar, and this polarity is only marginally revealed in rho-mutants (15). It is conceivable that the structures described are responsible for this polarity, but this will have to be shown by independent experiments. If the stem and loop-structures do act as termination signals, the sequences in the interior of IS4 cannot be transcribed by readthrough from the outside.

We have scanned the sequence of IS4 for palindromes, and for similarities with IS1 and IS2, without finding conspicuous features apart from the two 11 base pair palindromes mentioned above. We have also looked for open reading frames. The extent of such sequences, starting with either AUG or GUG and extending to a nonsense codon in the same reading frame is remarkable. In one orientation, a sequence of 1326 nucleotides is present in an open frame starting with AUG. It is not very probable, however, that this region is transcribed efficiently. The open frame ends a few nucleotides before the terminus of IS4, and there are no indications that in this direction a termination signal is present between the nonsense codon and the very end of IS4. Therefore, transcription of this DNA-region would most probably extend into adjacent DNA. Thus, at least one of the orientations of IS4 in galT should be no more polar than a nonsense mutation in approximately the same region. This is not observed. In accordance with this conclusion, the sequence in front of the open reading frame does not show features characteristic for a promoter.

It will be interesting to find out whether transcription and translation of this region can occur under certain circumstances. It is conceivable that functions associated with transposition are expressed very rarely. It will be interesting to see, whether our open frame encodes such a function and under which circumstances it can be expressed.

B. Insertion Specificity of IS4

At present, IS4 is known at three sites. The first of these is shared by all E.coli K12 strains.This has been shown by cloning a HindIII fragment carrying IS4 from one E.coli K12 strain in pBR322. It could be shown that not only IS4 DNA from this fragment, but also other DNA sequences from the cloned HindIII fragment hybridized to the IS4-containing HindIII fragments of other E.coli K12 strains, even if these fragments differed in size from the one cloned in pBR322. This indicated that the chromosomal region carrying IS4 has evolved after the integration of IS4. This integration site will be called the "common" IS4 site.

The second integration site of IS4 is in an unmapped restriction fragment in strain F165, which hybridizes to IS4 in addition to the one found in all other strains too. This fragment has been cloned. Its DNA apart from IS4 does not hybridize to the HindIII fragment carrying the "common"site. A fragment from the vicinity of the second IS4 has been isolated and hybridized to all other K12 strains investigated. In each case, a fragment has been obtained, which was smaller by 1.4 Kb than the one in strain F165. It can be concluded that the IS4 found on this fragment in strain F165 has been transposed there after the divergence of these K12 strains.

The third integration site of IS4 is located in deletion group 9 of galT. Approximately 4o% of all insertions found as polar mutations in galOP, galE, and galT are caused by the integration of IS4. Three of these carried on lambda dgal phages and representing both possible orientations were selected for sequence studies.

The junctions of IS4 with its adjacent sequences have been determined in all sites. In each case, a duplication is found flanking the termini. In the case of galT, all three mutants are found in the same site and are flanked by an 11 base pair duplication. The sequence of 11 base pairs is found once only both in the wildtype and in a Gal^+revertant (16).

In the other two integration sites, duplications of 12 base pairs flank IS4. In these cases, sequences without IS4 are not available for comparison.

The three sequences are very dissimilar. Therefore, they cannot have been brought along by IS4 and must have been created at the site of insertion, as has been also found for IS1 (17, 18, 19). From this result, the following can be concluded:

Grindley has suggested that the duplications flanking IS sequences are created by the introduction of staggered nicks, linking of the IS element to the protruding end of the recipient DNA and subsequent filling in the gaps thus created. If this is a general mechanism, the enzyme responsible for introducing of the staggered nicks is most probably different from those creating either nine bp or five bp duplications. If it is an enzyme specific for IS4, it will be interesting to see, where it is encoded.

In contrast to other IS elements, the number of nucleotides that can be doubled upon insertion is variable in the case of IS4. This is reminiscent of some of the restriction endonucleases of class II, e.g. HphI (2o). Both in the case of these restriction enzymes and in the case of the insertions, the variability is found in those cases, where the region presumably making contact with the DNA is relatively large. Whether this necessitates a greater flexibility in the protein involved remains to be determined.

The duplicated sequences found at the three known insertion sites of IS4 are so different from each other that the specificity of insertion cannot reside within them. Still, there must be specificity for the insertion of IS4 at least in galT. We have looked for similarities in the sequences surrounding IS4, and have found that in all three sites a short tetranucleotide palindrome, AAAC.....GTTT, is found, and that the duplicated region is located between the two tetranucleotides. The distance between the two tetranucleotides is variable. All three duplicated regions have the dinucleotide CA at one 5'terminus, and in two of the cases this CA is followed by a G. The distance between the AAAC and the start of the duplicated region is two nucleotides in the cases, where the duplication is 12 base pairs long, and three nucleotides in the case of the eleven base pair duplication. Thus, the distance between AAC and the end of the duplicated region is identical on one side of this region.

Whether this structure is sufficient to confer specificity to IS4 integration is not known presently. Neither is it known, which sequences of IS4 are making contact with the integration and/or recognition sites on the chromosome. The tetranucleotide palindrome AAAC.....GTTT is found in the stem and loop-structures near the termini of IS4 that were described above.

C. The Transposition Of IS4

It has been hypothesized that transposition is accompanied by a replication of the transposable element, resulting in two copies of this element, of which one resides in its old position, while the other is found in a new one (21, 22, 23). We have isolated mutants created by the insertion of IS4 in galT. The DNA of three such mutants was digested with various restriction enzymes and analyzed for fragments containing IS4 by Southern's blotting technique. In all three mutants, two IS4-containing fragments are detected. One of these fragments is expected from the restriction map of the gal operon, while the other one is the same fragment that is also present in the wildtype strain. The same observation is made in strain F165, where a new fragment containing IS4 is observed in addition to the fragment present in the wildtype. These experiments support the hypothesis of replication during transposition, but cannot disprove the possibility of a suicidal transposition, which eliminates the chromosome arm participating in the transposition event.

In an attempt to study transposition frequencies from various positions more easily, we have constructed derivatives of IS4 (pKS59), which carry a chloramphenicol-transacetylase gene from plasmid NR1 in the PstI site of IS4 (carried on

P1Cm phages, generously given to us by S.Iida, Basel). We hoped that such a structure would still be transposable, because any function inactivated should be provided by another copy of IS4 present on the chromosome. We had previously constructed a plasmid carrying IS2 with a fragment carrying a gene for kanamycin-resistance in the HindIII site of this IS element. The artificial transposon was capable of transposition. In the case of IS4, however, no transposition was observed. In our experiments, the plasmid pKS59, carrying IS4 from a chromosomal site was harboured by a cell, which was lysogenic for either lambda b515, b519 or lambda dgal. The bacteriophage was induced, and the resulting lysate was tested for Cm^rtransducers. No such transducers were observed among $1o^{11}$ lambda or lambda dgal phages. The latter phage was employed because we considered the possibility that the insertion specificity of IS4 might be so high as to preclude the insertion into lambda. The insertion into galT should have been possible, however. The reasons for this failure are not yet known. We are now testing, whether the presence of several copies of IS4 enhances the level of an IS4-encoded transposition repressor and thus decreases transposition frequency.

D. Bidirectional Deletions

Upon isolation of lambda dgal phages carrying IS4 in galT, we observed a small additional band of phages of decreased density. These phages were isolated and their DNA investigated by heteroduplex techniques. The DNA carries a deletion which removes IS4 and sequences adjacent to it. The frquency of these deletions is considerable as judged from the amount of phages carried in the extra band in CsCl. In order to exclude the possibility that an unknown lambda function is involved in the production of these deletions, we have cloned the gal operon and several kb up- and downstream of it in pBR322. These plasmids were used to transform cells carrying a mutation in galE. Antibiotic-resistant transformants with Gal^--phenotype were isolated. Colonies of these cells were subsequently hybridized to a DNA probe containing IS4, and approximately 1 kb up- and o,5 kb downstream. The Gal^-cells failed to hybridize to this DNA fragment, indicating that the Gal^-cells carry a deletion removing IS4 and adjacent sequences. In different plasmids carrying IS4, the frequency of these deletions is o,1-3/1ooo. No deletions were obtained in plasmids carrying a reversion of the galT mutations to Gal^+. (There is one exception, in which a plasmid not carrying IS4 also yields deletions. This plasmid carries part of IS1, and we are presently investigating, whether this is responsible for the deletion formation.) Bidirectional deletions were also found, when Gal^+cells were selected from galT::IS4 mutants which carried the gal operon on F'8gal, a deletion of the gal operon in the chromosome, and were lysogenic for a lambda dgal prophage providing a functional transferase but carrying a deletion of galK. This selection detects, in addition to true revertants, mutations relieving polarity, but not restoring galT. This latter class is 1o times more frequent than reversions and some of them do not recombine with point mutations on either side of the site of IS4 in appropriate crosses. These may also be bidirectional deletions.

From these observations we conclude the following:
1) At least the frequent deletions must be stimulated by the presence of IS4 between the endpoints of the deletions, because no such deletions are detected, when IS4 is present only on the chromosome in its "common" site but not in galT.
2) The endpoints of the deletions must be highly non-random, because the deletions detected on lambda dgal phages are found in a discrete phage band in CsCl gradients, and also because the frequencies of such deletions are so different, depending on the selection procedure. Selection for galE cells, followed by hybridization detects the deletions at frequencies up to $1o^{-3}$ while selection for polarity relief leads to Gal^+ cells at $1o^{-8}$/cell plated. These latter mutations must have their endpoints in a restricted region of the gal operon, but this cannot explain by itself a difference in the frequency of detection of several orders of magnitude.

This work was supported by Deutsche Forschungsgemeinschaft through SFB 74.

REFERENCES

(1) Cohen, S. N. (1976), Nature 263, 731.
(2) Starlinger, P., and Saedler, H. (1976). Curr. Top. Microbiol. Immunol. 75, 111.
(3) Bukhari, A., and Shapiro, J., and Adhya, S. (eds.) "DNA Insertion Elements, Plasmids and Episomes". Cold Spring Harbor Laboratory, Cold Spring Harbor, N. Y.
(4) Kleckner, N. (1977). Cell 11, 11.
(5) Starlinger, P. (1980). Plasmid, submitted.
(6) Heffron, F., Bedinger, P., Champoux, J., and Falkow, S. (1977). Proc. nat. Acad. Sci. 74, 702.
(7) Chou, J., Casadaban, M., Lemaux, P., and Cohen, S. (1979). Proc. nat. Acad. Sci. 76, 4020.
(8) Ohtsubo, H., and Ohtsubo, E., (1978). Proc. nat. Acad. Sci. 75, 616.
(9) Johnsrud, L. (1979). Molec. gen. Genet. 169, 213.
(10) Ghosal, D., Sommer, H., and Saedler, H. (1979). Nucl. Acid. Res. 6, 1111.
(11) Chadwell, H. A., Fritz, H.-J., Habermann, P., Klaer, R., Kühn, S., and Starlinger, P. Cold Spring Harbor Symp. quant. Biol. 43, 1187.
(12) Shimada, K., Weisberg, R. A., and Gottesman, M. (1973). J. mol. Biol. 80, 297.
(13) Pfeifer, D., Kubai-Maroni, D., and Habermann, P. (1977). In: "DNA Insertion Elements Plasmids and Episomes" (A. I. Bukhari, J. Shapiro, S. Adhya, eds.). Cold Spring Harbor Laboratory, Cold Spring Harbor, N.Y., 31.
(14) Adhya, S., and Gottesman, M. (1978). Ann. Rev. Biochem. 47, 967.
(15) Besemer, J., and Herpers, M. (1977). Molec. gen. Genet. 151, 295.
(16) Habermann, P., Klaer, R., Kühn, S., and Starlinger, P. Molec. gen. Genet. 175, 369.
(17) Calos, M. P., Johnsrud, L., and Miller, J. H. (1978). Cell 13, 411.
(18) Grindley, N. D. F. (1978). Cell 13, 419
(19) Kühn, S., Fritz, H.-J., and Starlinger P. (1979). Molec. gen. Genet. 167, 235.
(20) Kleid, D., Humayun, Z., Jeffrey, A., and Ptashne, M. (1976) Prov. nat. Acad. Sci. 73, 293.

(21) Grindley, N. D. F., and Sherrat, D. J. (1979). Cold Spring Harbor Symp. quant. Biol. 43, 1257.
(22) Shapiro, J. A. (1979). Proc. nat. Acad. Sci. 76, 1933.
(23) Arthur, A., and Sherrat, D. (1979). Molec. gen. Genet. 175, 267.

DISCUSSION

C. HERSCHBERGER: The number of IS4 sequences is held at one per chromosome even though transposition occurs without loss of the IS4 sequence from the donor site. The low frequencies of excisions and deletions seem to be at odds with these observations. One would expect a build-up in the number of copies of IS4 per chromosome. How do you explain the conflict?

P. STARLINGER: The reason for the relatively constant number of IS elements in the E. coli K12 chromosome is not clear. If transposition increases this number, another process should decrease it to keep the number in equilibrium. Precise excision of IS elements does occur and could provide such a mechanism. How this can explain that a single copy of IS4 is retained in the same site for a time long enough to allow sequence rearrangements to occur in the vicinity I do not know.

C. HERSCHBERGER: Your EM showing your heteroduplexes, your loops in the site of the deletion seem to have an internal loop where you have regions or internal homology in the denaturation loop. Does this represent ends of the IS element or does it represent another internal homology, a coincidence that was present in just those two examples? Is that coincidence or is there a significance to that loop within the loop?

P. STARLINGER: I do not think there is a loop within the loop because we know the DNA structure. Of course, if the loop within the loop were due to some surrounding DNA, I could not say it. Such things are seen relatively often and one would have to have them in a very reproducible way in each picture. I do not think this is the case.

A. BUKHARI: I do not understand your deletion studies. First of all, the evidence does not seem to be clear that you have specific end points of deletions. Your evidence only shows that deletions are generally large. Secondly, I did not understand the mechanism you propose for generating those deletions.

P. STARLINGER: To answer the first question, of course we do not yet really know that any of these deletions have exactly the same end points because these days this would have to be proven by sequence studies. I can only say from the statistical experiment there are other regions in which the end points are much higher. These regions must be relatively small because phages carrying these deletions tend to form a band in the CsCl gradient rather than a smear. This means that the DNA content of these phages must be very similar, but, of course, I know that this does not go to the nucleotide level. It may go to the kilobase range - it is not better than that.

To answer your second question, the mechanism that was proposed by Norman Davidson was that if you form a Cairns structure by a bi-directional replication and at a certain time cleave out the eye of this structure, you may get a structure which has a head-to-head rather than a tandem duplication of a certain sequence. This is observed in the formation of lambda dv. Now, if you assume that there is a mechanism which cleaves out the eye of the Cairns structure, then you could hypothesize that if this mechanism ligates the ends of the non-replicated part of the chromosome, this would be bi-directional deletion. But, of course, this is sheer speculation and just an extension of another speculation by Norman Davidson.

R. NUTTER: I was wondering if you could tell me, do you know if your bi-directional deletions are rec A dependent?

P. STARLINGER: They are not rec A dependent.

DNA SEQUENCE REARRANGEMENTS WITHIN IS2 AND THEIR INFLUENCE ON GENE EXPRESSION

Heinz Saedler, Hans Sommer, Debabrota Ghosal, John Cullum and Peter Peterson

Institut für Biologie III
Universität Freiburg
Freiburg i.Br.
West Germany

INTRODUCTION

The insertion sequence IS2 in *Escherichia coli* K12 is 1327bp long (Ghosal et al., 1979a). It is a natural component of the *E.coli* chromosome where there are about 5 copies (Saedler and Heiß, 1973) and of some plasmids such as F (Hu et al., 1975). IS2 can transpose into a variety of operons in the chromosome or on phages (Starlinger and Saedler, 1976) but seems to have a preference for insertion into control sequences i.e. operator-promoter sequences. Most and perhaps all IS2 insertions studied so far have been insertions into non-coding DNA sequences.

When IS2 inserts in one orientation (orientation II) downstream genes are expressed constitutively from a promoter in IS2 (Saedler et al., 1974). Recently IS2, orientation II insertions have lead to the expression of foreign genes that have been cloned into *E.coli*, but were previously poorly expressed (Walz *et al.*, 1979).

IS2 insertions in the opposite orientation (orientation I) have a strong polar effect on downstream genes due to a *rho*-dependent transcriptional termination signal on IS2 (Besemer and Herpers, 1977).

ISBN 0-12-633360-2

DNA Rearrangements in IS2

When IS2 inserts in orientation I into the galOP region, gal operon expression is switched off due to the polar effect of IS2 (Fig. 1). Precise excision of IS2, which occurs at a frequency of about 10^{-7}, can restore the structure of the gal-operon and, therefore, inducible gal operon expression (Fig.1). However, as well as these inducible Gal^+ revertants, there are also constitutive Gal^+ revertants that are often unstable (Saedler et al., 1974, Hill and Echoli, 1966). Recently,

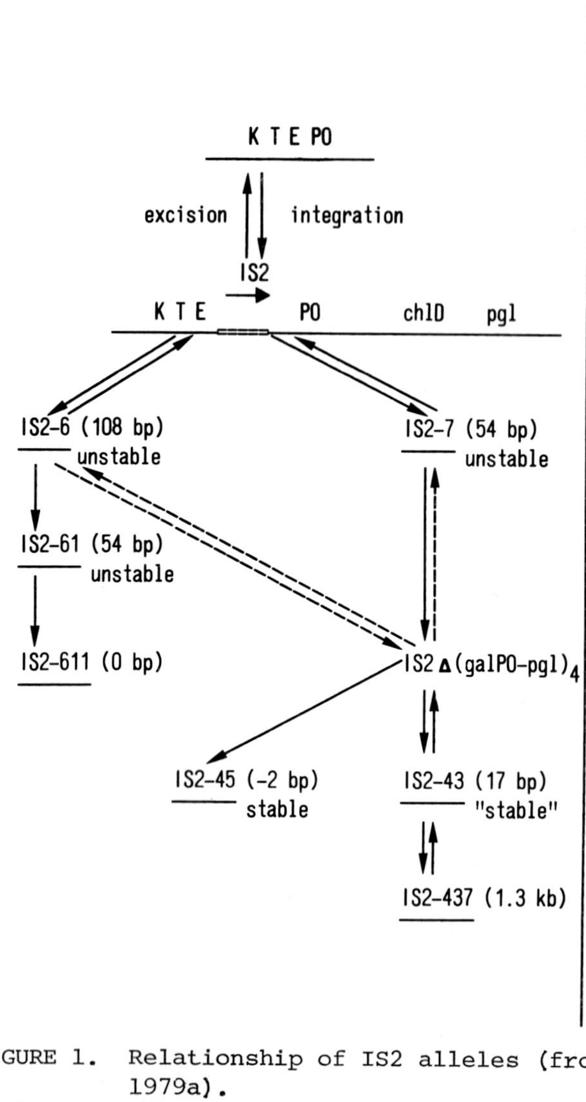

FIGURE 1. Relationship of IS2 alleles (from Sommer et al., 1979a).

cloning of the gal operon on a plasmid has allowed the DNA structure of the constitutive revertants to be studied (Ghosal and Saedler, 1977). All of the constitutive revertants studied so far have retained IS2 and have various DNA rearrangements within IS2. The revertants can be grouped into three classes with respect to the stability of the Gal^+ phenotype:

Very unstable - the "mini-insertions". These revertants gave more than 1% Gal^- segregants in an overnight culture grown without selection. Restriction analysis showed that they had extra DNA near the galE end of IS2 (Ghosal and Saedler, 1977). Sequencing of the "mini-insertion" IS2-6 showed that the extra DNA was due to a duplication of the IS2 sequence of rather a complicated structure to give an extra 108bp (Fig. 2, Ghosal and Saedler, 1978).

A model was put forward for the formation of these duplications (Ghosal and Saedler, 1978). The region where the duplications occur is AT-rich and has several inverted repeats. If melting out of the newly synthesized strand occurred during replication it might fold back on itself at an inverted repeat (Fig. 3b, c). The polymerase might then resume synthesis using the newly synthesized strand as a template, thus producing a duplication in invers orientation (Fig. 3d). In order to complete replication, it is necessary for the polymerase to switch back to the original template and the heteroduplex section of DNA will segregate after the next round of replication to give a wild type sequence and the mini-insertion (Fig. 3 f, g). IS2-6 needs two such cycles for its formation. This model predicted the sequence of IS2-7, which was confirmed by sequencing (Ghosal et al., 1979b).

IS2-6 and IS2-7 segregate Gal^- clones at a high frequency. This process is recA-independent (Ghosal and Saedler, 1979). Restriction analysis showed the loss of the extra DNA to restore the original size of IS2. IS2-6 is extremely unstable (4% Gal^- in an overnight culture). A derivative, IS2-61, was isolated that was still Gal^+, but was more stable (1% Gal^-) (Ghosal and Saedler, 1979). IS2-61 has 54bp more DNA than IS2 and the Gal^- segregants from it are the same size as IS2 (IS2-611). Fig. 2 shows that IS2-61 seems to arise from IS2-6 by recombination between 9bp direct repeats and IS2-61 gives rise to IS2-611 by recombination at direct repeats also. IS2-611 is of slightly different sequence than IS2 (Fig. 2).

Slightly unstable alleles - the "super minis". Other Gal^+ constitutive revertants were fairly stable yielding only about 0.01% Gal^- cells in an overnight culture. Sequencing of two such revertants (IS2-43 and IS2-44) showed that they were

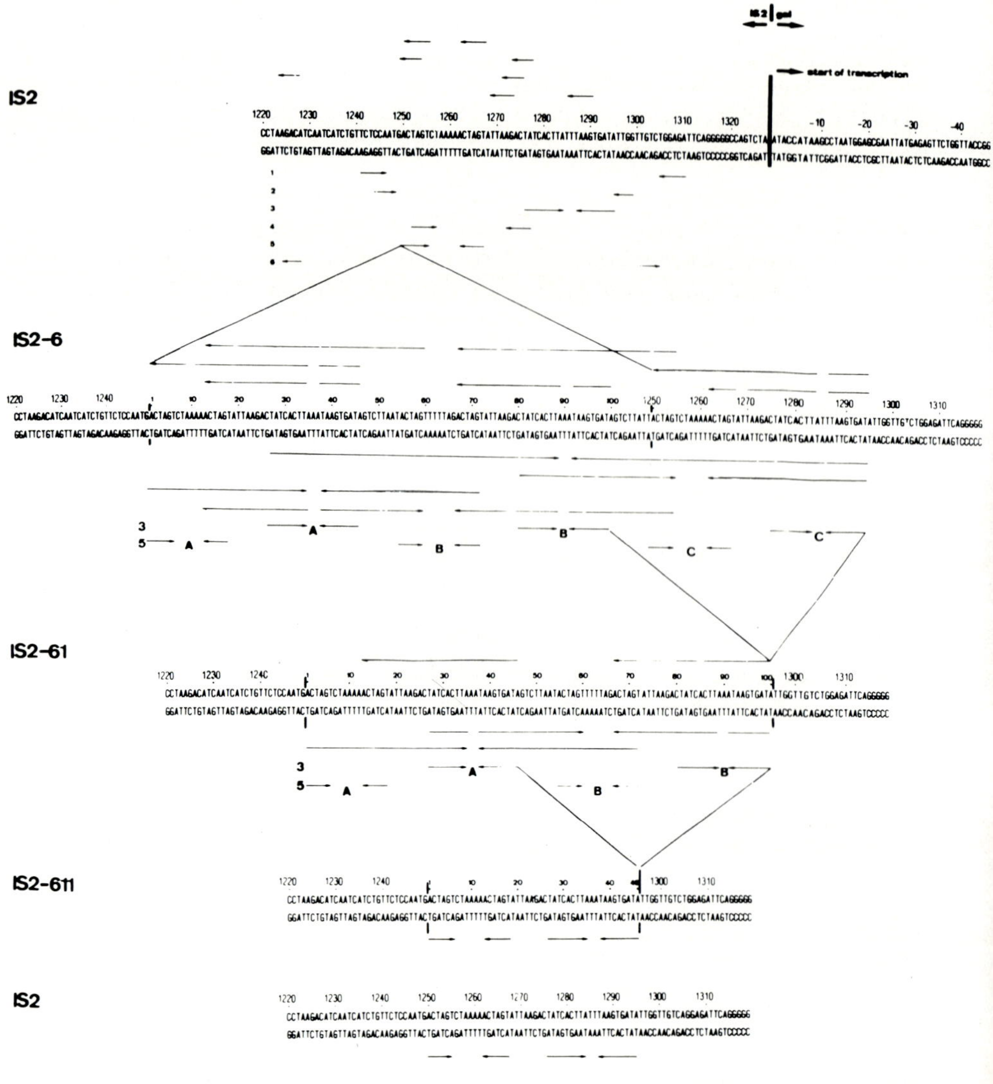

FIGURE 2. The arrows above the sequences indicate direct sequence repetitions, while the arrows below the sequences show the inverted repeats. For IS2 the small inverted repeats are numbered top to bottom, for the other alleles only the small inverted repeat pairs 3 and 5 are drawn separately and if recurrent they are marked by A, B, C respectively.

identical and had a 17bp direct duplication (Fig. 4) near the galE end of IS2 (Sommer et al., 1979). A Gal^- segregant from IS2-43 was sequenced (Sommer et al., 1979) and had the same sequence as IS2; this "precise excision" can be viewed formally as a recombination event between the 17bp tandemly repeated sequences (or some shorter subsequence of the 17bp).

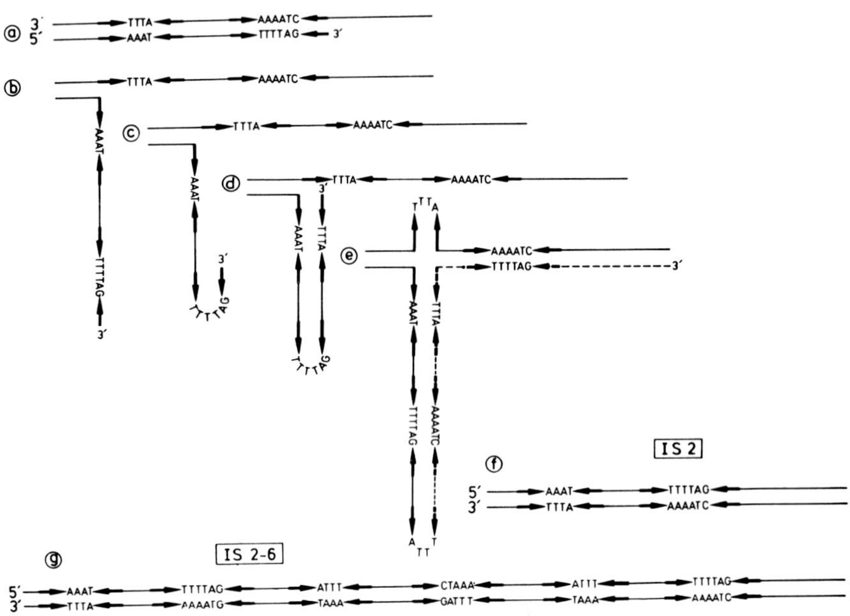

FIGURE 3. Model for the formation of mini-insertions.

Usually there were about equal numbers of "mini-insertions" and the more stable "super-mini" revertants. However, a spontaneous chromosomal mutation was found that increased the frequency of the 17bp "super-mini" revertants about one hundred-fold, whilst not increasing the frequency of "mini-insertions" (Sommer et al., 1979a). This suggests, as does the structure of the duplications, that the mini-insertions IS2-6 and IS2-7 and the 17bp duplications IS2-43 and IS2-44 arise by different mechanisms. It is possible that the 17bp duplications are caused by slippage of the DNA polymerase; such slippage has been observed in vitro with DNA polymerase I on synthetic AT templates (Kornberg et al., 1964).

A stable allele - IS2-45. A stable Gal$^+$ allele (IS2-45) was isolated; to date, no Gal$^-$ segregants have been observed (frequency less than 10^{-5}). Sequencing showed that IS2-45 had a 2bp deletion (Fig. 4).

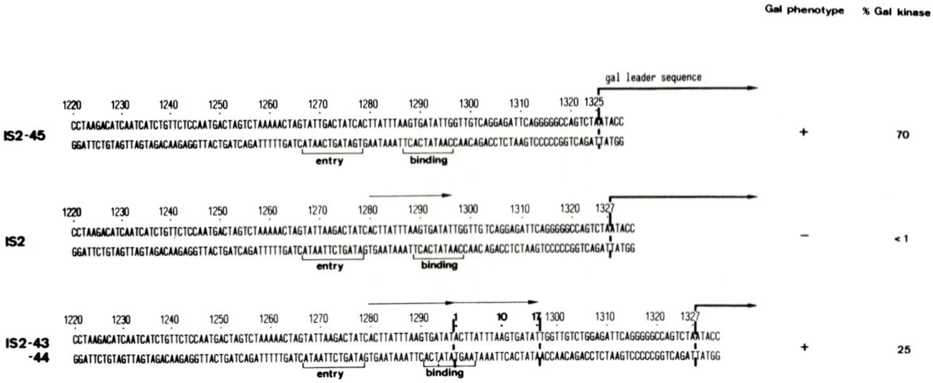

FIGURE 4. Alleles IS2-43, IS2-44 and IS2-45 showing "entry" and "binding" sites for RNA polymerase suggested by computer analysis.

This type of event seems to be rare compared to the duplication events; less than 5% of the constitutive Gal$^+$ revertants are stable.

The different constitutive Gal$^+$ revertants show a range of expression of the gal-operon enzymes. In the cases of the mini-insertions the promoter sequences are unknown; computer analysis of the sequences of IS2-6 (50% wild type galactokinase activity) and IS2-7 (25% activity) failed to reveal any close homology to the sequences of known promoters (Ghosal et al., 1979a). However, the sequences of the "super-mini" IS2-43 (25% wild type activity) and the 2bp deletion, IS2-45 (70% activity) closely resemble those of known promoters (see Fig. 4).

INSERTIONS OF ONE INSERTION SEQUENCE INTO ANOTHER

Insertion of IS3 into IS2. The major class of Gal$^-$ segregants from the "super-mini" allele IS2-43 was due to loss of the 17bp duplication. However, 1/11 Gal$^-$ segregants examined carried an insertion of IS3 into the 17bp duplication (Sommer et al., 1979b), thus preventing expression of the gal operon. This allele (IS2-437 could revert to Gal$^+$ by precise excision of IS3 at a frequency of about 10^{-7} (Sommer et al., 1979a). It seemed likely from the sequence that IS3 insertion had disrupted the new promoter present in IS2-43.

Insertions of IS2 into IS1.IS1 insertion into the galOP region leads to loss of gal operon expression (Saedler et al., 1972). Subsequent deletion of the gal promoter can prevent the occurrence of Gal$^+$ revertants by precise excision of IS1. Two such strains were constructed that carried IS1 in either orientation and Gal$^+$ revertants were selected. Among the Gal$^+$ revertants were several independent clones that carried a 1400bp insertion into IS1. Restriction analysis suggested to us that the insertion was IS2 in orientation II, which would provide a promoter for the expression of the gal operon. The identification of IS2 was confirmed by "southern transfer" experiments. For both orientations of IS1, the IS2 insertion was into the same end of IS1 (the end nearer the unique Pst1 cut) and was very close to the end of IS1. This suggested that this end of IS1 is a "hotspot" for IS2 integration. Thus, in one case IS2 was inserted very near the galE gene, whereas in the other case most of IS1 was between IS2 and the gal operon genes. In the latter case, operon expression was relatively weak (4% of wild type, compared to 25% in the former case) - probably due to transcriptional termination within IS1.

DISCUSSION

Genetic evidence (above and Saedler et al., 1974, Walz et al., 1979, Pilacinski et al., 1977) indicates that IS2 must carry a promoter in orientation II. However, computer comparison of the IS2 sequence with known promoter sequences (Ghosal et al., 1979a) did not reveal any good candidates for a promoter in either orientation. This might mean that the copy of IS2 that was sequenced (which came from an orientation I insertion) differs from IS2 insertions in orientation II, for there are about five copies of IS2 in the E.coli chromosome (Saedler and Heiß, 1973). However, it may just reflect the present lack of knowledge about promoter sequences, because, also in the case of the mini-insertions, the sequences corresponding to the new promoters could not be identified by computer analysis. The study of new promoters found in such systems may help to clarify the essential properties of promoters; the "natural" promoters usually studied are subject to complex control systems and, thus, it is not clear which of their properties reflect control requirements rather than the requirement to initiate transcription.

The small duplications in IS2 that we observed probably reflect a more general process in bacteria; in most systems such rearrangements might not lead to a promoter structure or might turn on a gene that was not easily assayed and would thus not be detected. We suggested models for the formation of the duplications based on replication errors. Further isolation and characterisation of mutations affecting these processes should clarify this question.

The instability of the constitutive promoters appears to be due to a mechanism that allows recombination between short directly repeated sequences; this process is not a function of the normal host homologous recombination process because it is *recA*-independent. Similar processes seem to be responsible for some spontaneous deletion formation in the *lac* region (Farabaugh et al., 1978). It is possible that such events occur between any directly repeated sequences of sufficient length; however, an intriguing possibility is that specific sequences are involved. In the case of all the IS2 unstable promoters, the nine base pair sequence TAAGTGATA (which is at about position 1290, Fig. 2) is part (or all in the case of IS2-6) of the directly repeated sequence at which excision takes place. This sequence is also the inverted repeat involved in the generation of the mini-insertions (Fig. 3) and also provides the integration site for IS3 in IS2-437 (Sommer *et al.*, 1979b). The instability of the alleles decrease as the repeats become closer together, being 4% for IS2-6 (108bp apart), 1% for IS2-7 and IS2-61 (54bp apart), and 0.01% for IS2-43 (17bp apart). These figures show the relative stability of the alleles, but it is difficult to compare the frequencies of reversion to Gal^+ with these, because the genes are carried on a multicopy plasmid. It would be expected that a Gal^+ allele would be detected relatively rapidly after formation whereas a Gal^- allele would only be detected after it had segregated from all its Gal^+ sister plasmids. It might be possible to obtain a more absolute measure of the frequencies using equations such as those found in Cullum and Broda (1979).

The work described in this paper shows that IS2 can have a variety of effects on the expression of a bacterial operon.

It can switch off an operon or switch it on with promoters of a variety of strengths and stabilities. Interaction with another IS sequence can lead to still more subtle effects. This ability to modulate gene activity and the ability to promote DNA rearrangements in neighbouring sequences (Peterson et al., 1979) is probably vital in the evolution of bacteria.

ACKNOWLEDGEMENTS

This work was supported by the Deutsche Forschungsgemeinschaft. JC was supported by a Royal Society European Programme fellowship.

REFERENCES

Besemer, J. and Herpers (1977). Molec.gen.Genet. 151, 295-304.

Cullum, J. and Broda P. (1979). Genet.Res., Camb. 33, 61-79.

Farabaugh, P.J., Schmeissner U., Hofer M. and Miller J.H. (1978). J.Mol.Biol. 126, 847-863.

Ghosal, D. , Gross, J. and Saedler, H. (1979). Cold Spring Harbor Symp.Quant.Biol. 43, 1193-1196.

Ghosal, D. and Saedler, H. (1977). Molec.gen.Genet. 158, 123-128.

Ghosal, D. and Saedler, H. (1978). Nature 275, 611-617.

Ghosal, D. and Saedler, H. (1979). Molec.gen.Genet. 176, 233-238.

Ghosal, D., Sommer, H. and Saedler, H. (1979). Nucleic Acids Res. 6, 1111-1122.

Hill, C.W. and Echols, H. (1966). J. Mol.Biol. 19, 38-51.

Hu, S., Ohtsubo, E., Davidson, N. and Saedler, H. (1975). J.Bacteriol. 122, 764-775.

Kornberg, A., Bertsch, L.L., Jackson, J.F. and Khorana, H.G. (1964). Proc.Natl.Acad.Sci., USA 51, 315-323.

Peterson, P.A., Ghosal, D., Sommer, H. and Saedler, H. (1979). Molec.gen.Genet. 173, 15-21.

Pilacinski, W., Mosharrafa, E., Edmundson, R., Zissler, J., Fiandt, M. and Szybalski, W. (1977). Gene 2, 61-74.

Saedler, H., Besemer, J., Kemper, B., Rosenwirth, B. and Starlinger, P. (1972). Molec.gen.Genet., 115, 258-265.

Saedler, H. and Heiß, B. (1973). Molec.gen.Genet. 122, 267-277.

Saedler, H., Reif, H.J., Hu, S. and Davidson, N. (1974). Molec.gen.Genet. 132, 265-289.

Sommer, H., Cullum, J. and Saedler, H. (1979a). Molec.gen.Genet. 175, 53-56.

Sommer, H., Cullum J. and Saedler, H. (1979b). Molec.gen.Genet. (in press).

Starlinger, P. and Saedler, H. (1976). Current Topics in Microbiol.Immunol. 75, 111-152.
Walz, A., Ratzkin, B. and Carbon, J. (1978). Proc.Natl.Acad. Sci.,USA 75, 6172-6176.

DISCUSSION

W. SZYBALSKI: As we both know, there are two general possibilities. Either you have IS2 in orientation "I" (and this has a polar effect) or in orientation II (and that in many cases acts as a new promoter. However, the IS2-6 and IS2-7 insertosomes (with rearrangements described by you) are both in orientation "I". Is that correct?

H. SAEDLER: Yes.

W. SZYBALSKI: Therefore your minor modifications which result in a new promoter did not result in the orientation "II". I wonder, however, why all the normally isolated promoter-generating IS2's are in the orientation "II"? Moreover, there must be some sequence modifications in such IS2-II, because we have seen many IS2 in orientation "II" which did not have promoter activity in lambda. (Only when there is a selective pressure for promoter activity (like for int-c mutations or nut L reversions) we seem to get IS2-II with promoter activity. Do you know sequence difference between promoter$^+$ and promoer$^-$ IS2-II?

H. SAEDLER: No. There has not been a complete sequence reported for IS2 in orientation II - at least to my knowledge. A portion of the sequence of IS2 in orientation II has been done by Waltz and no sequence deviation from an IS2 in orientation I is seen. By computer scanning of IS2 for promoter structures there is not an obvious promoter in either orientation of that particular IS2 sequence. So it is not known whether there is sequence divergence among the five copies of IS2 which naturally occur in the *E. coli* chromosome. It might very well be that the sequences are identical and turn-on signals exist on IS2 not resembling known promoters. I think we have to wait for an answer until complete sequences are available of IS2's in orientation II. But, I should stress that, as you pointed out already, the sequence rearrangements described here occurred in IS2 in orientation I and did not affect that orientation of the elements.

F. HAUROWITZ: In one of your slides you showed a cruciform arrangement of RNA or DNA. Such an arrangement had been suggested earlier by Gierer. In a paper published a few years later in the PNAS by Csech et al., the presence of cruciforms was denied. Did you indeed prove the presence of cruciforms of DNA or RNA?

H. SAEDLER: There is no direct evidence yet, and these are structures which you can form on paper assuming hydrogen bonding between complementary bases. I think that in a few of these cases where only a few hydrogen bonds are formed, these structures certainly would fall apart if not stabilized by proteins. But since we do not know anything about the proteins involved in these things I think it is premature to make a guess whether cruciforms exist or not.

I. HERSKOWITZ: G. Streisinger and colleagues have done experiments on the mechanism of acridine-induced frameshift mutagenesis in phage T4 which may be of interest. In this model deletion or addition mutations occur when bases on one DNA strand flip out during DNA repair or replication. Acridines are proposed to act not by intercalation into DNA but by forming a sandwich with the flipped out bases in the middle. Acridines thus stabilize the flipped out bases. If these bases are from the template strand, a deletion will be produced. Since you have invoked DNA strand slippage as a way to produce deletions and additions in IS2 and its derivatives, it would be interesting to know whether acridines affect the frequency of these events.

H. SAEDLER: Well, we have not tested it yet.

P. STARLINGER: I just wanted to remind you that the very first detection of these IS elements was by M. Malamy and he at that time found that they reverted under the influence of acridine and this has never really been cleared up. Usually the insertion of IS elements does not occur under the influence of acridine, but may be some kind of rearrangement in his situation may have been responsible. It might be worthwhile to look into this again.

P. BERG: I am curious whether the formation of deletions by recombination across these small duplications is a common phenomenon. If it is, why doesn't one see analogous recombinational deletions between restriction sites that have homology one to another? Three sites separated by 100 nucleotides might provide a target for recombination between them to excise segments of DNA just as these types of duplications, deletions, occurred between these direct repeats.

H. SAEDLER: Yes. I think it is not just a matter of having sequence duplication but the sequence might also be important and we have here the situation where there is a peculiar sequence of nine base pairs which is very effective in this process, leading up to 50% of segregation. Normal recombination between such small repetitive sequences should occur at a much lower frequency, and as a matter of fact Farabaugh described in lac i small sequence duplications being separated some 100 bases. The frequency of recombination between these is at least 2 to 3 orders lower than what was described here. Therefore, in the case of the insertions a special enzyme might be involved in recognizing that particular sequence.

M. BESSMAN: Did I understand you correctly in saying that Pol A does not seem to be involved in this stuttering type synthesis, and that you have a mutant which does seem to make a lot more of these repetitive sequences? If so, do you know which mutant protein is involved in this?

H. SAEDLER: The Pol A is a bit more complicated. We tested two alleles as a matter of fact. One temperature sensitive allele that turned out to effect the formation of the very unstable mini-insertions, even at the permissive temperature, while another pol A allele did not effect the formation at all. So we suspected that this might in effect be due to the bacterial background in which that ts mutation is - we did not succeed in crossing out that particular pol A ts allele into a related background. Now the other question concerning that mutation which apparently showed an increased incidence of stuttering - that is generally the 17 base pair duplication. That certainly does not map at the position where the pol A gene is located, but we do not know currently where it maps.

A. BUKHARI: Have you tested the formation of these rearrangements in IS2 which is carried on multicopy plasmid which replicates at a much higher frequency. What would be the frequency of these types of rearrangements?

H. SAEDLER: Are you asking whether that is occurring on a multicopy plasmid?

A. BUKHARI: Yes, that is one part of the question. If the mechanism involves replication, these rearrangements would also occur at a very high frequency. Is this correct?

H. SAEDLER: That is right. As a matter of fact they are on a pSC101 derivative which also occurs in multiple copies in the cell.

A. BUKHARI: If you take the type of strain in which you have a rearranged sequence such that it has a promoter in it, does that strain give rise to spontaneous promoter mutations at a high frequency at random? Let us say if you demand a promoter type of mutation in some other gene, then do you get that at a higher frequency?

H. SAEDLER: We have not tested that.

G. FINK: The large duplications were shown by John Roth to be resolved by the Rec A gene. How do your small duplications behave in the Rec+ and Rec-?

H. SAEDLER: The frequency is identical in both. They are not Rec A dependent.

G. FINK: The formation?

H. SAEDLER: Both formation as well as segregation.

G. FINK: Isn't that strange - because you do not think of these as IS2 mediated.

H. SAEDLER: They are sequence mediated and the important thing is not the IS2 part, but just the fact that there is a region rich in AT and having these symmetrical elements in it. That is the important part.

G. FINK: But for the resolution of the duplications one might have imagined that all duplications simply because of the fact that they are duplicated DNA segments would be sensitive to the Rec A gene product at least for large duplications this appeared to be the case in Roth's studies. Could it be that these are very small and therefore not recognized?

H. SAEDLER: That is actually what one thinks. Accurate excission of IS elements flanked by three to 12 base pair dupliates at the ends as well as those small ones up to 30 base pairs internal sequence duplications within IS2 are all independent of Rec A. The homology provided might not be large enough for synapsis to be informed.

G. FINK: So what this brings up then it seems to me is a question which has troubled me a great deal in all the studies of so-called homologous and non-homologous recombination is that critical question, "How large a region is a region of homology?"

H. SAEDLER: In order to be recognized by the normal recombination system?

G. FINK: Or by another recombination system.

H. SAEDLER: Well, I should stress that point here that I have shown you some data of recombinational events which either use small sequence homologies or else no homology whatsoever, like the IS2 driven deletion. Both processes are Rec A dependent. So there are a couple of ways, enzymatic pathways, to rearrange or to delete DNA, due to recombinational events.

A GENETIC ANALYSIS OF THE TRANSPOSON TN3: EVIDENCE FOR COINTEGRATES AS INTERMEDIATES IN TRANSPOSITION

Ronald E. Gill
Stanley Falkow

Department of Microbiology and Immunology
University of Washington
Seattle, Washington

Hisako Ohtsubo
Eiichi Ohtsubo

Department of Microbiology
SUNY, Stonybrook, N.Y.

Magdalene So
Fred Heffron[1]

Department of Biochemistry and Biophysics
University of California
San Francisco, CA

I. ABSTRACT

A large number of deletion and insertion mutations were made by *in vitro* mutagenesis in the transposon Tn3. From a genetic analysis of these mutations as well as by direct DNA sequencing and identification of the encoded polypeptides we have determined the overall organization of this transposon

[1]Supported by NSF grant PCM-7923858. Present address: Cold Spring Harbor Laboratory, Cold Spring Harbor, N.Y.

ISBN 0-12-633360-2

(see Figure 1). This transposon encodes at least three polypeptides, 1) the β-lactamase (bla), (2) a transposase; a high molecular weight protein required for transposition and 3) an autogenously regulated repressor which regulates expression of the transposase. Both ends of the transposon are required in *cis* for transposition. The most surprising result of these studies was the identification of another site which is encoded within this transposon. Deletions within one narrow region at the amino terminus of the transposase (between mutations 11 and 5 in Figure 1) invariably result in the formation of cointegrates (see Figures 3 and 4). Experiments described here and others reported elsewhere have led us to propose that cointegrates may be obligatory intermediates in transposition and that the function deleted between mutations 11 and 5 is absolutely required for their resolution.

II. INTRODUCTION

A bacterial transposon is a DNA sequence, often encoding antibiotic resistance, which is capable of transposition as a discrete DNA sequence from replicon to replicon independent of the host recombination system. Bacterial transposons have been observed carrying a wide variety of phenotypically identifiable genes, such as those for drug resistance. IS sequences, in contrast to transposons, are smaller and do not code for any readily observable phenotype. Some transposons contain IS sequences: Tn9 and Tn1681 are flanked by direct and inverted repeats of IS1, respectively (1,2). The structure of these two transposons is chracteristic of many other transposons and may reflect the way in which transposons evolved. One could imagine that two IS sequences at sometime became inserted at either side of a useful gene. The flanking IS sequences then picked up the intervening DNA and a new transposon was formed. Bacterial transposons provide a means for the rapid evolution of bacteria by the addition of "modules" having some transient selective advantage. Bacterial transposons account for the recent outbreaks of antibiotic resistant *Hemophilus* and penicillin resistant *Neisseria gonorrhea*. (3,4).

Transposons range in size from 2058 bp to over 15000 bp. All IS sequences and transposons in which the actual termini have been sequenced are flanked by an inverted repeat. Most likely the inverted repeats serve as recognition sites for enzymes involved in transposition. Since the same 38 bp sequence is present in inverted orientation at the ends of Tn3, a single enzyme could recongize and make breaks at either end of the transposon. In the case of Tn3, we have shown that both inverted repeats are required _in cis_ for transposition and that an internal mutation in one of these inverted repeats is _cis_ dominant transposition negative (5,6,7).

Transposons take part in a number of recA independent events in bacteria besides transposition. These include precise excision (8), deletion formation (8-11), inversions (8), and nearly precise excision (12). In the case of Tn10, it has been shown that neither precise excision (where a small fragment originally present at the ends of the transposon is left behind) depend on a functional transposon (i.e., one itself able to transpose) while inversion formation is strictly dependent on an intact transposon (N. Kleckner, personal communication).

The genetic organization of Tn3 appears to be quite different from Tn10. Tn10 is terminated by 1500 bp inverted repeated sequences which encode functions necessary for its transposition while Tn3 is flanked by only 38 bp inverted repeats (13) and the functions necessary for its transposition (aside from host encoded functions) are encoded in a 3.5 kb contiguous region on one side of the transposon. The genetic and biochemical identification of these functions is the subject of this report.

III. RESULTS

Analysis of mutants of the Tn3 element has resulted in a relatively simple map of the genetic organization of the Tn3 functions and structures necessary for its transposition and those related to the regulation of transposition frequency (Figure 1). This genetic map of Tn3 has been formulated from the analysis of two sets of Tn3 mutants. Both were the result of _in vitro_ manipulation of the plasmid RSF1050, a small derivative of ColE1 into which Tn3 has been transposed. The first of the RSF1050 mutants contained various size deletions having one or both endpoints within the transposon. A second collection was constructed by random _in vitro_ insertion of an oligonucleotide containing an _Eco_RI restriction site into the Tn3 element. Both kinds of mutations were scored for their affect on transposition frequency according to a genetic assay described in Figure 2. This assay takes advantage of the fact that RSF1050 like the parent ColE1 has a strict dependence on DNA polymerase 1 for its maintenance and replication. Following _in vitro_ manipulation of the DNA, the plasmid is transformed into an F+ strain. In a mating between these F+ transformants and a polA recipient, transconjugants resistant to penicillin can arise only by transposition of Tn3 onto F and subsequent transfer of F to the polA reci-

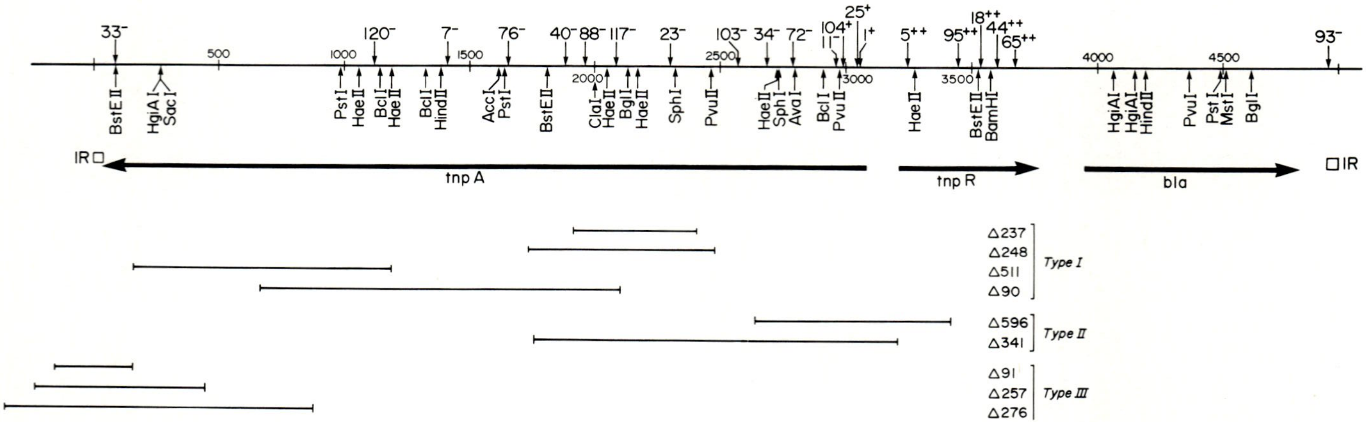

Fig. 1. Map of the transposon Tn3. This figure shows many of the mutations of Tn3 used in this study, and their phenotypes. The EcoRI insertion mutations and their phenotypes are shown above the line representing the transposon. The transposition negative deletions are shown at the bottom. Type I deletions all remove a portion of the transposase. Type II deletions remove the internal resolution site. Type III deletions remove the left-hand inverted repeat. The location and direction of transcription of genes encoded by the transoson are shown by the heavy lines. IR represents the 38 bp inverted repeats which are located at either end of the transposon. A number of useful restriction sites are shown in the figure.

pient. Mobilization of RSF1050 by F is precluded because RSF1050 cannot be maintained in the polA recipient. For a wild type transposon the frequency of Tn3 transposition is 10-3 penicillin resistant transconjugant/F+ transconjugant and at least 1000 fold lower for transposition negative mutants.

Three kinds of experiments have been carried out to identify functions encoded by this transposon as described in detail below. First genetic analysis by complementation of the various mutants which affect Tn3 transposition. Second, we have determined the complete DNA sequence of the transposon. Third, direct identification of the polypeptides encoded by Tn3 in minicells.

Table I. Test of Dominance for Derepressed Mutants

Plasmid(s)	Freq. of Transposition $\left(\frac{Ap^r Kan^r \text{ Transconjugant}}{Kan^r \text{ Transconjugant}}\right)$
RSF1050	10^{-3}
pFH5[1]	10^{-2}
pFH5 + pREG115[2]	10^{-3}

[1]Results shown here for pFH5, but pFH95, 18, 44 and 65 behave identically.

[2]pREG115 is a plsamid compatible with RSF1050 which contains a tetracycline resistant derivative of Tn3.

A. Genetic Analysis

Five mutations in a 500 bp region of Tn3 bisected by the BamHI restriction site showed an increase in transposition frequency (Figure 1). We have carried out a dominance test between these mutants and wild type Tn3 (Table I). As shown in Table I, these mutants show at least a 10-fold increase in transposition frequency relative to the wild type. When complemented by a second copy of Tn3, their transposition frequency decreases to the wild type level. These results show that these derepressed transposition mutants all lie within a recessive function which regulates transposition frequency.

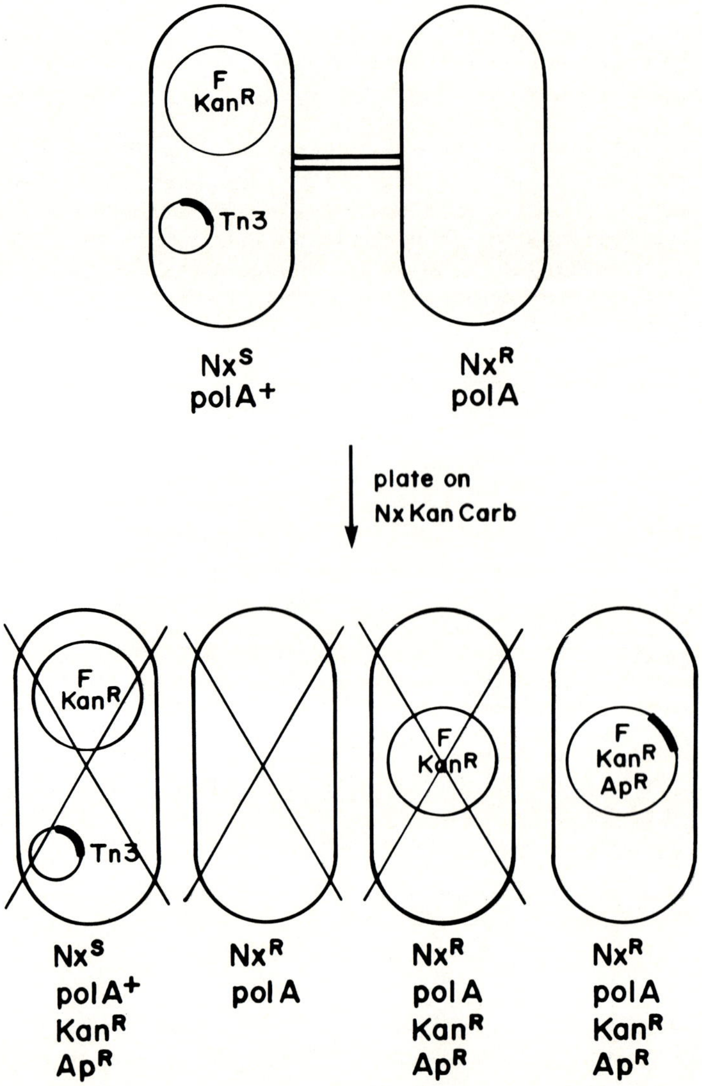

Fig. 2. Detection of transposition. This assay uses a kanomycin resistant derivative of the sex factor F as a recipient for transposition. The assay takes advantage of the strict dependence of ColE1 on polymerase I for its replication in order to prevent its mobilization by F. We use a polA mutant which is resistant to naladixic acid. The frequency of transposition is determined as the ratio of ampicillin resistant transconjugants to kanomycin resistant transconjugants. The bottom of the figure shows the four products of the mating.

To the left of this region as shown in Figure 1 is a 3 kb contiguous region in which all mutants were transposition negative. As shown in Table II they also were recessive to wild type in a dominance test suggesting that one or more transposition functions, presumably proteins, were encoded here. We have attempted to determine the number of complementation groups encoded in this region by complementing one transposition negative mutant with another. The results (see Table II) can be summarized simply; no two transposition negative mutants which mapped within this region would complement each other. This suggests that a single complementation group, perhaps only a single protein, is encoded here.

All the mutants described above were recessive to wild type in a dominance test and therefore affect a _trans_ acting function, presumably a protein. Several other mutants were _cis_ dominant. Three deletions which removed the left-hand inverted repeat and one _Eco_RI insertion mutation which was found to lie in the right hand inverted repeat were complemented with wild type Tn3. As shown in Table III, all of these mutations were _cis_-dominant transposition negative suggesting that the terminal 38 bp inverted repeat acts as an essential site in transposition.

B. Identification of the Internal Resolution Site

In order to validate the complementation experiments we have examined the nature of the transposed sequences from a variety of the Tn3 deletions. The product of transposition of the type I deletions (see Figure 1) were examined in the electron microscope by DNA heteroduplexing. As expected the sequence transposed corresponded precisely to the complemented

Table II. Complementation of Transposition Negative Mutations

Plasmid(s)	Frequency of Transposition
pFH33[1]	10^{-6}
pFH33 + JCL4::Tn3-135[2]	$5 \times 10^{-5} - 5 \times 10^{-4}$
pFH72 + JCL4::Tn3-72	10^{-6}
pFH33 + JCL4::Tn3-72[3]	10^{-6}

[1]Listed for pFH33 but other derivatives give similar results.

Table III. Test of Dominance for Mutations of the Inverted Repeat

Plasmid(s)	Transposition Frequency Ap^{-r} Transconjugants	Transposition Frequency Tet^{-r} Transconjugants
Δ91, pFH93	10^{-6}	---
pFH93 + pREG115	10^{-6}	10^{-3}

[2]This is a positive control. JCL4::Tn135 is a derivative of F containing a Tn3 derivative which confers tetracycline instead of ampicillin resistance.

[3]This complementation experiment is representative of a number of others in which we tested a number of transposition negative mutations pairwise.

deletion. When the type II deletions (Figure 1) were examined a novel structure was observed. The transconjugant plasmids contained a fusion between the donor and recipient in which a direct repeat of the transposon joins the donor to recipient plasmid (see Figs. 3,4). Instead of a true transposition event, in type II deletions, cointegrates appear to be formed by transposition (this is an unproven assumption, of course). Two additional pieces of data lead us to propose that the sequence deleted in the type II deletions could be required for recombination of direct repeats of the transposon under rec^- conditions. We have constructed direct repeats of the transposon by cloning. These direct repeats could never be recovered when transformed into a rec^- recipient. On the other hand when the same experiment was performed with a type II deletion the recombinant was perfectly stable under rec^- conditions. These results taken together suggest that cointegrates could be intermediates in transposition and that type II deletions have removed some site or function essential to resolution of the cointegrate. In fact, as shown in Table IV, a dominance test of the type II deletions suggests that these deletions affect a site and not a function. We have called this site the internal resolution site (IRS) and more detailed mapping of its location has demonstrated that it lies between mutations number 5 and 11 (Carol Morita and Fred Heffron, unpublished observations).

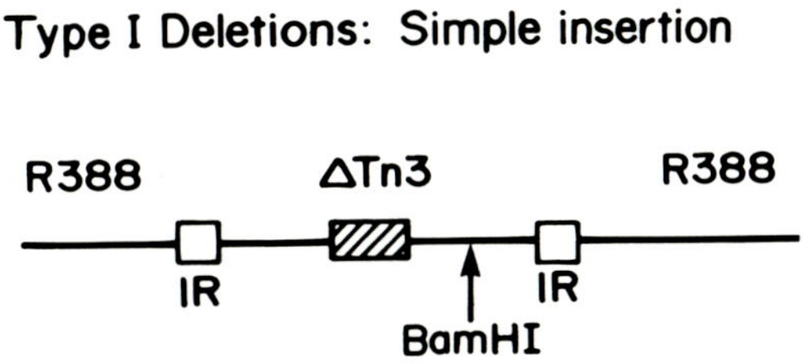

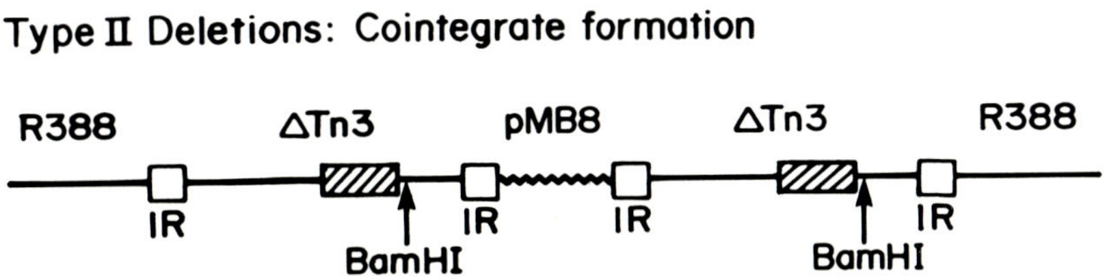

Fig. 3. Comparison of the sequences transposed by complementation of type I and type II deletions. The deletions have transposed into the transmissable plasmid R388.

Table IV. Dominance Test for Deletions of the Internal Resolution Site

Plasmid(s)	Freq. of Transposition	Cointegrates
Δ596 (All type II deletions)	10^{-6}	---
Δ596 + REG115	10^{-4}	90-100%[1]

[1]Cointegrates were scored by testing about 100 ampicillin resistant transconjugant colonies for colicin immunity.

C. DNA Sequence Analysis of the Mutants Affecting Tn3 Transposition

The complete DNA sequence of Tn3 and 20 mutations which map within it has been determined. In the last section we described five mutations all of which mapped within a 500 bp region bisected by the <u>BamHI</u> site which showed an increase

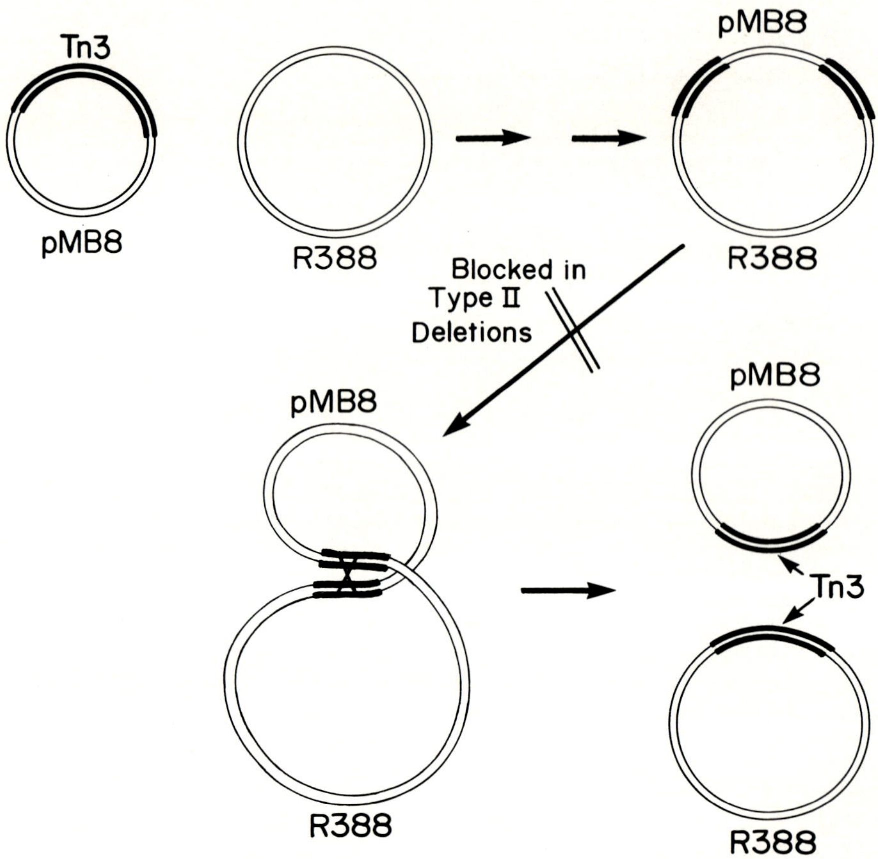

Fig. 4. Resolution of cointegrates by the internal resolution site. The first part of the figure shows the formation of cointegrates but leaves the mechanism undefined. The second part of the figure shows their resolution. We believe that type II deletions are missing a site essential for resolution and hence are blocked at this step.

in transposition frequency. This region together with these five mutations were sequenced and found to lie in the coding region for a polypeptide. The direction of transcription and initiation and termination sites we identified in the DNA sequence are shown in Figure 1. The DNA sequence would predict a 185 amino acid polypeptide.

The genetic data suggested that an adjacent three kilobase region was required for transposition but encoded only a single complementation group. This region of the transposon, together with 14 mutations which were located here, were sequenced. All these mutations were found to lie within the coding sequence for a single high molecular weight polypeptide which initiates in the center of the transposon and uses an ochre termination codon located within the terminal left hand 38 bp repeat. We refer to this high molecular weight protein as the transposase since it is absolutely required for transposition. In addition to the two proteins described above which were predicted from the DNA sequence there are several other open reading frames on either strand which are coencoded within the transposase. Whether these open reading frames encode proteins has not been determined.

D. Identification of Tn3 Encoded Peptides in Minicells

The genetics and DNA sequence suggested that there were at least two polypeptides encoded in Tn3; a high molecular protein absolutely required for transposition and a second protein which regulates transposition frequency. We have examined the proteins made in minicells on SDS slab gels from strains containing RSF1050 and many of the mutants which we have described.

The five mutations which showed an increase in transposition frequency were each examined in minicells. Minicells of RSF1050 show a polypeptide of about 20,000 daltons which corresponds to the molecular weight of the 185 amino acid polypeptide predicted from the sequence. Each of the five mutant derivatives above are missing this protein. From the DNA sequence of each of these mutations the molecular weight of the corresponding prematurely terminated polypeptide could be predicted. Each of the five mutations produced a polypeptide which corresponded closely to that predicted by their sequence. In fact, on the basis of DNA sequence we could predict that mutation 65 (see Figure 1) which lies near the carboxy terminus of this protein should bypass the normal translational stop and extend the protein from 185 to 197 amino acids. In fact, minicells containing this mutant do produce a protein which is slightly larger than 20,000 daltons. These results verify that a protein is encoded here and demonstrate that the direction of transcription agrees with that predicted from the DNA sequence. They do not explain why mutations in this polypeptide show an increase in transposition frequency. One surprising observation was that each of these prematurely terminated polypeptides was greatly overproduced compared to the wild type. This result suggests that the 20,000 dalton protein is autoregulated.

RSF1050 did not produce a polypeptide in minicells which would correspond to the high molecular weight transposase. We therefore examined the derepressed mutants for the appearance of new proteins. In mutants missing the 20,000 dalton protein a high molecular weight protein of 120,000 dalton is produced. This would suggest that the 20,000 dalton protein is in fact a repressor which regulates expression of the transposase. The reason why mutants in the repressor show an increased transposition frequency then becomes clear. Presumably, the transposase is the limiting enzyme in transposition and slight derepression of the transposase results in the increased transposition frequency observed. The direction of transcription of the transposase was verified by examining the size of truncated polypeptides produced from the sequenced EcoRI insertions in the tnpA gene. Since the transposase is not made in large enough amounts to be visualized in SDS-gels from minicells of a wild type transposon, it was necessary to construct derepressed mutants (repressor$^-$) of a number of the EcoRI insertion mutants in the transposase. The prematurely terminated polypeptides produced from these double mutants could be observed in minicells and in each case they corresponded closely to that predicted from the sequence. This result confirms the existance of the transposase and the direction of transcription predicted from the DNA sequence.

E. Genetic Demonstration of the Regulation of the Transposase by its Repressor

Mutants in the tnpR gene show an increase in transposition frequency and the appearance of the transposase in minicells. We have constructed a transcriptional fusion between the transposase and a second gene which specifies streptomycin phosphotransferase in order to directly test regulation of the transposase. Tn3 was transposed into an operon containing two genes; the proximal specifies resistance to sulphonamide and the distal to streptomycin. The insertion occurred into the proximal gene and was polar on expression of streptomycin resistance. From this original insertion a revertant was isolated which was no longer polar on streptomycin resistance. This revertant contained a deletion which removed the repressor (see Figure 5). From the DNA sequence and restriction digest information we know that the orientation of insertion is that indicated in Figure 5 and that the translational initiation site for the streptomycin gene occurs only about 300 bp downstream from the end of the transposon. The DNA sequence suggests that there is no transcriptional stop in the 31 bp between the end of the transposase and the end of the inverted repeat. It would therefore appear that we have created an artificial operon in which the transposase lies proximal and the streptomycin resistance gene is distal. The plasmid containing a deletion of the repressor is resistant to 100 g/ml of streptomycin. We can then ask "what is the effect on streptomycin resistance of supplying the repressor _in trans_. Supplying the repressor _in trans_ decreases the level of streptomycin resistance from 100 gm/ml to 10 gm/ml. However, supplying a mutant repressor _in trans_ does not change the level of streptomycin resistance. This result confirms the observation that the repressor regulates expression of the transposase. In addition, it suggests a new use for this transposon as a moveable regulated promoter. Exactly the same scenario, mentioned above, could be employed to put other genes under the regulation of the Tn3 repressor. By starting with a repressorless derivative of Tn3, it should be possible to select directly for transcriptional fusions in any gene in which a promoter minus mutation is available.

IV. DISCUSSION

Genetic, sequence and biochemical data all indicate the presence of three genes encoded by Tn3 as shown in Figure 1. These three genes use all but 350 bp of its total coding capacity and even these 350 bp encode sites and regulatory signals such as the two inverted repeats and the promoters and operators controlling expression of these genes. The _bla_ gene situated at the right hand end of the transposon and transcribed in the direction indicated was first sequenced by Sutcliffe (14). The remaining two genes form a unit of divergent transcription which begins near the center of the transposon and is transcribed in opposite directions.

A computer search of the 250 bp in which the IRS must lie revealed a sequence located at the amino terminus of the transposase which shows a striking homology with the inverted repeat (Figure 6). Assuming that the transposase can recognize the two inverted repeats it is logical to assume that it can also recognize this sequence as well. It would appear, under this hypothesis, that the transposase must have a

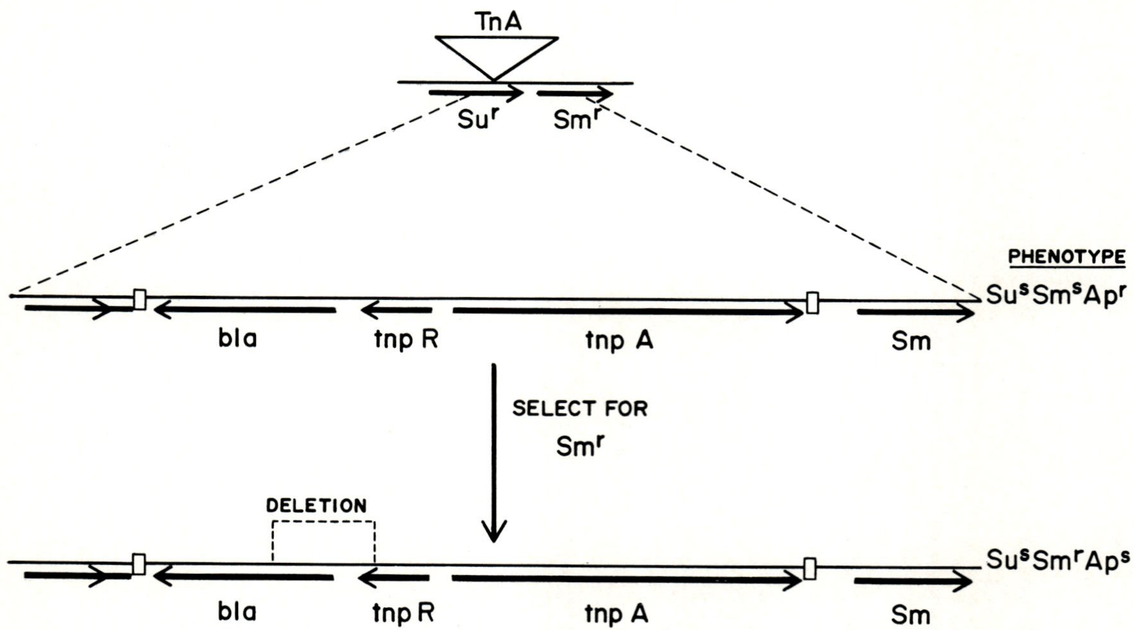

Fig. 5. Creation of transcriptional fusion between the transposase and a gene for streptomycin resistance. We isolated an insertion of Tn3 (here labled by its earlier name TnA) into a transcriptional unit containing the genes for sulfonamide and streptomycin resistance in which streptomycin is distal. The insertion had occurred into the gene for sulfonamide resistance. We then selected for revertants which releived the polar effect. One such revertant was mapped and found it contained a deletion of both the repressor and a portion of the *bla* gene

POSSIBLE INTERNAL RESOLUTION SITE

```
          3021                    A 3043
           GCTCAGTGGTCAAAAAATCACG
GGGGTCTGACGCTCAGTGGACGAAAACTCACGTTAG
1                   A               38
```

INVERTED REPEAT

Fig. 6. Comparison between a seqeunce located in the amino terminus of the transposase (nucleotides 3021-3043) and the inverted repeat (nucleotides 1-38). This sequence shows more homology with the inverted repeat than any other in the transposon.

number of different activities such as making specific breaks in the DNA at either end of the transposon and non-specific breaks in recipient DNA.

The work of Landy *et al*. (this volume) may suggest a strong parallel between activities which have been demonstrated for the lambda *int* protein and these hypothetical activities of the transposase. In the absence of the bacterial attachment site insertion of lambda appears to be very similar to insertion of transposons. In both cases a highly specific site is recognized in the donor DNA while insertion can take place at a number of sites in the recipient DNA (although at sequences which show some homology to the attachment site). Similarly, highly specific breaks must be made at each end of the transposon, yet the transposon can insert at many places in recipient DNA. The work of Cohen *et al*. (this volume) and other similar work by Kleckner (12) and Calos and Miller (personal communication) would suggest that insertion of a transposon, just as insertion of lambda, takes place at sites which contain sequences homologous to sequences at the ends of the transposon. In the case of lambda, donor and recipient DNA are distinguished by the presence of multiple *int* binding sites in the donor DNA. A 250 bp sequence surrounding the lambda att region contains at least four sites at which lambda int will bind. Landy has hypothesized a complex of proteins including host proteins on the donor DNA which takes part in the attack on the recipient DNA. An exactly analogous situation could be imagined for transposons. The presence of host proteins and multiple transposase binding sites may direct site specific recombination at the internal resolution site while other host proteins, together with the transposase must also be required for transposition.

REFERENCES

1. HacHattie, L.A. and Jackowski, J.B., in "DNA Insertion Elements, Plasmids, and Episomes" (A.I. Bukari, J.S. Shapiro and S. Adhya, eds.). Cold Spring Harbor Laboratory, New York (1977).
2. So, M., Heffron, F. and McCarthy, B.J., Nature 277, 453 (1979).
3. de Graafe, H., Elwell, L. and Falkow, S., J. Bacteriol. 125, 439 (1976).
4. Ellwell, Ll, Roberts, M., Meyer, L.W. and Falkow, S., Antimicrob. Agents and Chemother. 11, 528 (1977).
5. Heffron, F., Bedinger, P., Champoux, J.J. and Falkow, S., Proc. Natl. Acad. Sci USA 74, 702 (1977).
6. Gill, R.F., Heffron, F. and Falkow, S., Nature 282, 797 (1979).
7. Heffron, F., McCarthy, B.J., Ohtsubo, H. and Ohtsubo, E., Cell 18, 1153 (1979).
8. Botstein, D. and Kleckner, N., in "DNA Insertion Elements, Plasmids, and Episomes" (A.I. Bukari, J.S. Shapiro and S. Adhya, eds.), p. 185. Cold Spring Harbor Laboratory, New York (1977).
9. So, M., Heffron, F. and Falkow, S., J. Bacteriol. 127, 1520 (1978).
10. Weinstock, G.M. and Botstein, D., Cold Spring Harbor Symp. Quant. Biol. 117, (1978).
11. Nisen, P.D., Kopecko, D.J., Chou, J. and Cohen, S.N., J. Mol. Biol. 117, 975 (1978).
12. Kleckner, N., Cell 16, 711 (1979).
13. Ohtsubo, H., Ohmori, H. and Ohtsubo, E., Cold Spring Harbor Symp. Quant. Biol. 43, 1269 (1979).
14. Sutcliffe, J.G., Cold Spring Harbor Symp. Quant. Biol. 43, 77 (1979).

DISCUSSION

D. DRESSLER: Does the resolution of the co-integrate structure involve the E. coli gene that you referred to here, or is this a transposon coded function?

F. HEFFRON: That's a good question. I don't know the answer to that but what we would like to believe is that the transposase actually has two separate activities and that it can participate in transposition through the inverted repeat which is coded in its carboxy terminal end and also site specific recombination through the internal resolution site encoded in its amino terminal end. Whether it also requires host functions, I don't know.

D. DRESSLER: Is it known what goes wrong with the transposition process in the hip mutant of E. Coli?

F. HEFFRON: I'm, sorry, I missed the question.

D. DRESSLER: I believe that the transposition fails in the mutant of E. coli called hip, the question is where does it fail? Does it pile up co-integrates?

F. HEFFRON: I don't know the answer to that.

W. SZYBALSKI: You have two divergent operons between transposase and its repressor function, are there two promoters and how they are organized? Is there one operator which blocks both functions because you were saying that your repressor function affects both itself and the other one, because you have a sequence there, I guess?

F. HEFFRON: More studies will have to be done. We have not been able to identify two operators. The distance between the probable start of translation for both proteins is only around 250 base pairs.

A.O. SUMMERS: On stability. Did I hear you say at the end of your presentation that the induction of the transposes was lethal?

F. HEFFRON: Yes, it was constructed in such a way that it could be repressed. There is excess repressor supplied in the cell.

A.O. SUMMERS: So the stability of a cell that is producing excess transposes or the stability of the plasmid or the survival of the cell is very low. Is that right?

F. HEFFRON: On induction, it's lethal.

A.O. SUMMERS: What about the cells that are derepressed, the ones that you said were derepressed due to mutations in the repressor?

F. HEFFRON: In repressor minus cells, I can't say for sure, but I would guess that they are less viable than the wild type.

J. SCHELL: You were mentioning the lethality of the induction of the fusion product. Have you tested whether you can get survivals of this lethality and what kind of mutations they would be?

F. HEFFRON: That's a good question. You can get survivors, certainly, and you would assume therefore that you could get transposition negative mutants that way and that is being worked on, but I don't know the answer.

REGULATION OF Tn3 TRANSPOSITION AND SPECIFICITY OF ITS INSERTION SITES

Stanley N. Cohen, Malcolm J. Casadaban,
Joany Chou, Peggy G. Lemaux, Christine A. Miller
and C.-P. David Tu

Departments of Genetics and Medicine
Stanford University School of Medicine
Stanford, California 94305

INTRODUCTION

Studies carried out in our laboratory during the past several years have been directed at elucidating both the process of transposition and the specificity with which it occurs. Evidence obtained from these studies (1-5), as well as from investigations from other laboratories (6-11), has indicated that the **process** of transpostion of Tn3 is controlled by genes located within the element itself. Our studies of **specificity** of Tn3 insertion have shown that the nucleotide sequence of the recipient genome plays a role in determining which DNA sites receive Tn3 insertion most frequently.

USE OF LAC GENE FUSIONS TO IDENTIFY AND STUDY SIGNALS THAT REGULATE EXPRESSION OF TN3 GENES

Our approach to the study of control of transposition has been an outgrowth of earlier work involving *lac* gene fusions carried out *in vivo* (12-14) and *in vitro* (15). Plasmids containing both Tn3 and *lac* were constructed, and deletions that fuse *lac* genes to Tn3 sequences were introduced, bringing expression of *lac* under control of transcriptional and translational signals within the transposable element (5). In such constructs (Fig. 1), β-galactosidase synthesis is a function of Tn3 regulatory mechanisms and can be used to detect and analyze mutations that affect the regulation of Tn3 genes (16, 17).

In earlier investigations, fusion plasmids containing certain deletions were found to show an elevated frequency of transposition, suggesting that the gene interrupted by fusion to *lac* was involved in the regulation of transposition (5). Moreover, β-galactosidase production in these plasmids was reduced by the presence in *trans* of the wild type Tn3 element,

ISBN 0-12-633360-2

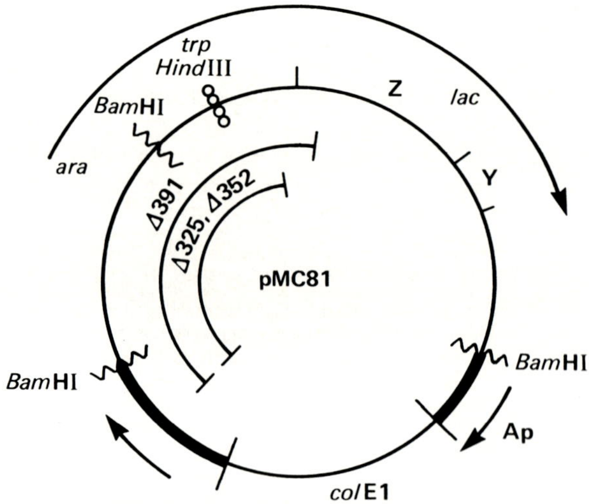

Figure 1. Map of prototype deletion plasmids that fuse lac structural genes to Tn3 promoters. pMC81 is ColE1::Tn3 into which an ara-trp-lac fusion has been inserted at the BamHI site of Tn3. Arrows indicate the direction of transcription. The thickened line segments indicate Tn3 sequences. Such fusions place β -galactosidase synthesis under the control of regulatory signals within Tn3.

leading us to suggest that Tn3 encodes a repressor of translocation, and that this repressor (whose gene had been fused to lacZ) is autoregulated.

Using one such lac fusion plasmid (pMC823), we isolated Tn3 mutants that fail to inhibit in trans phenotypic expression of β-galactosidase encoded by the lac genes fused to a Tn3 promoter on the plasmid, and identified both suppressible and nonsuppressible mutations among the nonrepressing clones (16). Mutants that rendered expression of the lac phenotype of the Tn3-lacZ fusion sensitive to repression by a coexisting wild type Tn3 (i.e., cis-dominant mutations) were also isolated. Figure 2 shows the effects of certain of the nonrepressing mutants on Tn3-encoded protein synthesis. In these experiments, plasmids carrying non-repressing Tn3 mutants or the parental Tn3 were individually introduced by transformation into a minicell producing E. coli strain, and their ^{35}S-labeled polypeptide products were examined by electrophoresis on SDS/polyacrylamide gels.

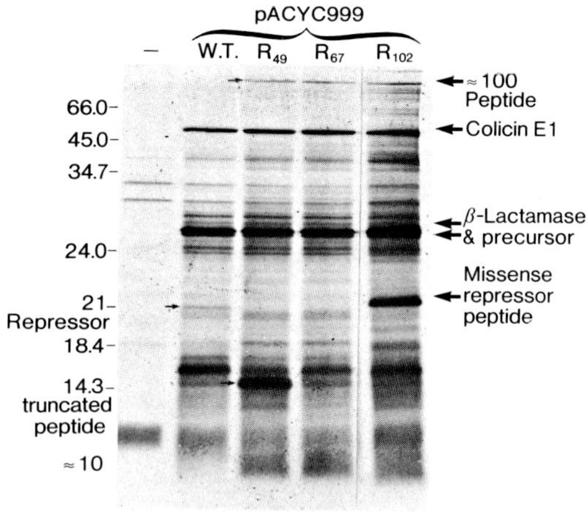

Figure 2. One-dimensional autoradiogram showing ^{35}S-labeled polypeptides synthesized by minicells containing a plasmid carrying Tn3 or its mutant derivatives. Experimental details have been published elsewhere (16). The locations of molecular weight standards are shown to the left of the gel. Arrows point to 21K polypeptide observed in extracts of wild type Tn3 (w.t.) and to related peptides made by Tn3 derivatives (R_{49} and R_{102}) carrying mutations in the repressor gene mutants. The positions of colicin El and of β-lactamase and its precursor are indicated. Numbers are x 10^{-3}. The track marked (-) is the minicell strain alone.

As seen in the figure, the amber-suppressible mutants (R_{49} and R_{67}) failed to synthesize a peptide that was produced in small amounts by the wild type Tn3 element, and which migrated at a rate corresponding to a molecular weight of 21,000. In one of the amber-suppressible mutants (R_{49}) a (truncated) 14K-peptide band was present in large amounts, consistent with premature termination of the 21K peptide as a consequence of amber mutation. A non-amber-suppressible mutation (R_{102}) showed an intense band migrating at the 21K position, suggesting that the mutation results in production of a missense peptide and that the failure of this missense peptide to function results in its overproduction - providing further indication that the repressor is self-regulated.

DNA sequence analysis of all the wild type and the amber-suppressible mutants showed the locations of the nonsense codons in these mutant Tn3 elements. In R_{49}, the mutation occurred at a position consistent with the length of the truncated peptide. In R_{67}, an amber termination codon (TAG) was located near the beginning of the gene at a position that would be expected to yield a truncated peptide too short to be detected in polyacrylamide gels. The translational reading frame and positioning of the amber codons, plus translational initiation and termination codons at positions consistent with the length of the 21K autoregulated peptide, defines the Tn3 repressor gene.

The genetic properties of the cis-dominant (putative operator-constitutive type mutations) which render the Tn3-lac fusion insensitive to expression in trans by a wild type Tn3 element, suggested that these mutations are located in the operator-promoter control region of the Tn3 gene that was fused to lac on pMC823. DNA sequence analysis of endonuclease-generated DNA fragments that include this region of the mutant and of wild type plasmids (Figure 3) showed a single-base change (G→A) at a position 67 nucleotides before the AUG translational start codon of the gene that contains the two amber mutations. The nucleotide changed by the mutation was within an 18-base pair incomplete palindrome of the type found previously to be associated with bacterial operators (18-19). The DNA sequence in this region includes a dA-dT rich region characteristic of an RNA polymerase recognition site (20), followed by other elements characteristic of a transcriptional and translational control signals: a candidate Pribnow box (21) and a segment (S.D.) partially complementary to the sequence at the 3' end of 16S ribosomal RNA proposed by Shine and Dalgarno (22) to be involved in the binding of messenger RNA to ribosomes.

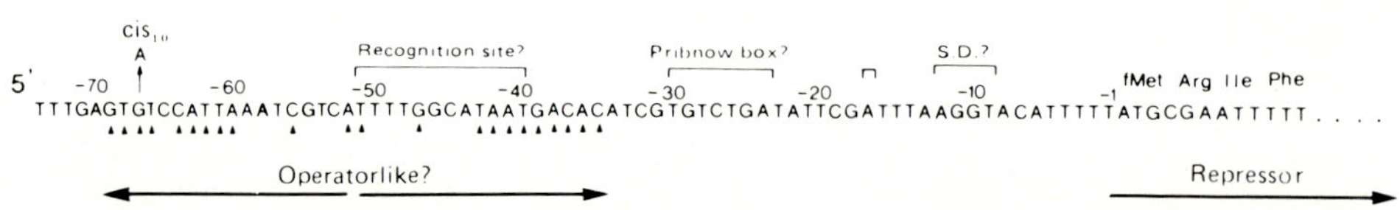

Figure 3. Schematic diagram of putative transcriptional and translational control regions of Tn3 repressor of translocation. The numbers refer to the nucleotide distance preceeding the ATG (AUG) translational start codon. Complementary nucleotides within the palindromic operator-like sequence are indicated. S.D. indicates the segment partially complementary to the Shine and Dalgarno sequence (see text).

The last of the 185 amino acids of the gene defined by the translational reasing frame of the start codon and the amber nonsense codon are followed by a translational termination codon. The amino acid compostion predicted from the experimentally determined DNA sequence indicates a molecular weight for the peptide that is in agreement with the size estimated for the repressor protein by gel analysis.

In the absence of functional repressor, we observed a prominent increase in the synthesis of another Tn3-encoded protein that has a molecular weight of 100,000. This result suggested that the gene specifying the peptide might also be regulated by the repressor. By fusing the _lac_ gene to other regulatory signals within Tn3, we have been able to characterize the gene encoding the 100K peptide and to demonstrate that this peptide is a repressor-controlled product required for transposition of Tn3 (17).

Using the amber suppressible and missense mutants of Tn3 described above, and the series of _lac_ fusion plasmids shown in Figure 4, it was possible to show directly that the Tn3 repressor controls synthesis of a transposase _(Tnp)_ protein, and to localize and determine the direction of reading of the repressor-controlled gene encoding the _Tnp_ peptide (17). To investigate genetic control of the _Tnp_ gene, and to identify more precisely where transcription and translation on this gene begin, we used _lac_ gene fusions to isolate promoter-containing fragments of Tn3 that are repressible by a wild type Tn3 _in trans_. These experiments utilized the pMC874 plasmid (Figure 5), which includes a gene encoding resistance to kanamycin (km) plus a _lac_ gene fragment in which the regulatory region for the _lacZ_ gene and the codons for eight amino acids at the amino terminal end of this gene have been removed and a _BamHI_ cleavage site has been introduced (17,23). The segment of _lacZ_ that remains determines sufficient primary structure to yield a functional β-galactosidase enzyme, which can be expressed when a DNA fragment carrying both transcriptional and translational start signals and the amino-terminal segment of a gene is introduced into the _BamHI_ site. In such a case, _lac_ gene expression is placed under control of the promoter, ribosomal binding site, and translational start codon on the inserted fragment, and the resulting fusion peptide contains an amino terminal end encoded by DNA sequences on the fragment.

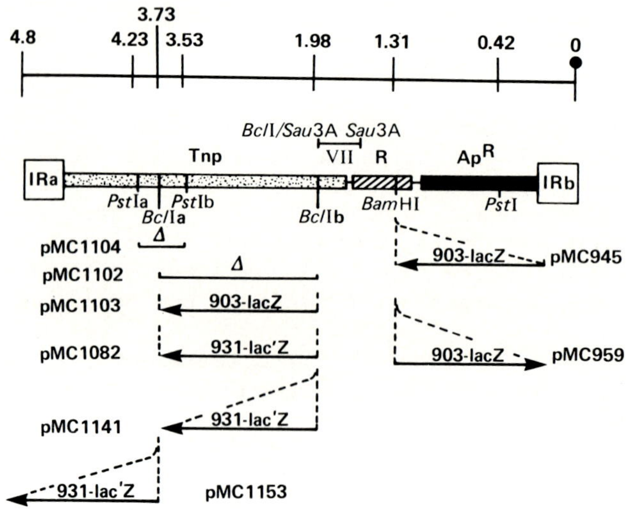

Fig. 4. Schematic map showing the structure of lac insertions into Tn3 that place lac expression under the control of Tn3 regulatory signals. Tnp is the transposase gene, R the repressor gene, and Ap the β-lactamase gene. The 38-base-pair long inverted repeat termini of Tn3 (IR_a and IR_b) are not drawn to scale. The arrows indicate the direction of insertion of 7-kilobase BamHI-generated fragments containing lac genes that lack promoters. For details of construction see reference 17.

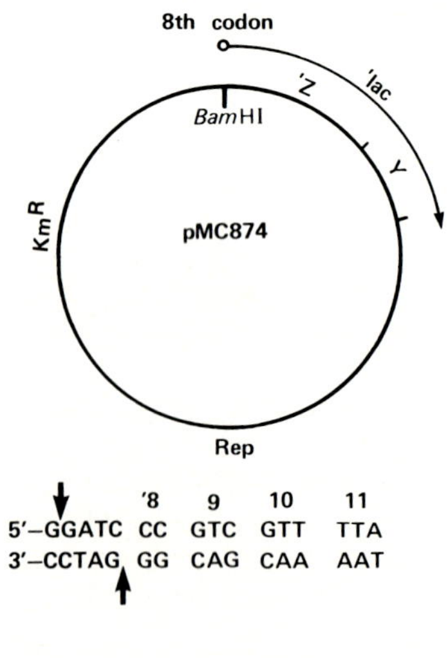

Fig. 5. Simplified map of the pMC874 plasmid, indicating the position and orientation of the lacZY gene and the locations of the Km^r determinant and the replication region (rep). This plasmid contains a unique BamHI site that results in cleavage before the eighth codon of the lacZ gene. DNA fragments that contain transcriptional signals and translational start codons yield a lac^+ phenotype when introduced in frame with the lacZ gene at the BamHI cleavage site. The addition of exogenously derived amino acids to the β-galactosidase protein does not affect its activity (17, 23).

By isolating a Sau3A endonuclease-generated DNA fragment that contains two promoters controlled by the Tn3 repressor, and determining the nucleotide sequence of this region, it has been possible to show that the Tnp and repressor gene promoters are located adjacent to each other and that the genes expressed from the two promoters are transcribed in divergent directions (Figure 6). Similar conclusions have been reached independently by Gill et al. (11) and by Heffron et al. (10) through the use of a different approach.

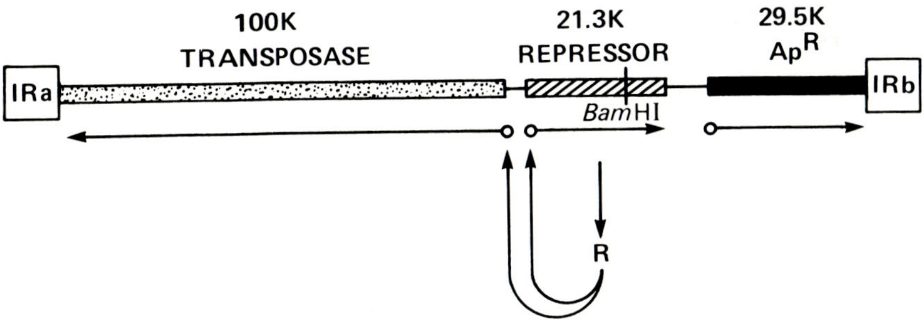

Figure 6. Schematic diagram showing locations of Tn3 genes in the regulatory signal; as shown, a 21.3K repressor peptide acts to reduce transcription from the promoters of the transposase and repressor genes --- which are expressed in divergent directions.

Nucleotide sequence analysis of the segment of Tn3 that contains the transcriptional control regions for both genes, plus the amino terminal sequences for both the repressor and transposase peptides is shown in Figure 7. The location of certain restriction endonuclease cleavage sites and the site of the cis-acting mutation in the putative operator for the Tn3 repressor gene are shown. The entire sequence of the repressor gene that has been defined by these fusion methods has been reported elsewhere (16).

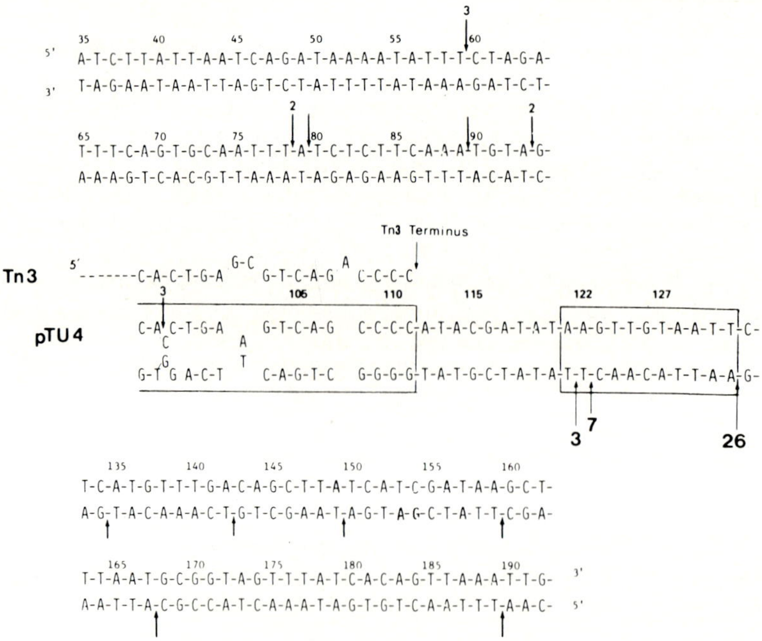

Figure 7. Partial DNA sequence of the transposase and repressor genes of Tn3 showing the inter-gene genetic control structures.

These results provide a structural and functional basis for the previously reported regulation of transposition frequency of Tn3 (1), and also explain the observation that structural integrity of the region to the left of the BamHI site of Tn3 is required for transposition to occur (6-8). Since transposition is thought to be a multi-step process that involves DNA nicking and ligation steps, plus replication related functions, the large size of the Tnp peptide may reflect an ability to carry out multiple steps in the transposition process.

PROMOTION OF REPLICON FUSIONS BY A TRANS-ACTING GENE PRODUCT OF THE F PLASMID

Recently, it was reported that the tetracycline (Tc) resistance plasmid pSC101 can undergo integrative recombination (i.e., fusion) with a plasmid or bacteriophage genome that lacks extensive DNA sequence homology with pSC101 (1, 24-25). Such fusion of pSC101 with a second replicon can occur in the absence of the bacterial recA gene function, and appears to involve DNA sequences on pSC101 that resemble closely the inverted repeat termini of the Tn3 element. In both of the observed instances of replicon fusion involving pSC101, the fertility plasmid F was present in the bacterial cell; in one case an autonomously replicating F plasmid was used to support the infection of the male-specific bacteriophage fl (25, 26), while in the other F was integrated into the chromosome of an Hfr bacterial strain being used to study mobilization of a non-conjugative plasmid (27). However, in numerous studies with the pSC101 plasmid carried out in the absence of F in our laboratory and elsewhere, pSC101 has not been observed to undergo fusion to a second concurrently present replicon. These observations suggested to us that the F plasmid might carry a function that enables pSC101 to undergo replicon fusion.

Recent evidence described elsewhere in detail (28) indicates that the F plasmid does in fact provide a *trans*-acting function that promotes *rec*A-independent site-specific recombination by pSC101, and that the function carried by F is located in a segment of the plasmid that includes the *tra* genes (Table 1). Current models for transposition propose that replicon fusion is an intermediate step in the transposition process (29-31); if this proposal is current, the function provided by F seems to be capable of carrying transposition through the intermediate step. Analogous fusions are also permitted by the *Tnp* protein of Tn3 (6); however, in the case of Tn3 an additional function accomplishes resolution of the fusion event to yield full transposition.

TRANSLOCATION SPECIFITY OF THE Tn3 ELEMENT

To investigate the specificity of translocation, we constructed a small plasmid (pTU4) having a defined structure and containing a DNA segment that previously had been sequenced in its entirety. Using this plasmid, we could effectively study independent insertion events; our analysis of almost 250 independent isolates of insertions of Tn3 into the pTU4 plasmid indicates that multiple insertions occur at specific nucleotides located within certain regions of the plasmid (32). Moreover, our findings indicate that the clustering of such insertions at certain DNA locations is determined by the nucleotide sequence of the recipient genome.

TABLE 1

Bacterial Strain and Plasmids	Tc^R Colonies at 42°/ Total colonies at 42°	Plasmid DNA isolated from Tc^R Colonies Grown at 42°
PM191 (pPM103, ColE1)	1.5×10^{-5}	16/17 two plasmids: 9kb, 6kb 1/17 one plasmid: 9kb
PM191 (F'-Km, pPM103, ColE1)	1.2×10^{-3}	19/21 one plasmid: 15kb 2/21 two plasmids: 9kb, 6kb
PM191 (R6-5, pPM103, ColE1)	1.5×10^{-6}	10/11 two plasmids: 9kb, 6kb 1/11 one plasmid: 9kb
PM191 (pPM103, pDPT234)	2.5×10^{-8}	3/3 two plasmids: 9kb, 5kb
PM191 (F'-Km, pPM103, pDPT234)	3.0×10^{-5}	3/3 one plasmid: 14kb
PM191 (R100.1, pPM103, ColE1)	1.0×10^{-1}	4/4 two plasmids: 89kb, 6kb
PM191 (pCM1891[1], pPM103, ColE1)	2.5×10^{-7}	2/2 two plasmids: 9kb, 6kb
JC1569 (pPM103, pDPT234)	9.0×10^{-8}	4/4 two plasmids: 9kb, 5kb
JC1569 (F'-Km, pPM103, pDPT234)	5.0×10^{-5}	4/4 one plasmid: 14kb
JC1569 (F Δ 446, pPM103, pDPT234)	4.0×10^{-7}	4/4 two plasmids: 9kb, 5kb

[1] pCM1891 is a pACYC189 (27) derived plasmid containing EcoRI fragment 1 of F (28). Similar results were obtained with EcoRI fragments 2,3,5,6,15 of F.

Table 1. Single colony isolates of the recA S6 E. coli strains PM191 carrying a ts plasmid derivative of pSC101 (i.e. pPM103) plus other plasmids as shown.

Cells grown overnight at 30° in the absence of selection were plated on media containing Tc and grown at 42°. Plasmid DNA isolated from cultures was examined by agarose gel electrophoresis to determine the nature and size of the plasmids present. The F Δ 446 plasmid has a deletion of the tra region of F, leaving the sequence between 16 and 58 minutes. Experimental details are provided in reference 28.

A description of the construction of the pTU4 plasmid and a map of the plasmid have been presented elsewhere (32). Our initial analysis of Tn3 insertions into this plasmid indicted that 25% of 247 independent insertion events occurred within the 222-base pair <u>HhaI</u> fragment 6 of the pTU4 plasmid. DNA sequence analysis of these insertions indicated the presence of several distinct hot spots (i.e., nucleotide positions that had received multiple insertions of Tn3): a total of 36 independent insertions (15% of those examined) occurred at three separate locations within an 11 nucleotide region. Moreover, 31 of these 36 insertions were in the same orientation. Determination of the DNA sequence in the vicinity of the hot spot region for Tn3 insertion (Figure 8) indicates that major homology

```
       PvuII/AluI                                                       HhaI
(A) 5' C AGC TGT GAA CGC TGT TCG GCT GAC GGT ATC AGT GCC AGT TTG TTC CAC AGG CGC AAC GTC GCC TTT TCC CTT ACC TCT GAA ATC AAC
    3' G TCG ACA CTT GCG ACA AGC CGA CTG CCA TAG TCA CGG TCA AAC AAG GTG TCC GCG TTG CAG CGG AAA AGG GAA TGG AGA CTT TAG TTG
         Ala Thr Phe Ala Thr Arg Ser Val Thr Asp Thr Gly Thr Gln Glu Val Pro Ala Val Asp Gly Lys Gly Lys Gly Arg Phe Asp Val
       HpaII                   HpaII                                                                     TaqI
(B)    CGG GTC AGC GTG GTG GCT CCG GGG AGA ATA ATA CGA TGT TGC ATA AGC CAC CCT GTC GCA GGT CGA AAA GCA GGC AGG ACG TTC GTT
       GCC CAG TCG CAC CAC CGA GGC CCC TCT TAT TAT GCT ACA ACG TAT TCG GTG GGA CAG CGT CCA GCT TTT CGT CCG TCC TGC AAG CAA
       Pro Asp Ala His His Ser Arg Pro Ser Tyr Tyr Ser Thr Ala Tyr Ala Val Arg Asp Cys Thr Ser Phe Cys Ala Pro Arg Glu Asn
                           HpaII                                         EcoRII,HaeIII   EcoRI
(C)    GCC TAT CCA GCT CCG GGT ATA TAA AAG ACG GGT AAG GCG AAA TGT CCA GGG CCA GGC AAA TTC ACG ATA CTG ATA GTG CTG ACG TAT
       CGG ATA GGT CGA GGC CCA TAT ATT TTC TGC CCA TTC CGC TTT ACA GGT CCC GGT CCG TTT AAG TGC TAT GAC TAT CAC GAC TGC ATA
       Gly Ile Trp Ser Arg Thr Tyr Leu Leu Arg Thr Leu Arg Phe Thr Trp Pro Trp Ala Phe Glu Arg Tyr Gln Tyr His Gln Arg Ile
        HaeII/HhaI                                                                          HinfI AvaI AluI
(D)    CAG CGC TGC ATG CTC ACG GCG GGT ATT TTC CCT CTG ACC GTA TTC TGC AAG AAC GGT GAT ATC ACG AAT CCC GAG CTG TCT GGC GGT
       GTC GCG ACG TAC GAG TGC CGC CCA TAA AAG GGA GAC TGG CAT AAG ACG TTC TTG CCA CTA TAG TGC TTA GGG CTC GAC AGA CCG CCA
       Leu Ala Ala His Glu Arg Arg Thr Asn Glu Arg Gln Gly Tyr Glu Ala Leu Val Thr Ile Asp Arg Ile Gly Leu Gln Arg Ala Thr
               HpaII   HpaII       HinfI
(E)    AAA ATG CCG GAC GCC GGA AGG AAT ATG ATT CAT GTC AGT AAG AAA AGT GCC CAG AAA ACG CAC ACA CCC GAT TTG CAG GGC AAT ACC
       TTT TAC GGC CTG CGG CCT TCC TTA TAC TAA GTA CAG TCA TTC TTT TCA CGG GTC TTT TGC GTG TGT GGG CTA AAC GTC CCG TTA TGG
       Phe His Arg Val Gly Ser Pro Ile His Asn Met Asp Thr Leu Phe Thr Gly Leu Phe Arg Val Cys Gly Ile Gln Leu Ala Ile Gly
           TaqI    BclI/Sau3A HpaII              EcoRI                                 TacI,HhaI PvuII/AluI
(F)    AAG TCG ATT GTG ATC ACC CCG GCT TTT CCC GAT AAA TTC TTT ATC CGC TTC ATC AAG ATG AAA ATA ACG CGC CAG CTG AAG TTC ATC
       TTC AGC TAA CAC TAG TGG GGC CGA AAA GGG CTA TTT AAG AAA TAG GCG AAG TAG TTC TAC TTT TAT TGC GCG GTC GAC TTC AAG TAG
       Leu Arg Asn His Asp Gly Arg Ser Lys Gly Ile Phe Glu Lys Asp Ala Glu Asp Leu His Phe Tyr Arg Ala Leu Gln Leu Glu Asp
                                                                                      HaeIII
(G)    GGG TTC GCC AGT GAA CCT GCC ATA ACT CTC AAC CTG CTC AGT GGT CAA AAA ATC AAC GGG CAT ATCGGCCTCCCTGCCTGACGGCTTTTTTAACA
       CCC AAG CGG TCA CTT GGA CGG TAT TGA GAG TTG GAC GAG TCA CCA GTT TTT TAG TTG CCC GTA TAGCCGGAGGGACGGACTGCCGAAAAAATTGT
       Pro Glu Gly Thr Phe Arg Gly Tyr Ser Glu Val Gln Glu Thr Thr Leu Phe Asp Val Pro Met    S.D. ?              P.B.?
                 TaqI                                         A                                               TaqI
(H)    CAACTGCAACCGTTCGAAATATTATAAATTATCAGACATAGTAAAACGGCTTCGTTTGAGTGTCCATTAAATCGTCATTTTGGCATAATGACACATCGTGTCTGATATTCGATTT
       GTTGACGTTGGCAAGCTTTATAATATTTAATAGTCTGTATCATTTTGCCGAAGCAAACTCACAGGTAATTTAGCAGTAAAACCGTATTACTGTGTAGCACAGACTATAAGCTAAA
                                                              T  operator-like   ?                   P.B.?
                         EcoRI*          HhaI/TacI                           TaqI      Sau3A
(I)    AAGGTACATTTTT ATG CGA ATT TTT GGT TAT GCG CGG GTC TCA ACC AGC CAG CAG TCC CTC GAT ATT CAG ATC AGA ... ... ... 3'
       TTCCATGTAAAAA TAC GCT TAA AAA CCA ATA CGC GCC CAG AGT TGG TCG GTC GTC AGG GAG CTA TAA GTC TAG TCT ... ... ... 5'
        S.D. ?       Met Arg Ile Phe Gly Tyr Ala arg Val Ser Thr Ser Gln Gln Ser Leu Asp Ile Gln Ile Arg
                      1                              10                                  20
```

Figure 8. DNA sequence homology between terminal nucleotides of Tn3 and DNA segments in the vicinity of insertional hot spots. Thirty-six multiple insertions of Tn3 into the recipient genome occur within an 11-base pair region near a segment of the recipient genome showing homology with the terminal nucleotides of Tn3. Sites of single insertions are indicated by arrows; in instances of multiple insertion the number of insertions observed at that site is indicated. For details, see reference 32.

with a sequence within Tn3 exists. Fifteen of 17 nucleotides (mismatch at 3 locations) in a DNA segment near the hot spot region are identical to 15 of 18 nucleotides within the sequence determined previously for the Tn3 inverted repeat termini (5, 33-35).

Analysis of multiple insertion sites also showed that the previously proposed regional specificity (1) of Tn3 insertion is associated with a strong preference A-T rich segments. Furthermore, we have identified a direct repeat of a 10-base pair sequence that is located within the Tn3 termini: it is of some interest that the preferred orientation of insertion of Tn3 brings this 10-base pair sequence into a position near the Tn3-like sequence previously identified on the recipient genome (35).

These findings indicate that the site and orientation of insertion of Tn3 are at least partly determined by the primary nucleotide sequence of the recipient genome, and suggest that insertional hot spots may result from the combined effects of A-T-richness plus homology of the recipient genome with terminal sequences of Tn3. Thus, despite the recA-independence of transposition Tn3 (36), DNA sequence homology may have an important role in such "illegitimate" recombinational events.

ACKNOWLEDGMENTS

These studies were supported by funds from the American Cancer Society and the National Institutes of Health. M.J.C., P.G.L., and C.-P. D.T. were recipients of postdoctoral fellowship awards from both organizations.

REFERENCES

1. P.J. Kretschmer and S.N. Cohen, J. Bact. 130 (1977) 888.
2. D.J. Kopecko, J. Brevet, and S.N. Cohen, J. Mol. Biol. 108 (1976) 133.
3. P.D. Nisen, D.J. Kopecko, J. Chou, and S.N. Cohen, J. Mol. Biol. 117 (1977) 975.

4. S.N. Cohen, J. Brevet, F. Cabello, A.C.Y. Chang, J. Chou, D.J. Kopecko, P.J. Kretschmer, P. Nisen and K. Timmis, In Microbiology (D. Schlessinger, ed.), American Society for Microbiology, Washington, D.C., 1978.
5. S.N. Cohen, M.J. Casadaban, J. Chou, and C.-P.D. Tu, Cold Spring Harbor Quant. Biol. XLIII (1979) 1247.
6. F. Heffron, P. Bedinger, J. Champoux, and S. Falkow, Proc. Natl. Acad. Sci. (USA) 74 (1977) 702.
7. F. Heffron, M. So, and B.J. McCarthy, Proc. Natl. Acad. Sci. (USA) 75 (1978) 6012.
8. R. Gill, F. Heffron, G. Dougan, and S. Falkow, J. Bact. 136 (1978) 742.
9. G. Dougan, M. Saul, A. Twigg, R. Gill, and D. Sherratt, J. Bact. 138 (1979) 48.
10. F. Heffron, B.J. McCarthy, H. Ohtsubo, and E. Ohtsubo, Cell 18 (1979) 1153.
11. R.E. Gill, F. Heffron, and S. Falkow, Nature 282 (1979) 797.
12. M. Casadaban, J. Mol. Biol. 104 (1976) 541.
13. M.J. Casadaban and S.N. Cohen, Proc. Natl. Acad. Sci. (USA) 76 (1979) 4530.
15. M.J. Casadaban and S.N. Cohen, J. Mol. Biol., in press.
16. J. Chou, M.J. Casadaban, P.G. Lemaux, and S.N. Cohen, Proc. Natl. Acad. Sci. (USA) 76 (1979) 4020.
17. J. Chou, P.G. Lemaux, M.J. Casadaban, and S.N. Cohen, Nature 282 (1979) 801.
18. W. Gilbert and A. Maxam, Proc. Natl. Acad. Sci. (USA) 70 (1973) 3581.
19. T. Maniatis, M. Ptashne, and R. Maurer, Cold Spring Harbor Symp. Quant. Biol. 38 (1974) 857.
20. J. Majors, Nature (London) 256 (1975) 672.
21. D. Pribnow, Proc. Natl. Acad. Sci. (USA) 72 (1975) 784.
22. J. Shine and L. Dalgarno, Proc. Natl. Acad. Sci. (USA) 71 (1974) 1342.
23. M.J. Casadaban, J. Chou, and S.N. Cohen, submitted for publication.
24. P.J. Kretschmer and S.N. Cohen, J. Bact. 139 (1979) 515.
25. S.V. Ravetch, M. Ohsumi, P. Model, G.F. Vovis, D. Fischhoff, and N.D. Zinder, Proc. Natl. Acad. Sci. (USA) 76 (1979) 2195.
26. M. Ohsumi, G.F. Vovis, and N.D. Zinder, Virology 89 (1978) 438.
27. A.C.Y. Chang and S.N. Cohen, J. Bact. 134 (1978) 1141.

28. C.A. Miller and S.N. Cohen, Nature, in press.
29. N.D.F. Grindley and D.J. Sherratt, Cold Spring Harbor Symp. on Quant. Biol. XLIII (1979) 1257.
30. J.A. Shapiro, Proc. Natl. Acad. Sci. (USA) 76 (1979) 1933.
31. A. Arthur and D. Sherratt, Molec. gen. Genet. 175 (1979) 267.
32. C.-P. D. Tu and S.H. Cohen, Cell 19 (1980) 151.
33. H. Ohtsubo, H. Ohmori, and E. Ohtsubo, Cold Spring Harbor Quant. Biol. 43 (1979) 1269.
34. T. Takeya, H. Nomiyama, J. Miyoshi, K. Shimada, and Y. Takagi, Nucleic Acid Res. 6 (1979) 1831.
35. C.-P.D. Tu and S.N. Cohen, Molec. gen. Genet. 177 (1980) 597.
36. K.J. Kopecko and S.N. Cohen, Proc. Natl. Acad. Sci. (USA) 72 (1975) 1373.

DISCUSSION

C. HERSHBERGER: Does the region of homology on the plasmid line up with the homologous sequence of Tn3 so that the end of the Tn3 is in precise register with the site of insertion in the recipient DNA?

S.N. COHEN: The sequence on pTU4 that is in the proximity of the 11 base pair segment that received multiple insertions is down a few nucleotides and in register.

A. LANDY: These homologies may not reflect any role of hydrogen bonding but may instead be recognition sites for the same protein(s). In the latter case the spacing would not have to be so critical. It would probably be better to avoid the term "sequence complementarily" for the present.

S.N. COHEN: I thought I had used the term homology but I agree with your statement.

N. FRANKLIN: Is there any possibility that a general recombination function exists on any of the plasmids?

S.N. COHEN: We know that the plasmids that are involved do not contain a function which complements rec^-, but whether or not there is some other generalized recombination function I can't answer.

P. STARLINGER: There are several transposons having very similar inverted repeats and all making a 5-base pair duplication. Upon insertion is there any indication that they are using the same transposase and therefore are under the same control?

S.N. COHEN: I have no indication of that. We are doing some experiments that relate to complementation of various TN3-like sequences by F but the answers are incomplete at the moment.

Transposable Genetic Elements: The Mu Paradigm

G. Chaconas[1], R.M. Harshey, M. S. DuBow
N. Sarvetnick and A.I.Bukhari[2]

Cold Spring Harbor Laboratory
Cold Spring Harbor, New York 11724

Integration of the bacteriophage Mu genome into the host genome involves transposition of Mu DNA from one set of host sequences to another. Mu DNA is flanked by host sequences even when packaged in virions (see Bukhari, 1976). During the process of Mu integration, Mu DNA apparently leaves these host sequences and inserts itself at a new site; hence the term transposition. This also occurs upon prophage induction in a Mu lysogen - that is, Mu DNA is found to be attached to new host sequences after induction. The behaviour of Mu DNA during its growth is extraordinary; not only does it integrate at a high frequency at different sites on the chromosome, but it causes many diverse rearrangements in the host DNA (Toussaint, et al., 1977). The genetic rearrangements that have been observed during Mu growth are listed in Figure 1.

The most striking property of Mu is the frequency of transposition. As indicated in Figure 1 there may be as many as 100 (or more) transposition events per cell. This frequency can be compared with other transposable elements that would generally undergo, under the best of conditions, 10^{-2} to 10^{-3} events per cell. Moreover, the process of transposition of Mu DNA can be triggered at will by simply infecting the cells with Mu particles or by inducing a lysogen. It is for this reason that Mu provides a good system for biochemical studies on the mechanism of transposition of DNA sequences.

[1]Supported by a postdoctoral fellowship from the Canadian Medical Research Council.
[2]Supported by grants from NSF PCM 7826710, NIH GM23566, and a Career Developmental Award from NIH.

ISBN 0-12-633360-2

FIGURE 1. DNA rearrangements involving Mu.

(i) Transposition of Mu CNA generates 5 bp duplications (Allet, 1979; Kahmann and Kamp, 1979). It involves integration of about 37 kb of Mu DNA at new sites.

(ii) Heterogeneous circles contain Mu sequences (Waggoner et al., 1974; Parker and Bukhari, 1976). These heterogeneous circles are thought not to carry 5 bp duplications.

(iii) Replicon fusion results in cointegrate formation in which two copies of Mu are directly repeated (see Toussaint et al., 1977).

iv) In the X mutants of Mu which contain a functional A gene, the Mu prophage can be cut out from the host DNA at a low frequency (Bukhari, 1975). This process is superseded by replication-integration during active growth of Mu.

(vi) Flip-flop of G refers to a specialized intramolecular recombination process that is responsible for inverting a 3 kb segment of Mu DNA, called the G segment. This inversion controls the host range functions of Mu (P. van de Putte and coworkers, personal communication; D. Kamp, personal communication).

(v) Deletions and inversions, occur during Mu growth (Toussaint et al., 1977). Large duplications have not been characterized so far.

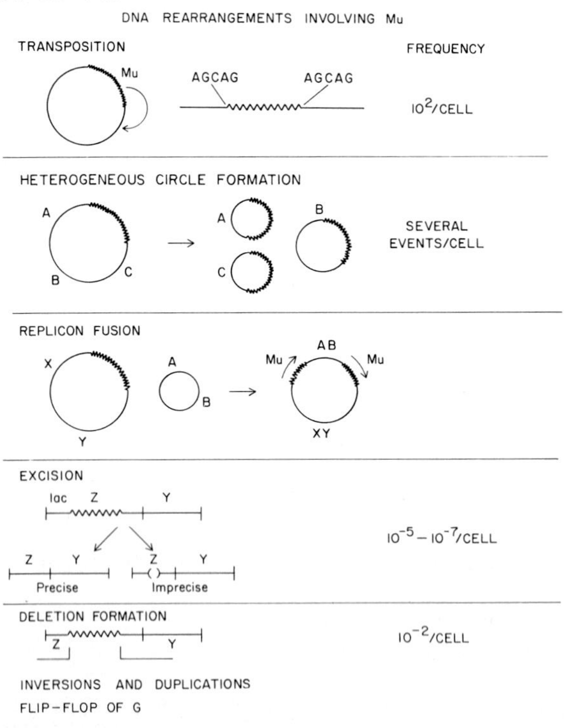

The transposable genetic elements in prokaryotes have been shown to cause duplications of short host sequences, ranging from 5 to 11 base pairs, at the sites of insertion (see Heffron, et al., this volume). It has recently been demonstrated that the phenomenon of host base pair duplications transcends the prokaryote-eukaryote boundary. Mu, Tn*3*, IS*2* and γδ sequences cause duplication of 5 base pairs of host sequences; G. Fink and coworkers have found that at Ty*1* a transposable element in yeast causes a 5 base pair duplication (this volume). G. Rubin and coworkers have reached a similar conclusion for an element in *Drosophila* (this volume). Thus, it would seem that the process of bacteriophage Mu DNA transposition has even wider applicability than was thought previously.

Based on our studies of Mu DNA transposition, we proposed that transposition of DNA sequences involves replication such that the transposable element becomes duplicated during transposition (Bukhari, 1977; Ljungquist and Bukhari, 1977; Bukhari, et al., 1977; Bukhari and Ljungquist, 1978). This model of integrative recombination has gained support from studies on other transposable elements and various structures have been proposed to explain the replication-integration model (Grindley and Sherratt, 1978; Shapiro, 1979). One of the difficulties in studying the DNA structures that may be intermediates of transposition is that transposable elements associate with host DNA during replication and transposition. This in particular has been shown in bacteriophage Mu (Ljungquist and Bukhari, 1979; Ljungquist, et al., 1978). To circumvent this problem we have focussed our efforts on dissecting the process of Mu transposition by examining the behaviour of small plasmids containing either whole Mu or internally deleted Mu (called mini-Mu). These plasmids can be readily separated from the host DNA, allowing us to examine their behaviour during transposition.

Construction of mini-Mu Plasmids

All of the plasmids used for studies on Mu replication and transposition were derived from a 9.09 kb plasmid pSC101. We removed from a pSC101 plasmid containing the whole Mu genome (constructed by Malcolm Casadaban) the central 27 kb *PstI* fragment of Mu DNA (see Figure 2). This prototype mini-Mu contains 1.6 kb from the left end of Mu and about 7.8 kb from the right end, including the invertible G segment. Within this mini-Mu, we cloned various selectable markers such as the entire *lac* operon of *Escherichia coli*, a gene for ampicillin resistance (encoding β-lactamase),

FIGURE 2. Schematic representation of pSC101::Mu and the mini-Mu plasmid.

The heavy lines denote pSC101 sequences, the thin lines phage Mu DNA, and the left and right ends of the Mu genome are indicated by c and S, respectively. The shaded box within the plasmid indicates the position of the transposable element carried by pSC101 (G.F. Vovis, personal communication).

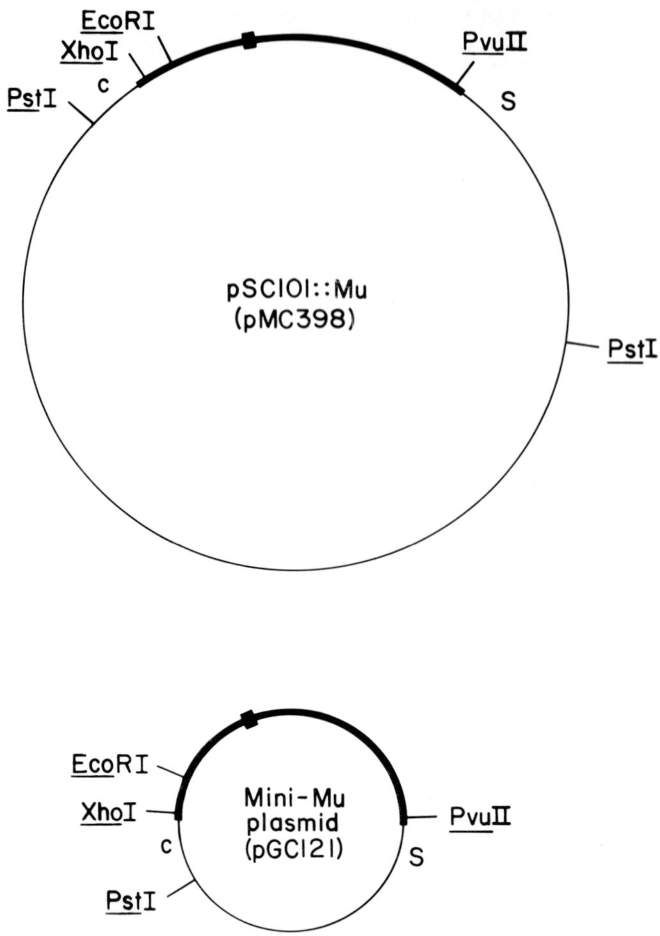

and a gene for kanamycin resistance. The prototype mini-Mu was further manipulated by first recombining the gene for ampicillin resistance (Bla) at the right extremity of the G segment. This was done by a genetic cross involving the Mu amp previously isolated by Leach and Symonds (1979), Since the Bla gene contains a PstI cleavage site, we were able to cut out the G segment, removing all but the 1.6 kb segment of Mu DNA at the right end. In addition, we cloned the lac operon and the Bla gene within this 3.2 kb mini-Mu. Many of the important plasmids constructed by us during the past year are shown in Table 1.

These plasmids provide us with valuable tools with which to dissect the process of Mu transposition and replication. We can study the various events during the life cycle of Mu by monitoring the markers we have inserted within the ends of Mu.

Replication and Transposition of mini-Mu Sequences

By genetic, biochemical, and electron microscopic methods, we have shown that the mini-Mu sequences carried in our plasmids undergo Mu-mediated transposition, replication, and packaging when provided with Mu helper functions in trans. When a phage lysate is prepared from a strain containing mini-Mu and a helper Mu phage, 5-20% of the phage particles carry mini-Mu sequences, indicating that mini-Mu sequences increase in number during Mu growth. Figure 3 shows a packaged mini-Mu that was heteroduplexed with wild-type Mu DNA.

The replication of mini-Mu is shown directly by assaying the activity of β-lactamase for mini-Muamp and the activity of β-galactosidase for mini-Mulac. When these mini-Mu's are provided with normal Mu functions, the levels of the enzymes begin to increase. This increase is in good agreement with the previously reported kinetics of wild-type Mu DNA replication. The replication of mini-Mu requires both functional A and B genes of Mu.

High-efficiency transposition of mini-Mu sequences has been shown by a genetic assay that involves integration of the mini-Mu sequences into a conjugally transferable plasmid such as F'prolac. Following induction of a strain containing a Mu prophage and mini-Muamp, mini-Mulac, or mini-Mukan, the mini-Mu sequences appear in about one in ten transferred episomes. Many of these mini-Mu sequences still have the plasmid attached to them; the integrated DNA has the structure mini-Mu-plasmid-mini-Mu. This type of structure has been

TABLE 1 Plasmids Containing Mu DNA

Plasmid	Number	Mu DNA	Mu G region	Selectable marker	Transposition
pSC101::Mu	pMC321	entire Mu	+	--	+
pSC101::Mu*kan*	pMC398	entire Mu	+	kanamycin^r	+
Mini-Mu	pGC121	c+S	+	--	+
Mini-Mu*kan*	pGC102	c+S	+	kanamycin^r	+
Mini-Mu*amp*	pMD861	c+S	+	ampicillin^r	+
Mini-Mu*lac*	pCL198	c+S	+	Lac$^+$ phenotype	+
Right end	pCL151	S only	+	--	-*
Left end	pGC302	c only	-	--	-*
Mini-Mu	pCL222	c+S	-	--	+
mini-Mu*lac*	pGC401	c+S	-	Lac$^+$ phenotype	+
Mini-Mu*amp*	pGC501	c+S	-	ampicillin^r	+

All mini-Mu's contain the immunity gene *c* and only the beginning of the *A* gene. The left and right ends of Mu are denoted by *c* and S, irrespective of the extent of DNA present at the ends.

*Transposition not detected genetically by conjugation experiments.

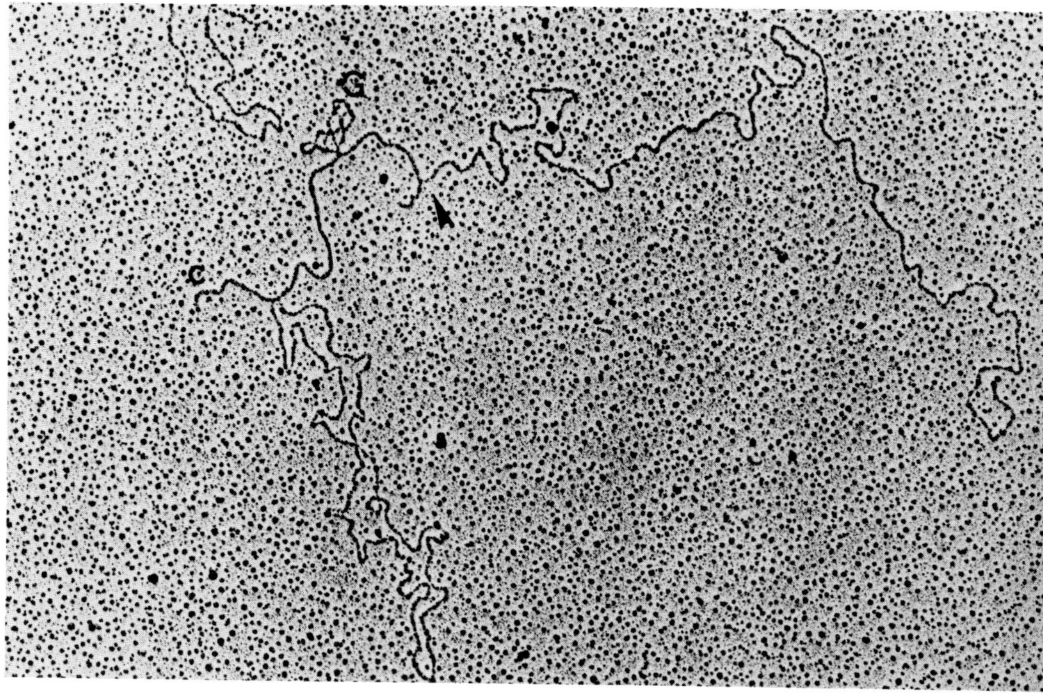

FIGURE 3. A package mini-Mu sequence. Mini-Mu DNA in a transducing lysate is heteroduplexed with wild-type Mu DNA. A large loop near the left end (c end) represents the deletion in mini-Mu. An underwound G segment can be seen. The arrow indicates a normal split end and a very long split end, indicating the pack aging of extra-long host DNA attached to the mini-Mu sequence.

seen with all transposable elements and is referred to as a cointegrate (Toussaint and Faelen, 1973; Gill et al., 1978).

Behaviour of Mu- and mini-Mu-Containing Plasmids During Transposition

A singularly important advantage of using small plasmids with Mu insertions is that the plasmid forms can be separated both from one another and from chromosomal DNA by agarose gel electrophoresis. In a series of experiments, we extracted the DNA from strains containing the Mu or mini-Mu plasmids at various times after prophage induction and subjected it to electrophoresis in low-percentage-agarose gels followed by transfer of the DNA to nitrocellulose paper. Separate hybridizations with ^{32}P-labeled pSC101 and Mu DNA revealed the position of the plasmids and the replication of Mu DNA. We found that the plasmids are not cleaved to generate either a linear form or smaller products; instead, they migrate with *E. coli* DNA. This interaction of the plasmids with the host chromosome is of two types: (1) formation of cointegrates of the structure Mu-plasmid-Mu and (2) association of circular plasmids with the chromosomal DNA as shown by electron microscopy.

Electron Microscopy of Plasmid-Host Association

Electron microscopy of mini-Mu plasmids after Mu induction showed that many circular plasmids were in contact with the host DNA. About 20% of the observable plasmids were seen associated with the host chromosome in this manner. This association requires functional *A* and *B* gene products of Mu. Figure 4 shows examples of such association. These structures apparently represent intermediates in the transposition process. Analysis of the nature of the contact between the plasmids and the host DNA promises to provide important clues to the mechanism of transposition.

Analysis of Cointegrates

That some of the plasmid DNA migrating with the host DNA after induction represents cointegrates has been shown by restriction endonuclease digestion followed by blotting hybridization. When the DNA was digested with *KpnI*, which cleaves only once within Mu or mini-Mu sequences, linear

FIGURE 4. Electron microscopic visualizationof mini-Mu plasmids associated with *E. coli* DNA.

The DNA was extracted after induction of strains containing mini-Mu plasmids and a Mu*cts* prophage (a) or a Mu*cts* *Ats* prophage where the culture was incubated at the permissive temperature for 50 minutes prior to extraction of the DNA (b) and (c). The schematic diagrams depict the plasmids as a heavy line and the *E. coli* DNA as a fine line.

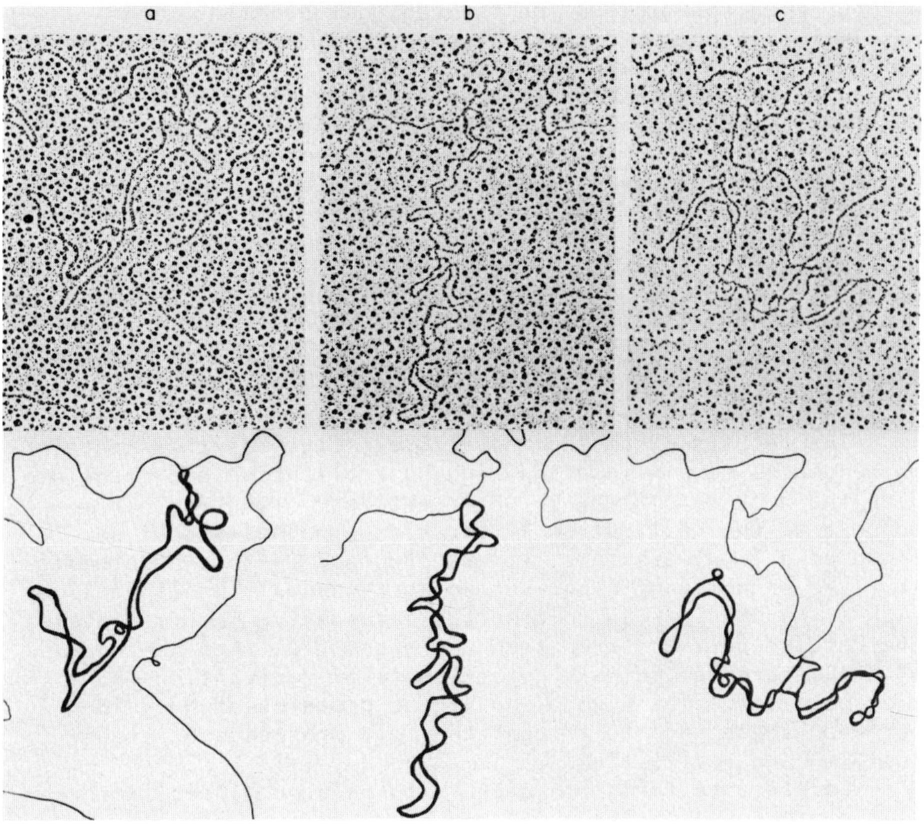

plasmids of the original size were released. When the DNA was cut with *XhoI*, which cuts the plasmid DNA once but does not cut the Mu sequences, a smear of plasmid DNA was generated. This would be expected if Mu-plasmid-Mu sequences had integrated at different sites in the host DNA.

As mentioned above, we have studied the cointegrates genetically by inserting them in conjugally transferable plasmids. As much as 100% of the integrated Mu sequence can be in the form of cointegrates. The cointegrates are resolved in *rec*$^+$ cells, generating the original Mu plasmid and a copy of Mu (or mini-Mu) on the conjugal plasmid. This resolution is much reduced in *recA*$^-$ cells. We have been unable to detect a Mu function that will resolve cointegrates in a *recA*$^-$ background. We have also studied the cointegrates by transduction experiments. When Mu lysates are used for transduction of plasmid markers, we can recover the original plasmid in the recipient cells. This can only happen if cointegrates are packaged in the phage particles and are then resolved in the recipient cells. The packaging of cointegrates is being examined carefully.

The Mu Transposition Problem in 1980

The results discussed above, i.e., plasmids that contain Mu sequences remain intact during Mu replication and associate with the host chromosome in their entirety, lend further support to the replication-integration hypothesis that we proposed a few years ago. It was proposed in this hypothesis that the predominant event during Mu transposition is replication of Mu DNA *in situ* to generate integrative intermediates. The results further strengthen the assumption that the ends of Mu DNA are the primary determinants of replication and transposition. The *A*-gene and *B*-gene proteins of Mu evidently act upon these ends to trigger the twin processes of transposition and replication.

Cointegrate formation appears to be a prominent feature of the process of Mu transposition and, indeed, of all other prokaryotic transposable elements. Occurrence of cointegrates in response to Mu induction has been seen in at least two other laboratories working on Mu. Certainly, cointegrate formation (i.e., replicon fusion) is an important consequence of transposition of an element from one replicon to another. It is not clear though whether the cointegrates are obligatory intermediates in all transposition events.

REFERENCES

Allet, B. (1979). Cell 123, 123.

Bukhari, A.I. (1975). J. Mol. Biol. 96, 87.

Bukhari, A.I. (1976). Annu. Rev. Genet. 10, 389.

Bukhari, A.I. (1977). Brookhaven Symp. Biol. 29, 218.

Bukhari, A.I., Ljungquist, E., deBruijn, F., and Khatoon, H. (1977). In "DNA Insertion Elements, Plasmids and Episomes". (A.I. Bukhari, J.A. Shapiro and S. Adhya, eds.). p. 249. Cold Spring Harbor Laboratory, New York.

Bukhari, A.I., and Ljungquist, E. (1978). In "Microbiology 1978. (D. Schlessinger, ed.). p. 52. American Society for Microbiology.

Gill, R., Heffron, F., Dougan, G., and Falkow, S. (1978). J. Bact. 136, 742.

Grindley, N.D.F., and Sherratt, D. (1978). Cold Spring Harbor Symp. Quant. Biol. 43, 1257.

Kahmann, R., and Kamp, D. (1979). Nature 280, 247.

Leach, D., and Symonds, N. (1979). Molec. Gen. Genet. 172, 179.

Ljungquist, E., and Bukhari, A.I. (1977). Proc. Natl. Acad. Sci. USA 74, 3143.

Ljungquist, E., and Bukhari, A.I. (1979). J. Mol. Biol. 133, 339.

Ljungquist, E., Khatoon, H., DuBow, M., Ambrosio, L., deBruijn, F., and Bukhari, A.I. (1978). Cold Spring Harbor Symp. Quant. Biol. 43, 1151.

Parker, V., and Bukhari, A.I. (1976). J. Virol. 19, 756.

Shapiro, J.A. (1979). Proc. Natl. Acad. Sci. USA 76, 1933.

Toussaint, A., and Faelen, M. (1973). Nature New Biol. 242, 1.

Toussaint, A., Faelen, M., and Bukhari, A.I. (1977). In "DNA Insertion Elements, Plasmids and Episomes". (A.I. Bukhari, J.A. Shapiro and S. Adhya, eds.). p. 275. Cold Spring Harbor Laboratory, New York.

Waggoner, B.T., Gonzalez, N.S., and A.L. Taylor (1974). Proc. Natl. Acad. Sci. USA 71, 1255.

DISCUSSION

C. HERSHBERGER: Would you please explain your evidence of the association between Mu-plasmid DNA and the chromosome? If you have fragmentation of the circular plasmid DNA so that the fragments approach the same molecular weight range as the chromosomal fragments, it would be difficult to establish a true association.

A.I. BUKHARI: First of all, it is clear that something happens to the plasmid in response to Mu induction. This event makes the plasmid migrate with the chromosomal DNA on gels. This migration with the chromosome is dependent upon Mu functions. The <u>A</u> gene and <u>B</u> gene functions appear to be required for this effect. Thus, association with the chromosome is not merely fragmentation. It is a Mu specific effect. Secondly, restriction enzyme analysis and electron microscopy indicate that there is very little fragmentation. Fragmentation will result in a smear, something smaller than the plasmid.

D. MERLO: I did not get your interpretation of the last electron micrograph, could you please restate that?

A.I. BUKHARI: There is no real interpretation of the last electron micrograph. We would like to interpret it if anybody can help. In some cases we can measure the length of the plasmid and can be sure that we are looking at a plasmid in contact with the "chromosomal" DNA. In this case, the plasmid and the chromosome appear as a jumble, so it is difficult to tell what is going on. In general, we think these structures represent Mu the undergoing transposition and replication. It is a difficult problem. Our hope is to dissect it further.

THE FUSION OF DNA MOLECULES AND GENETIC RECOMBINATION

Huntington Potter and David Dressler

The Biological Laboratories, Harvard University,
Cambridge, Massachusetts 02138

DNA recombination is among the cell's most significant activities. It is important both for the generation of individuals with new genotypes and for the maintenance of a constant communication between individuals, preventing over-rapid speciation and devolution.

While numerous models have been proposed that can logically lead to the formation of recombinant chromosomes, the precise molecular mechanisms involved in the recombination process remain a topic of current experimentation.

A convenient place to begin a discussion of the molecular mechanisms involved in recombination is with a suggestion made by Robin Holliday in 1964. Holliday's essential contribution was to propose a mechanism for fusing two DNA molecules together so that they could ultimately undergo a reciprocal exchange of genetic information (1-3).

The Prototype Holliday Model

A prototype model that uses the Holliday recombination intermediate is shown in Figure 1. Two homologous double helices are aligned, and in each the positive strands (or, alternatively, the negative strands) are nicked open in a given region. The free ends thus created leave the complementary strands to which they had been hydrogen-bonded and become associated instead with the complementary strands in the homologous double helix (Figure 1 A→D). The result of this reciprocal strand invasion is to establish a tentative physical connection between the two DNA molecules. This linkage can be made stable through a process of DNA repair, which in this case can be as simple as the formation of two phosphodiester bonds by the enzyme ligase (Figure 1-E).

The structure shown in Figure 1-E is the Holliday recombination intermediate. Although the intermediate is

ISBN 0-12-633360-2

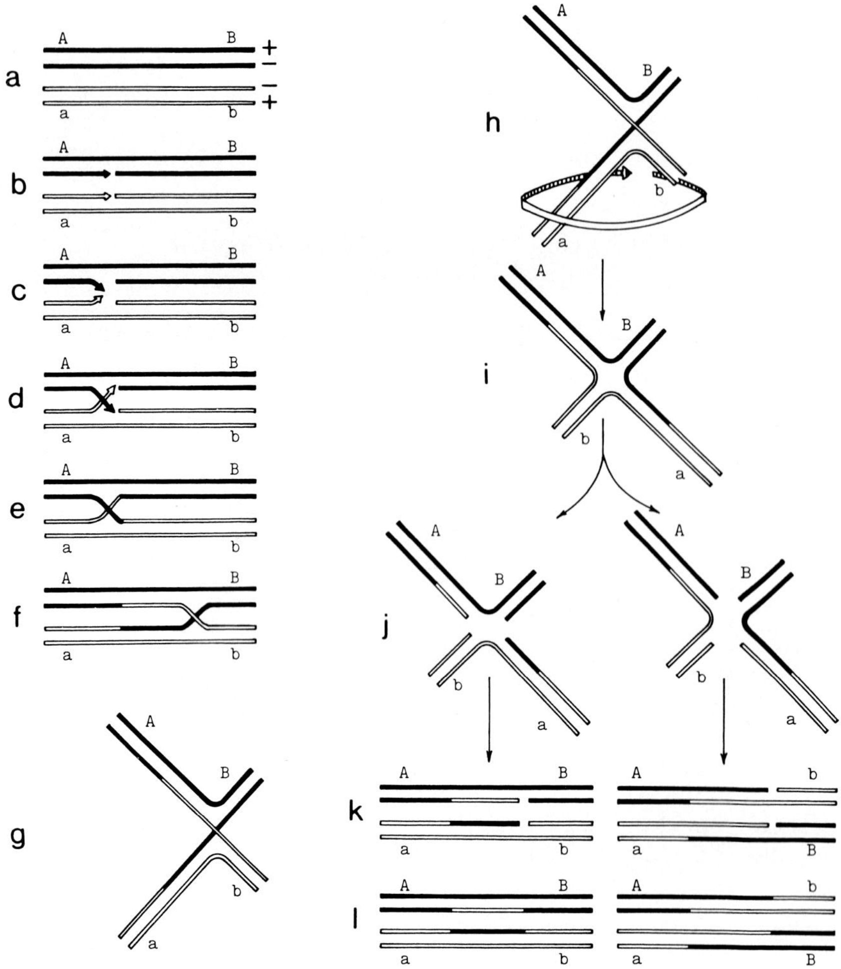

Figure 1. A prototype model for genetic recombination using the Holliday intermediate.

expected to be stable, it need not be static. A continuing reciprocal strand invasion by the two polynucleotide chains involved in the crossover can occur (in conjunction with the rotation of the four double-helical arms around their cylindrical axes), allowing the point of linkage between the two DNA molecules to move to the right or to the left (Figure 1 E→F). This dynamic property of the Holliday structure can lead to the development of regions of heteroduplex DNA during recombination, and when mismatched bases occur, to the formation of areas of double helix that are genetically heterozygous. It was, in fact, the observation of heterozygosity in the recombinant chromosomes of haploid fungal spores that motivated the proposal of this model by Holliday.

The remainder of Figure 1 shows the proposed maturation mechanism for the Holliday intermediate. Because of its structural symmetry, it is expected that the intermediate can be processed in either of two related ways, to give rise to two different pairs of recombinant chromosomes (3-5). The symmetric structure of the Holliday form, and the dual maturation potential that results, are most easily appreciated if one rotates the intermediate into another planar configuration, as in Figure 1-I. Then, cleavage on an east-west axis, or a north-south axis, leads to the release of unit size DNA molecules in which potentially heterozygous regions exist, and in which the flanking parental alleles either remain in their original linkage, or, with equal probability are reciprocally exchanged (Figure 1-L). Genetic distances are measured from the 50% of the cases in which the flanking markers are exchanged; the remaining maturation events are silent from the point of view of traditional recombination, except for the region of heterozygous DNA which remains as a footprint of a former crossover connection.

The reason for preferring a recombination model of this type comes from experiments that biochemists might not have thought to do, but geneticists did.

This model or one related to it, is attractive as a general mechanism for recombination because it accounts for the genetic properties of recombinant chromosomes that emerge from the most highly ordered form of gene exchange known---the eukaryotic meiosis (6-10). As studied in fungi, where all four meiotic products are preserved together and can be comparatively analyzed, the following general conclusions have emerged: (1) recombination proceeds with a net conservation of genetic material: for every two chromosomes that enter the recombination process, two emerge. (2) Recombinant chromosomes are produced in reciprocal pairs---not just in the population as a whole, but as the result of individual crossover events. (3) A region of heterozygosity is frequently found in the immediate area of the chromosome where recombination has occurred. And (4) when one examines a chromosome that shows genetic heterozygosity, it is found that the genes on either side of the heterozygous region either appear in their original linkage or, with equal probability, emerge in a recombinant linkage. The prototype recombination mechanism shown in Figure 1 accounts for all of these genetic findings. Other models for recombination differ in that they produce only one recombinant at a time (for instance, the model of Boon and Zinder, (11)), or piece together one recombinant chromosome while discarding the rest of the participating DNA in the form of fragments (for instance, the model of Broker and Lehman, (12)).

It is taken as axiomatic by workers in the field that the genetic data obtained with fungi, and the resulting implications about reciprocity and heterozygosity in recombination, will be applicable to a wider range of plant and animal systems, where the dispersion of the immediate products of recombination precludes a similar genetic analysis.

Physical Evidence for the Existence of the Holliday Recombination Intermediate

The general pattern of reciprocal recombination inherent in the prototype Holliday mechanism is supported by the data which we will now consider. These data deal with the observation of fused DNA molecules from *E. coli*---plasmid molecules that appear to represent the physical counterparts of the structure proposed on genetic grounds by Holliday (13-15, see also 16-21).

Observation of figure-8s. To search for intermediates in the recombination process, plasmid-containing cells were chosen as an experimental system. The plasmid DNA molecules are small and easy to isolate without breakage---and because the plasmids do not encode their own recombination system but utilize the recombination apparatus of the host cell, it is expected that they can provide a window onto a larger view.

When plasmid DNA molecules were recovered from cells, treated lightly with DNase to remove superhelical twists, and examined in the electron microscope, they were found to consist of a set of monomer-size DNA rings and related multimers. A small percentage of the DNA rings appeared to be touching each other and thus were candidates for intermediates in recombination. These molecules had the appearance of a figure-8---which is the topology that is expected if two DNA circles were to undergo the type of strand nicking and exchange events proposed by Holliday (see Figure 2).

An example of a plasmid figure-8 molecule is shown in Figure 3A.

The figure-8, although a highly suggestive structure, is inherently ambiguous. This geometry could represent two genomes covalently held together at a region of DNA homology (as in Figure 2C), but two alternative interpretations are also possible. The figure-8 could result from two monomer circles interlocked like links in a chain, or from a double-

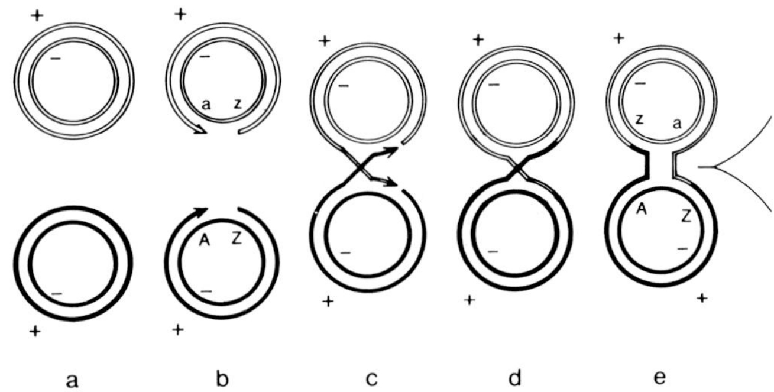

Figure 2. Circular DNA molecules undergoing a reciprocal strand invasion to form a figure-8 structure.

length circle that accidently overlies itself in the middle. To remove this ambiguity, one may open the plasmid DNA molecules with the restriction enzyme Eco R1 prior to electron microscopy (13, 19). This enzyme cuts monomeric plasmid DNA rings once, at a unique site, generating unit-size rods. Interlocked monomer rings or double-length circles should be cleaved into two separable, unit-size rods upon enzyme digestion. On the other hand, if the figure-8s represent two plasmid circles covalently connected at a region of DNA homology, then the enzyme is expected to leave the fusion point intact and to convert the figure-8 into a bilaterally symmetric dimer shaped like the Greek letter chi.

Observation of chi forms. Figure 3B shows an electron micrograph that is representative of more than 1500 chi-shaped molecules found among plasmid DNA after linearization with Eco R1. Such dimeric forms are observed at a frequency of about 1% amidst a simple background of unit-length plasmid rods.

The chi forms could not result from the accidental overlapping of two monomer rods during spreading of the DNA for electron microscopy. This follows from the fact that the crossed molecules always have a special symmetry. The point of contact between the unit-size plasmid genomes occurs so as to divide the molecule into a structure with two pairs of equal-length arms (Figure 3B). Because the fusion point (the potential crossover) is doubly equidistant from a defined base sequence (the Eco R1-cut ends), it must almost certainly occur at a region of DNA homology.

In the chi forms, the fusion point is observed with equal probability at numerous and perhaps all locations along

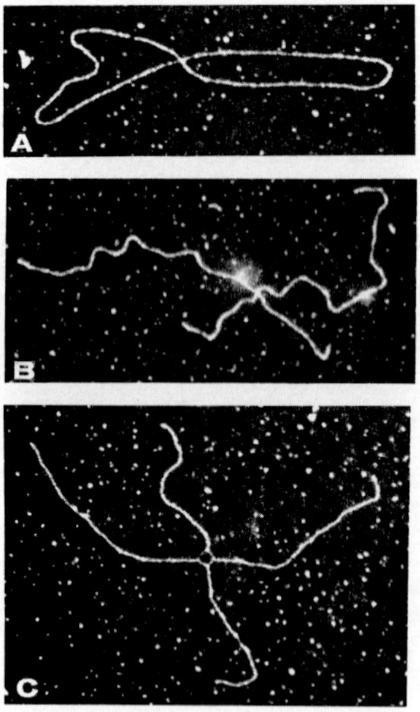

Figure 3. Electron micrographs showing (A) a plasmid figure 8 form, (B) a chi form, and (C) a chi form in which the single-stranded connections can be seen in the region of the crossover.

the plasmid DNA molecule---that is, at varying distances from the Eco R1 cutting site.

The existence of chi forms indicates that one can recover from intact cells genomes fused together by an interaction that occurs at a region of DNA homology.

The crossover point. Of course what one would really like to observe is something of the nature of the connection at the crossover point. This would allow a direct comparison of the experimentally observed chi forms with the diagrammatic structures shown in Figure 1. However, when the Holliday form occurs in the planar configuration shown in Figure 1-E, it is expected that the two polynucleotide chains forming the crossover will overlie each other, obscuring the nature of the crossover connection. But if the intermediate were to assume the planar configuration shown in Figure 1-I, then the region of the crossover would occur in a potentially more open state with a more analyzable geometry. In this case one would expect to see four double helical segments emerging from a ring of connecting single stranded DNA (see Figure 1-I).

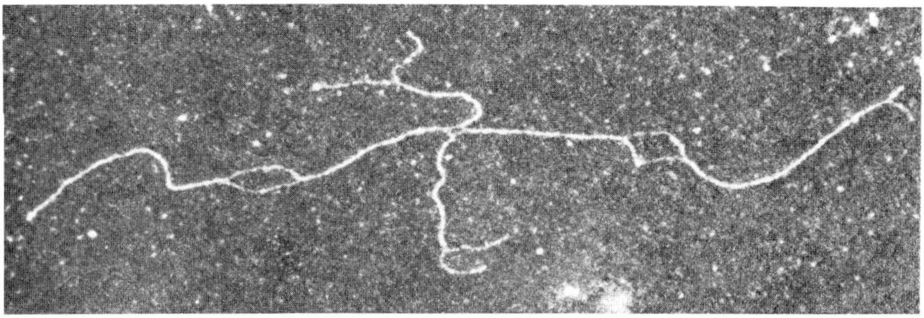

Figure 4. A chi form that has been prepared for electron microscopy in the presence of a high concentration of formamide to cause partial denaturation of the DNA double helix at regions rich in A-T base pairs. The single-stranded connections in the crossover region can be seen. Compare with Figure 1-I.

By pushing the electron microscope technology toward its limit, we have been able to observe about 100 molecules in which the nature of the polynucleotide strand connections in the crossover region is visible. Figure 3C shows a molecule in which the double-helical DNA has become partially unwound in the crossover region during spreading for the electron microscope. The single strands connecting the four arms of the recombining molecules are easily distinguished, showing that the point of contact is indeed a crossover connection involving DNA strands and not, for instance, the result of a synaptic protein holding the two DNA molecules together. The observation of this DNA strand substructure in the crossover region allows the molecule to be correlated exactly with the planar representation of the Holliday intermediate shown in Figure 1-I.

It is a predicted that when the intermediate has assumed the planar configuration in which the crossover region is open, the equal-length arms will always be oriented in the trans configuration. This prediction has been fulfilled in all of the open molecules we have observed.

Figure 4 shows another example of a chi form that has become partially denatured during spreading for the electron microscope, in this case because the DNA was prepared in a high concentration of formamide. It is again possible to see the covalent single-strand connections in the region of the crossover. Furthermore, the homologous arms can be identified explicitly by their characteristic partial denaturation patterns, and can be seen to occur in the trans configuration.

To summarize the results obtained from examining DNA molecules recovered from intact cells, we interpret the molecules shown in Figures 3 and 4 to represent the two planar configurations of the Holliday intermediate as diagramed in Figure 1, Panels G and I.

<u>How is the Holliday Structure Formed?</u>

While the finding of chi forms offers physical evidence in support of the existence of the Holliday structure, the data do not bear directly on the question of how the recombination intermediate is formed. Several alternative initiation mechanisms, distinct from the one proposed by Holliday and shown in Figure 1 A→E, can be considered. Two of these alternatives are shown in Figures 5 and 6.

Because genetic studies in fungi indicate that the regions of heterozygous DNA in a finished pair of recombinant chromosomes are not perfectly reciprocal (8-10), Meselson and Radding were motivated to propose a two-step initiation mechanism for the formation of the Holliday structure (22). They sought to explain the asymmetric heterozygous regions by suggesting that, initially, only one of the participating double helices is nicked. After nicking, DNA synthesis occurs and leads to the displacement of a single-stranded DNA tail (Figure 5). The displaced tail then invades a

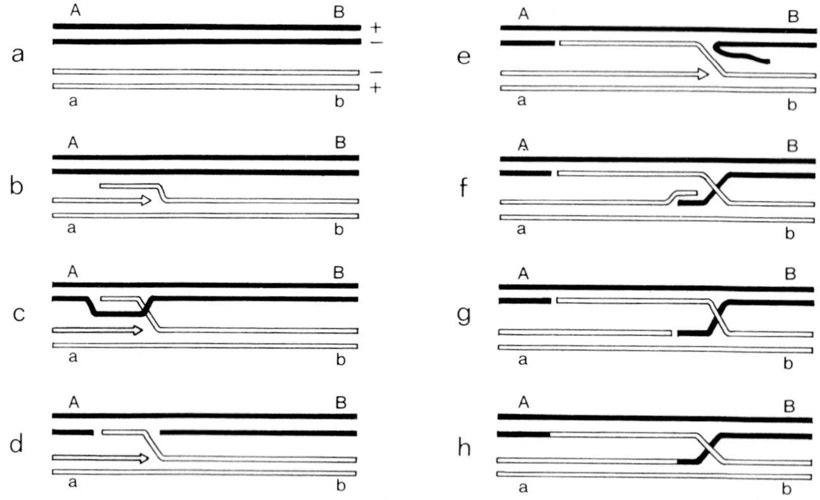

Figure 5. The Meselson-Radding initiation mechanism which couples DNA synthesis to the formation of the Holliday recombination intermediate.

second double helix at a region of homology and provokes a reciprocal nicking and strand invasion which occurs at a slightly different place. After the filling-in of any resultant gaps and the trimming of any remaining tails, this would yield asymmetric heteroduplex regions at the outset of recombination (Figure 5).

Another type of initiation mechanism is also possible. In the mechanism shown in Figure 6 genome fusion occurs when two DNA molecules undergo a localized denaturation at homologous places and the exposed pairs of positive and negative strands become interwrapped (23-27). After the initial contact, strand interwrapping is extended for several hundred base pairs, with transient nicks being introduced by a nicking-closing enzyme (25,26) to allow the necessary interwrapping. The structure shown in Figure 6-D results and is characterized by two crossover connections of the type shown in Figure 1, one at either end of the wrapped segment (compare Figure 6-D with Figure 1-E). If the two crossover connections are close together, such a wrapped structure would appear to have an extended region of pairing when examined in the electron microscope. Indeed figure-8s and chi forms sometimes have this appearance.

Although the wrapped structure is twofold more complex, it is not fundamentally different from the single crossover structure shown in Figure 1. One imagines that such an intermediate may be matured by the same type of strand-nicking events that have been proposed for the Holliday-type recombination intermediate. A pair of maturation nicks are introduced independently at each of the crossover positions (Figure 6-G→I). After one crossover is matured, a structure with a single crossover connection results (compare Figure 6-H with Figure 1-I). Maturation at the second crossover finally separates the two recombinant genomes. These finished molecules will each have a region of potentially heterozygous DNA in the area of the crossover, which, as discussed earlier, is known from genetic studies to be a property of recombinant chromosomes. And, depending on the type of cutting that occurs at the two crossovers, there will be either a retention of the flanking genes in their parental linkage or, with equal probability, the production of a recombinant linkage for the flanking genes (again in accord with genetic findings about recombinant chromosomes).

How can the genetic data about asymmetric heterozygous regions be accommodated without relying on the DNA synthesis invoked by the Meselson-Radding model at the outset of recombination? Perhaps the formation of asymmetric heterozygous regions arises not at the time of the formation of

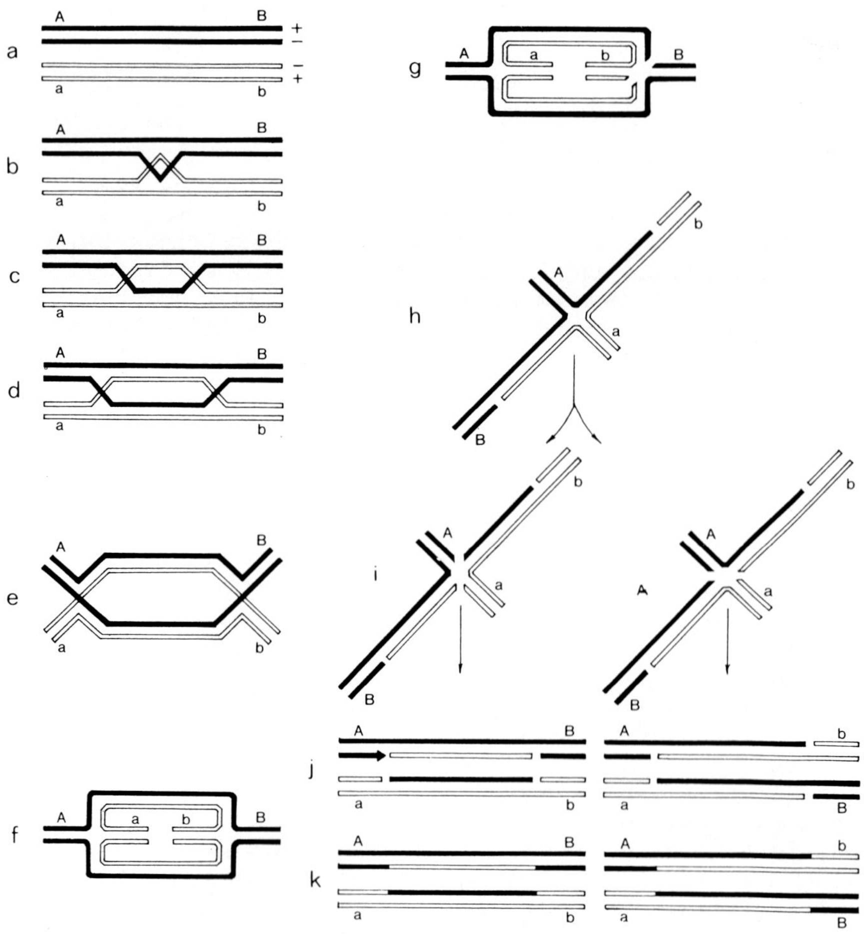

Figure 6. A prototype model for DNA recombination in which the recombination intermediate is formed by the interwrapping of complementary DNA strands after localized denaturation of two double helices at a region of homology. DNA synthesis is not involved in this initiation.

the recombination intermediate, but rather at the stage of maturation, when the wrapped intermediate is broken apart (15,27). Nicks are then introduced, which necessarily occur at the borders of the heteroduplex DNA (see Figure 6-J). If nick translation, or exonuclease digestion followed by resynthesis, occurs at this time (prior to sealing of the recombinant chromosomes), this will generate dissimilar regions of heterozygous DNA in the finished pair of recombinant chromosomes, and thus account for the genetic data. Even if a recombination intermediate is formed with a single crossover, as shown in Figure 1 C→D, nick translation at the time of maturation is still able to give asymmetric heteroduplex regions and remains an alternative to the model of Meselson and Radding.

Because several reasonable alternative mechanisms can be proposed for the formation of a recombination intermediate, to choose between the mechanisms, or to discover the true one, it will be necessary to purify recombination enzymes from cells and study their mode of action.

Genome Fusion Enzymes

Thus far, two enzymes have been purified, both from E. coli, which can carry out genome fusion reactions. In our laboratory, we have purified an enzyme called DNA synaptase, which is capable of fusing DNA molecules at a region of homology (15,28,29).

DNA synaptase was detected when monomer-size plasmid duplex rings were added to an unfractionated cell extract derived from E. coli and incubated at 37°C, with aliquots being withdrawn over a period of hours for analysis by electron microscopy. During the incubation, an increasing number of monomer rings were converted to multimeric circles (Figure 7). Routinely, the percentage of plasmids appearing as multimers rises from a background of less than 0.25% at 0 hours to about 20% at 20 hours.

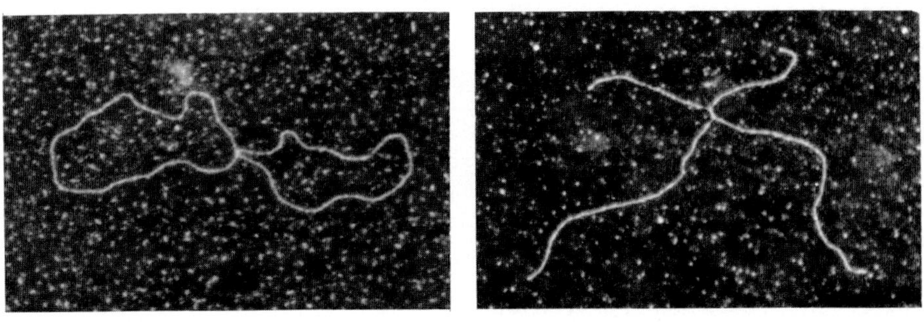

Figure 7. Electron micrographs of (A) a representative figure-8 and (B) a derived chi form made in vitro. One often observes (as in the chi form shown here) that the two genomes in the fused structure appear to be connected over a rather long region of homology, ranging up to about 500 base pairs in length.

As was the case for the in vivo figure-8s, we were able to show that the molecules formed in vitro by the unfractionated cell extract consist of two genomes covalently held together at a region of DNA homology. That is, one can convert the figure-8s to chi forms by cleavage with Eco R1.

Plasmid molecules linearized with a restriction enzyme are also substrates for the fusion reaction. In this case, the reaction products appear directly as chi forms (15).

Purification of DNA Synaptase

Although the electron microscope provided an excellent method for detecting genome fusion in cell extracts---because of the specificity with which the products of an incubation could be analyzed---it would not have been easy to use this technique for enzyme purification. This results from the fact that the electron microscopic analysis of a reaction mixture requires about an hour. Therefore to process the many samples that would arise as the proteins of the cell are fractionated---for example, across chromatographic columns, we developed a radio-biochemical assay for detecting genome fusion (28). In this assay one incubates a cellular protein fraction with two distinct but related plasmid DNA molecules. The first plasmid is heavily labeled with radioactive thymidine. The second plasmid is essentially homologous to the first but is non-radioactive and has been supplemented through recombinant DNA techniques to contain multiple copies of the _lac_ operator. The presence of the _lac_ operator DNA allows the second plasmid to be bound to a membrane filter in the presence of the _lac_ repressor protein.

To assay for genome fusion, the two plasmids are incubated together with a protein fraction, then exposed to repressor and passed through a nitrocellulose filter. The radioactivity of the first plasmid will only be retained on the filter if it becomes fused to an operator-containing plasmid.

This assay for genome fusion in vitro is (i) rapid, (ii) quantitative, (iii) applicable to the many parallel samples that arise during enzyme purification and, (iv) because it does not destroy the fused structures, allows their subsequent recovery from the filter and further analysis, for example by electron microscopy.

The radio-biochemical assay has been used to purify the genome fusion activity initially detected in the cell extract (29). The data in Figures 8-11 show how the DNA synaptase is purified to apparent homogeneity by the application of (1) ion-exchange chromatography on DEAE-sephacel, (2) molecular sieve chromatography on Bio Gel P-200,

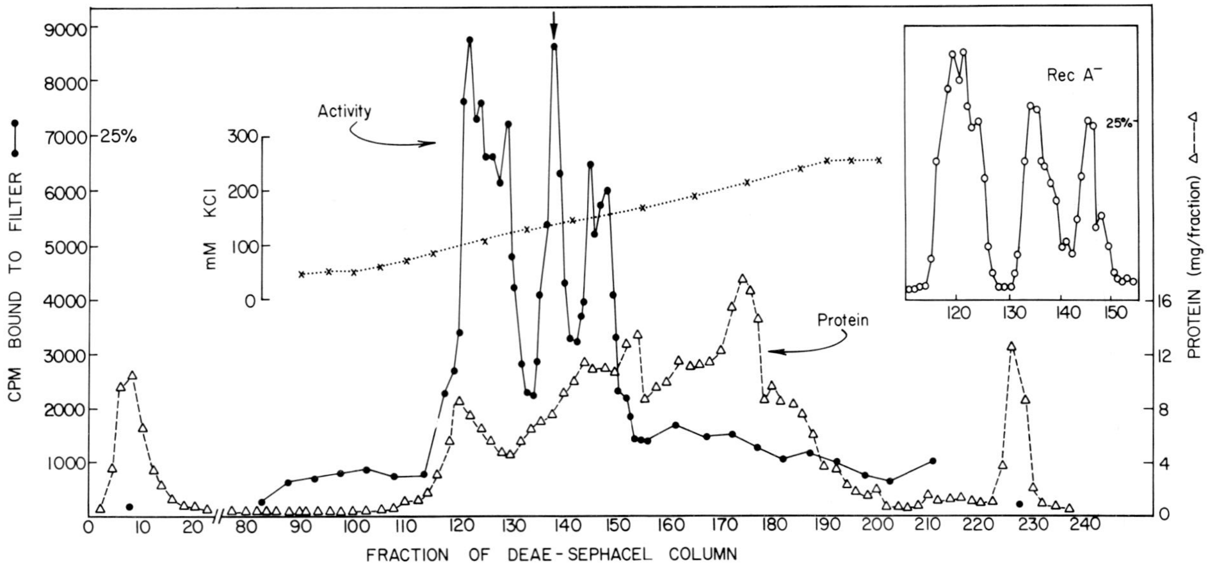

Figure 8. Purification of DNA synaptase by ion-exchange chromatography on DEAE-sephacel.

(3) affinity chromatography on DNA-cellulose, and (4) affinity chromatography on phenyl-sepharose. The purified enzyme without the aid of other proteins present in the cell extract is still able to fuse double-stranded genomes at a region of homology.

DEAE Ion-exchange Chromatography. Figure 8 shows the purification of the fusion activity by ion-exchange chromatography on DEAE-sephacel. Following an initial ammonium sulfate fractionation, the fusion activity was absorbed to the DEAE column in low salt and eluted with a KCl gradient. Each column fraction was then assayed for its ability to fuse radioactive and operator-containing plasmids. The fusion activity eluted at about 100 mM KCl, and showed evidence for a distribution into a small number of discrete peaks. The profile in Figure 8 shown was typical of five independent preparations.

The design of the assay leads one to expect that the retention of radioactivity on the filter represents the formation of figure-8 structures. This can be tested directly. An electron microscopic analysis of selected column fractions confirmed the results of the filter-binding assay: throughout the active areas of Figure 8, degradation of the input DNA did not occur and monomer plasmids were fused to form figure-8s and higher multimers. These structures could be cleaved by Eco R1 to yield chi forms.

The results of the DEAE-sephacel chromatography are somewhat unexpected in that they indicate that cells may contain three genome fusion activities. Alternatively, a single activity may occur in multiple forms.

The inset of Figure 8 shows another result one might not have anticipated. The fusion activity is clearly detectable in a preparation from Rec A^- cells. As shown in the inset, DEAE chromatography of an extract prepared from the Rec A^- strain MM 152 yields essentially the same activity profile as that obtained from Rec^+ cells. This result has encouraged us to broaden our view of genome fusion reactions to consider the possibility that such fusions may also play a role in cellular processes other than recombination.

For example the repair of damaged DNA could involve the reversible fusion of double helices at a region of homology. In this case the pairing would allow homologous DNA sequences to intertwine and serve as correctional templates for lesions in complementary DNA strands (30,31).

One may also consider the terminal segregation of two daughter DNA molecules as potentially involving a recombination-type event, in which fusion at a region of homology (the terminus of replication in a Cairns form) is reversed so as to achieve separation of the progeny chromosomes.

Lastly the as-yet not understood catenanes, or interlocked DNA circles, are structures whose formation or resolution might require a DNA fusion reaction, proceeding in an intermediate stage through a figure-8 form.

Because the DEAE-sephacel column appeared to resolve the fusion activity into several peaks, it was decided to focus first on one well-defined peak (marked with the arrow in Figure 8) for further purification. The fusion activity in this peak is associated with about 3% of the protein originally loaded on the column (compare the symbols ● and Δ in Figure 8).

Molecular Sieve Chromatography. The genome fusion activity recovered from the ion exchange column (Figure 8, arrow) was next purified on the basis of size using molecular

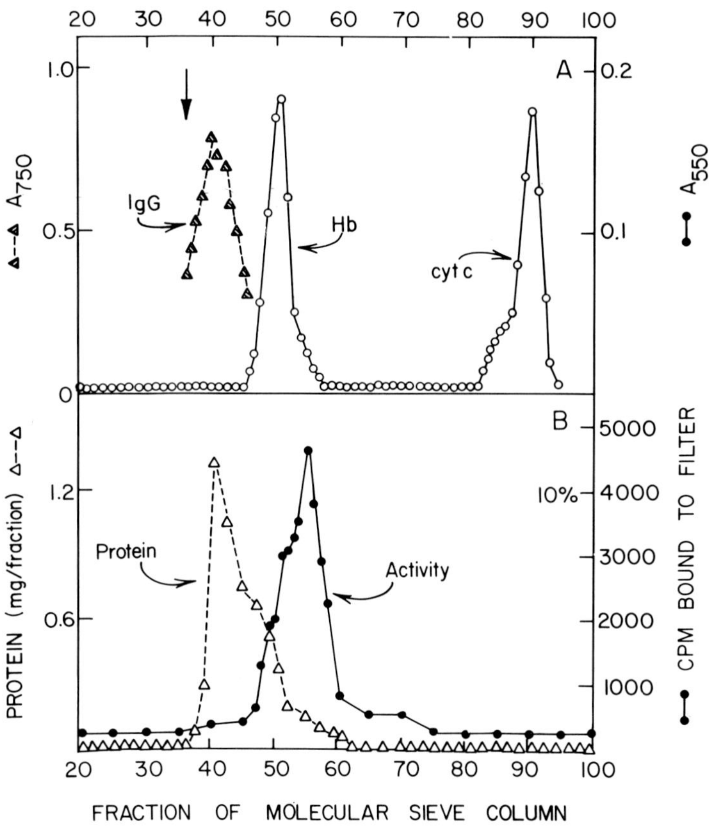

Figure 9. Purification of DNA synaptase by molecular sieve chromatography on BioGel P-200.

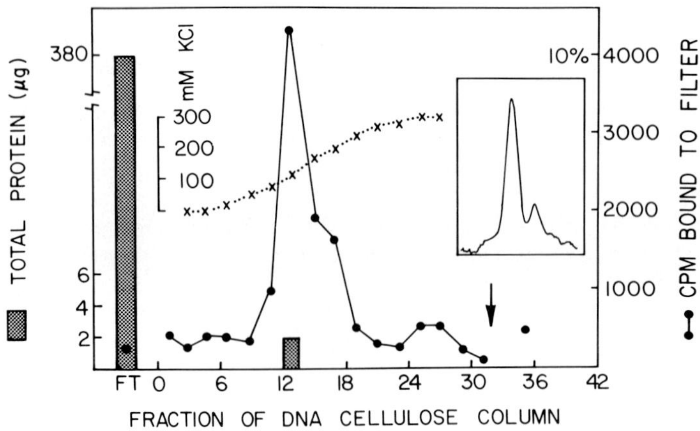

Figure 10. Purification of DNA synaptase by affinity chromatography on DNA-cellulose.

sieve chromatography. The column resin chosen was Bio Gel P-200 which, as shown in Figure 9A, was capable of resolving three marker proteins into distinct peaks: immunoglobulin G (150,000), hemoglobin (63,000), and cytochrome c (12,500).

An assay of the biological activity across the molecular sieve column gave the profile shown by the solid dots in Figure 9B. The peak of fusion activity was associated with about 5% of the protein originally loaded on the column and eluted at a position corresponding to a molecular weight of less than 60,000 daltons.

Affinity Chromatography on DNA-cellulose. The third column used for the purification of the fusion activity was DNA-cellulose. As shown in Figure 10, more than 95% of the protein loaded on the column (vertical bars) was recovered in the flow-through, while the fusion activity (●) bound to the DNA-cellulose column and eluted at about 120 mM KCl.

At this stage of the purification, polyacrylamide gel analysis showed only two proteins, with molecular weights of 33,000 and 30,000 daltons (Figure 10, inset). The fusion activity quantitatively co-purified with the 33,000 molecular weight protein when chromatographed on the final column, phenyl-sepharose.

Affinity Chromatography on Phenyl-sepharose. The last step in the purification of the fusion activity involved chromatography on phenyl-sepharose. This resin retains proteins on the basis of hydrophobic areas on their surfaces.

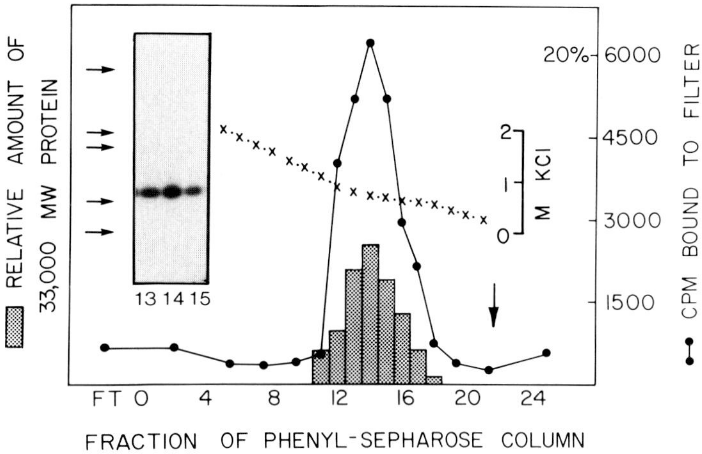

Figure 11. Purification of DNA synaptase by affinity chromatography on phenyl-sepharose.

The material recovered from the DNA-cellulose column was applied to phenyl-sepharose, and the bound proteins were eluted with a decreasing salt gradient. The fusion activity was recovered in a single peak (Figure 11).

At this stage the enzyme had been purified to apparent homogeneity. This was determined by iodination and polyacrylamide gel analysis of the phenyl-sepharose column fractions. As shown in the inset of Figure 11, a single protein appears whose amount (vertical bars) rises and falls coincident with the peak of biological activity. The monomer molecular weight of the fusion activity is judged to be about 33,000 daltons, by comparison with the gel profile of four marker proteins run in parallel (arrows).

Summary of the Purification and Yield. Four chromatographic columns were used in the purification of the DNA synaptase: DNA-sephacel, molecular sieve Bio Gel P-200, DNA-cellulose, and phenyl-sepharose. The recoveries from the four columns were estimated to be 92%, 45%, 40% and 74%.

The most highly purified material (from the phenyl-sepharose column) gave rise to a single band upon polyacrylamide gel analysis (Figure 11) and contained about 0.5 μg of protein. Correcting for the fact that during purification only the centers of the activity peaks were pooled, and that less than the entire pooled fraction derived from any one column was used on the subsequent column, it can be estimated that about 20 to 50 μg of the fusion protein may be obtained from about 100 liters of cells. This corresponds to about 10 to 20 molecules of DNA synaptase per logarithmically growing bacterial cell.

The DNA Fusion Reaction

The DNA synaptase was purified on the basis of its ability to join radioactive and operator-containing plasmids. However an electron microscopic analysis of Eco R1 cut molecules is necessary to establish the existence of fusion points and show that they occur at a region of DNA homology. Therefore, the most highly purified synaptase preparation, from the phenyl-sepharose column, was assayed with the electron microscope for its ability to form figure-8s which could be cleaved into chi forms. With the purified enzyme, the percentage of monomer circles that were fused to give multimers rose to 16% over a period of 20 hours. When

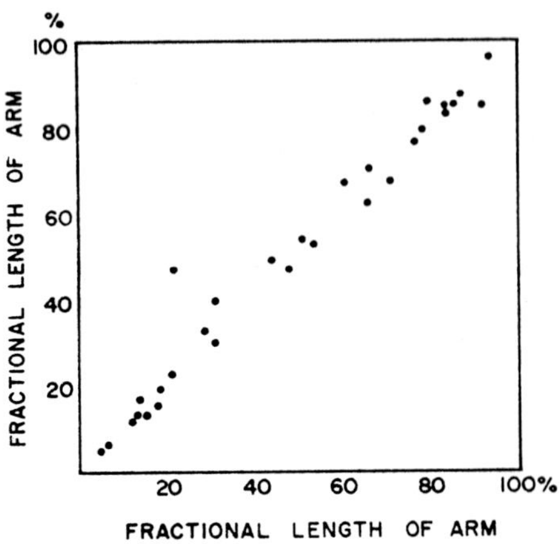

Figure 12. An analysis of the lengths of the arms in 15 randomly chosen chi forms derived from figure-8s made in vitro by the purified DNA synaptase. For each molecule the lengths of the four arms were measured, summed, and divided by two to obtain the unit genome length. The proportional lengths of the two shorter arms were then plotted, one as the abcissa and the other as the ordinate of a single point. Similarly, the two longer arms were used to produce a single point for the curve. The fact that the points generate essentially a straight line of slope 1 establishes that the chi forms contain pairs of equal length arms, and thus that fusion has occurred at a region of DNA homology. Furthermore, the finding that pairs of arms have different lengths indicates that the point of contact between the two genomes can occur at many locations, perhaps randomly.

an aliquot of the DNA was cleaved with Eco R1, a roughly equivalent number of chi forms were found, as in Figure 7.

By measuring the lengths of the pairs of arms in different chi forms, it was possible to conclude that the points of fusion were distributed at numerous sites along the length of the plasmid genome (Figure 12).

In sum, the purified DNA synaptase appears to catalyze the same reaction as was originally observed when the enzyme was present with numerous other proteins in the crude cell extract. The reaction with the purified enzyme proceeds in the presence of Mg, spermidine, and β-mercaptoethanol and does not require a high energy cofactor (29).

While the synaptase was detected and purified on the basis of its ability to fuse fully duplex rings to form figure-8 structures, it has also proved possible to observe an apparent partial reaction (29). If the purified protein is incubated with circular duplex plasmid molecules, and linear positive strands derived after Eco R1 digestion of the plasmid, the single strands become partially assimilated into the duplex rings. An example is shown in Figure 13A. Cleavage of these fusion products with Eco R1 linearizes the duplex ring element and shows that the single strand has become associated at a region of homology (Figure 13B). In this partial reaction the fusion enzyme shows a functional similarity with the Rec A protein of E. coli as will be discussed below.

Using the purified enzyme and DNA substrates appropriately labeled with radioactive and density isotopes, in future

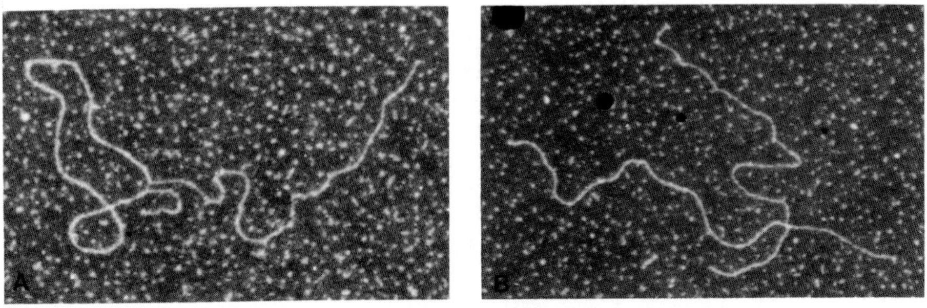

Figure 13. (A) The purified DNA synaptase has fused a plasmid duplex ring to a unit-length single strand, derived from the plasmid by Eco R1 digestion. (B) Upon Eco R1 digestion of the composite structure, the duplex ring is linearized and one sees that the single-stranded rod had been fused to the duplex ring at a region of homology.

experiments it should be possible to explore the range of substrates the synaptase can work upon, and to determine the biochemical mechanism by which the enzyme promotes genome fusion. Thus far our initial results with the purified enzyme indicate that intact superhelical DNA rings are not efficient substrates, and that a nick in the participating molecules is highly stimulatory.

The Rec A Protein

The other genome fusion enzyme that has been isolated is the product of the Rec A gene of *E. coli*.

The Rec A gene is pleiotropic with effects extending into 1) genetic recombination, 2) sensitivity to ultraviolet light. 3) inducibility of the λ prophage and 4) ability to support error-prone DNA repair (32-35).

Normally, the Rec A protein is present in small amounts in actively growing cells, but it becomes a major protein if the cells are exposed to DNA damaging agents such as ultraviolet light or nalidixic acid (36-39). This inducibility has been a great aid in the purification of the protein. Beginning with cells in which Rec A constitutes 5-10% of the cellular protein, several effective purification protocols have been developed (40-43). The Rec A gene product has been obtained in milligram amounts, and shown to be a 40,000 MW monomer that can exist in multimeric forms.

Roberts and Roberts discovered the first enzymatic activity to be associated with the Rec A protein---its ability to inactivate the λ phage repressor, by cleaving the repressor into two parts (44,40). This result strengthened the earlier suggestion that the Rec A protein might exert its pleiotropic effects by functioning as a regulatory protein---in this case by using its protease activity to activate pro-enzymes, or inactivate the repressors for various genes including those involved in recombination and repair.

More recent results, however, have suggested ways in which the Rec A protein might also participate directly in such processes as recombination. For instance, Ogawa et. al found that the purified Rec A protein has the ability to bind co-operatively to single-stranded DNA (45). Weinstock, McEntee, and Lehman then demonstrated that the separated single strands of phage P-22 DNA will reanneal more rapidly in the presence of large amounts of Rec A protein (39).

What is perhaps most striking in the reannealing reaction is that the amount of purified Rec A protein needed to effectively catalyze the reannealing is very high---one 40,000 MW monomer for every 100 bases of DNA. In more recent experiments, one monomer for every five bases of DNA has been found to be optimal. At this concentration, one might expect that DNA-binding and oligomerization of the rec A protein could serve to bring two DNA strands together anywhere along their length (rather than only at regions of homology). Once linked by protein-protein interactions, the DNA would exist at a high effective concentration, allowing an 'intramolecular' hybridization reaction to proceed very rapidly.

Further studies on Rec A-mediated DNA annealing by Shibata, DasGupta, Cunningham, and Radding and by McEntee, Weinstock and Lehman (42,46) have detected an allied reaction which potentially even more directly related to recombination. These investigators have shown that stoichiometric amounts of Rec A protein allow single-stranded DNA fragments to be locally taken up into homologous duplex DNA to form D-loops.

The current model for the action of the rec A protein assumes that the protein binds to single-stranded DNA, then denatures an area of target duplex DNA to form a non-specific, triple-stranded complex. This process would occur reiteratively, powered by the hydrolysis of ATP, until the single strand finds a region of homology in the duplex DNA and forms a D-loop. This could be the first step in a recombination process. But given the pleiotropic nature of the rec A gene, this reaction could also be a step in DNA repair for instance recombinational or 'post-replication' repair (30,31).

In that they can both promote the assimilation of a single strand into a duplex DNA molecule, the DNA synaptase and the rec A protein show a certain functional similarity. Nonetheless, the proteins are structurally and genetically distinct. Moreover DNA synaptase can fuse fully double-stranded DNA whereas the rec A protein does not. Another difference is that the synaptase appears to function catalytically whereas the rec A protein is needed in stoichiometric amounts. And finally, the requirement for ATP by rec A also serves to distinguish the two proteins.

Given the early state of knowledge about enzymes that can catalyze genome fusion reactions, discussions about recombination models must necessarily be speculative. One may expect, however, that the development of in vitro systems that carry out recombination-type reactions will make it possible to choose between the various models that have been proposed or to discover the true one.

REFERENCES

1. Holliday, R. 1964. Genet. Res. 5: 282-304.
2. Holliday, R. 1968 in Replication and Recombination of Genetic Material, eds. Peacock, W. and Brock, R. (Austrialian Acad. Sci., Canberra), pp. 157-174.
3. Holliday, R. 1974. Genetics 78: 273-287.
4. Emerson, S. 1969. Genetic Organization (Academic Press, New York), Vol I, pp. 267-360.
5. Sigal, N. and Alberts, B. 1972. J. Mol. Biol. 71: 789-793.
6. Pascher, A. 1918. Ber. Dtsch Bot. Ges. 36: 163-168.
7. Kitani, Y., Olive, L. and El-ani, A. 1962. Am. j. Bot. 49: 697-706.
8. Hurst, D., Fogel, S. and Mortimer, R. 1972. Proc. Natl. Acad. Sci. USA 69: 101-105.
9. Fogel, S., R. Mortimer, K. Lusnak and F. Tavares 1978. Cold Spring Harbor Symp. Quant. Biol. 43: 1325-1341.
10. Rossignol, J.L., N. Paquet and A. Nicolas 1978. Cold Spring Harbor Symp. Quant. Biol. 43: 1343-1352.
11. Boon, T. and N. Zinder 1971. J. Mol. Biol. 58: 133.
12. Broker, T. and Lehman, R. 1967. J. Mol. Biol. 60: 131-149.
13. Potter, H. and D. Dressler 1976. Proc. Natl. Acad. Sci. USA 73: 3000.
14. Potter, H. and D. Dressler 1977. Proc. Natl. Acad. Sci. USA 74: 4168.
15. Potter, H. and D. Dressler 1978. Cold Spring Harbor Symp. Quant. Biol. 43: 969.
16. Thompson, B., Escarmis, C., Parker, B., Slater, W., Doniger, J., Tessman, I., and Warner, R. 1975. J. Mol. Biol. 91: 409-419.
17. Thompson, B., M. Camien and R. Warner 1976. Proc. Natl. Acad. Sci. USA. 73: 2299.
18. Benbow, R., Zuccarelli, A., Shafer, A. and Sinsheimer, R. 1974. Mechanisms in Recombination, ed Grell, R. (Plenum Press, New York), pp. 3-18.
19. Benbow, R., Zuccarelli, A. and Sinsheimer R. 1975. Proc. Natl. Acad. Sci. USA 72: 235-239.
20. Valenzuela, M. and Inman, R. 1975. Proc. Natl. Acad. Sci. USA 72: 3024-3028.
21. Warner, R., R. Fishel, and F. Wheeler 1978. Cold Spring Harbor Symp. Quant. Biol. 43: 957-968.
22. Meselson, M. and Radding, C. 1975. Proc. Natl. Acad. Sci. USA 72: 358-361.
23. Cross, R. and Lieb, M. 1967. Genetics 57: 549-560.
24. Sobell, H. 1972. Proc. Natl. Acad. Sci. USA 69: 2483-2487.

25. Champoux, J. 1977. Proc. Natl. Acad. Sci. USA 74: 5328-5332.
26. Kirkegaard, K. and J. Wang 1978. Nucleic Acids Res. 5: 3811.
27. Potter, H. and D. Dressler 1978. Proc. Natl. Acad. Sci. USA 75: 3698.
28. Potter, H. and D. Dressler 1979. Proc. Natl. Acad. Sci. USA 76: 1084-1088.
29. Potter, H. and D. Dressler 1980. Proc. Natl. Acad. Sci. USA Vol. 77: March.
30. Rupp, W.D., C. Wilde, D. Reno, and P. Howard-Flanders 1971. J. Mol. Biol. 61: 25-44.
31. Clark, A. and M. Volkert 1978 in DNA Repair Mechanisms eds. Hanawalt P., E. Friedberg and F. Fox (Academic, New York) pp. 57-72.
32. Clark, A.J. and A.D. Margulies 1965. Proc. Natl. Acad. Sci. USA 53: 451.
33. Howard-Flanders, P. and L. Theriot 1966. Genetics 53: 1137.
34. Brooks, K. and A.J. Clark 1967. J. Virol. 1: 283.
35. Witkin, E. 1966. Bacterial. Rev. 40: 869.
36. Gudas, L. and Pardee, A.B. 1975. Proc. Natl. Acad. Sci. USA 72: 2330-2334.
37. McEntee, K. 1977. Proc. Natl. Acad. Sci. USA 74: 5275-5279.
38. Gudas, L. and Mount, D. 1977. Proc. Natl. Acad. Sci. USA 74: 5280-5284.
39. Emmerson, P.T. and West, S.C. 1977. Mol. Gen. Genet. 155: 77.
40. Roberts, J. and Roberts, C. and Craig, N. 1978. Proc. Natl. Acad. Sci. USA 75: 4714-4718.
41. Weinstock, G.M., McEntee, K., and Lehman, I.R. 1979. Proc. Natl. Acad. Sci. USA 76: 126-130.
42. Shibata, T., DasGupta, C., Cunningham, R., and Radding C. 1979. Proc. Natl. Acad. Sci. USA 76: 1638-1642.
43. West, S., E. Cassuto, J. Mursalim, and P. Howard-Flanders 1980. Proc. Natl. Acad. Sci. USA 77: March.
44. Roberts, J. and Roberts, C. 1975. Proc. Natl. Acad. Sci. USA 72: 147-151.
45. Ogawa, T., Wabiko, H., Tsurimoto, T., Horii, T., Masukata, H., Ogawa, H. 1978. Cold Spring Harbor Symp. Quant. Biol. 43: 909-915.
46. McEntee K., Weinstock, G., and Lehman, R. 1979. Proc. Natl. Acad. Sci. USA 76: 2615-2619.

ACKNOWLEDGMENTS

The work of our laboratory is made possible by research grants from the National Institutes of Health and the National Science Foundation, an award from the Dreyfus Teacher-Scholar Fund, and a Postdoctoral Fellowship from the American Cancer Society.

DISCUSSION

I. HERSKOWITZ, University of Oregon: If you try to fuse two different plasmid DNAs which are partially homologous and plot the lengths of the arms in the chi forms to test for equivalence, do you find places where the synaptase cannot promote synapsis and do these correspond to the places where there are no homologies?

D. DRESSLER: We have used the genome fusion enzyme to connect molecules which are only partially homologous, for instance pBR-322 and pMB-9, and when we make the plot (as in Figure 12) to see where the crossover connections occur, we find that they are occurring at the regions of homology.

P. STARLINGER, Universitat zu Koln, FRG: I have two questions. One is, did you say that this enzyme, though it works much better in principle than the Rec A enzyme, is not involved in recombination because it occurs in the Rec A^- cell?

D. DRESSLER: One could consider it a 'good' feature of this enzyme that it functions in catalytic amounts, as opposed to stoichiometric amounts. But how should we interpret the fact that DNA synaptase can be recovered from Rec A^- cells (as can two potentially similar synaptase activities). Ought not all recombination enzymes to be absent in Rec A^- cells? This is a difficult question for us, but we would argue this way. The finding of an enzyme that carries out a genome fusion reaction in Rec A^- cells may indicate that the Rec A^- cell still contains some of the enzymes that participate in physiological recombination---perhaps because they are also used in other areas of cell DNA metabolism. We are not ready to accept the assumption that all enzymes involved in recombination must be absent in Rec A^- cells.

A second point to keep in mind is that the synaptase and the Rec A protein might work synergistically. Perhaps they give a better genome fusion reaction together than either one separately. We are testing this possibility.

Basically, the point that Peter has brought out is a good one. We want to emphasize that we do not yet have a gene associated with the synaptase, and therefore we do not know what the physiological consequence of the absence of the enzyme would be. Without such genetic data, we lack a basis for saying that this genome fusion enzyme is necessarily to be thought of in terms of general genetic recombination. Nonetheless, the very existence of the synaptase implies that there are genome fusion events in the cell which, as yet, we know nothing about. Besides traditional recombination, they could be involved in a kind of recombinational DNA repair, or even in transposition-type movements. *E. coli* is still full of mysteries.

P. Starlinger: I have a second question. Your reaction goes to a remarkable extent, 25% or so fusion. Did you ever see in the electron microscope triple structures?

D. DRESSLER: Absolutely. The reaction goes through progressive stages. Initially, you build up figure-8s and then as time goes by, you have trimers, and then structures that look like the Olympic Seal with four or five circles fused together. And if you let the reaction go long enough, hundreds of DNA molecules become involved; by this stage one is basically adding circles onto pre-existing multimers. When we showed you figure-8s and chi forms, these were obtained by stopping the reaction early, before the larger multimers had built up.

G.R. FINK, Cornell University: Have you carried out the reaction, or tried to carry out the reaction, with completely non-homologous molecules?

D. DRESSLER: From studies with homologous plasmids, it is clear that at the end the reaction one has genomes fused at a region of homology. The thrust of Gerry's question is could an initial interaction occur at a region of non-homology such that a preliminary fusion occurs, with the strands then sliding relative to one another until a homologous area is found and the fusion product becomes stabilized. Does the fusion reaction occur only after you have found homology, or does the synaptase interact DNA molecules first and then let them find the homology because they are now so close together. That is certainly a possibility and it is an important one to explore. Now that we have the purified enzyme, we can examine the various kinds of substates it is able to work upon. We are testing the synaptase with completely non-homologous molecules. We do not yet have a clear result, and so we don't want to take a position on this question. We are using SV-40 DNA and a plasmid, two sequenced molecules which have no obvious extended homology. It should certainly be possible to obtain a clear result as to whether fusion can occur.

J.W. LITTLE, University of Arizona: Do you know if the reaction products will give rise to recombinants if you put them back into the cell?

D. DRESSLER: Yes the reaction products will give rise to recombinants if we transfect them back into cells with the calcium-heat-shock treatment. We are looking to see whether this resolution is blocked in Rec A^- cells. However such an experiment may have a potential pitfall. The problem is that replication of the two elements in the fused structure might occur within the recipient cell and would regenerate the two parental molecules. These could then undergo a full cycle of recombination in the recipient cells. If this were to happen, then one would not be studying the maturation of the fused structure at all. Once you put the fused molecule back into a cell, it could be hard to know whether the resolution of the structure occurs by recombinational maturation or by replication to give recombinant chromosomes indirectly.

A. BUKHARI, Cold Spring Harbor: Have you looked at those cells which overproduce the Rec A protein to see if DNA synaptase goes up or down in those cells? Can you look at the amount or activity of DNA synaptase to see whether the regulation of its synthesis is in any way connected with the regulation at the Rec A protein.

D. DRESSLER: That's a good point. There appear to be only about 10 to 20 molecules of the synaptase protein per cell. This is on the order of the lac repressor protein, and one of the next things we will do is treat cells with reagents like nalidixic acid and ultraviolet light to see if the level of synaptase is increased. It would certainly be interesting if the level of the fusion enzyme is increased by the same reagents that turn on the Rec A protein. That will make life easier in preparing more of this enzyme, and it will also tend to imply a relationship between the two in terms of function. They would both come into play under the same conditions of stress to the cell. That is a good suggestion, and we're working on that.

EXCISION AND REPLICATION SEQUENCES IN THE MITOCHONDRIAL GENOME OF YEAST

Giorgio Bernardi
Giuseppe Baldacci
Francoise Culard
Godeleine Faugeron-Fonty
Claire Gaillard
Regina Goursot
Francois Strauss
Miklos de Zamaroczy

Laboratoire de Genetique Moleculaire,
Institut de Recherche en Biologie Moleculaire
2, Place Jussieu, F-75005 Paris, France

I. INTRODUCTION

It is now well established (Bernardi, 1979; Faugeron-Fonty *et al.*, 1979) that the mitochondrial genomes of spontaneous cytoplasmic petite mutants of *Saccharomyces cerevisiae* originate by a mechanism involving: a) the excision of a segment from one of the 50-100 mitochondrial genome units of the parental wild-type cells; b) the amplification of the excised segment into tandem repeat units to form a defective mitochondrial genome unit (Fig. 1). In general, the latter will: a) segregate into one of the buds originating from the parental wild-type cell; b) give rise, in a few generations time, to the mitochondrial genome of petite mutant cells. The genome units of the petite mutant may in turn undergo further excision-amplification processes leading to secondary petite genomes having shorter tandem repeats.

We report here recent results on the nucleotide sequences involved in the excision process giving rise to the mitochondrial genomes of petite mutants. In addition, we present data on sequences which appear to correspond to the origins of replication of the mitochondrial genome.

In order to put the present work in a proper perspective, it will be useful to mention two points: a) previous work from our laboratory had shown that the mitochondrial genome units

ISBN 0-12-633360-2

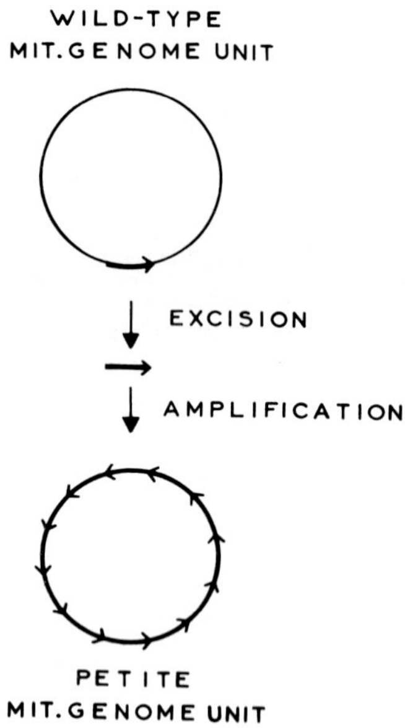

Fig. 1. Scheme of the process leading to the formation of spontaneous petite genomes (see Text).

of wild-type yeast cells are characterized by the presence of long AT spacers (which form at least half of each 50.10^6 genome unit and are formed by short alternating and non-alternating A:T sequences) and of many (a hundred or so) short GC clusters which are embedded in the AT spacers and consist, in most cases, of CCGG sequences clustered with GGCC or other GC-rich sequences; b) ten years ago, we proposed a deletion model as the basic mechanism underlying the petite mutation and suggested that excisions were due to illegitimate, site-specific recombination events taking advantage of homologous sequences which we supposed to occur frequently in the spacers of the mitochondrial genome units and to account for the extreme instability of the latter, the spontaneous petite mutation having a rate of 1-5% per generation (Bernardi, 1979).

II. RESULTS

A. *Excision Sequences Belong in Two Classes*

Restriction mapping of the repeat units of the mitochondrial genomes from six spontaneous petites (Fig. 2) and hybridization of these genomes on restriction fragments from the parental wild-type genomes have shown that these petite genomes and their excision sequences belong into two classes (Faugeron-Fonty *et al.*, 1979; Bernardi *et al.*, 1980). The repeat units of petite genomes of the first class appear to have been excised at (CCGG, GGCC) clusters, since: a) they are formed by Hpa II and Hae III fragments having the same sizes as Hpa II and Hae III fragments from the parental wild-type

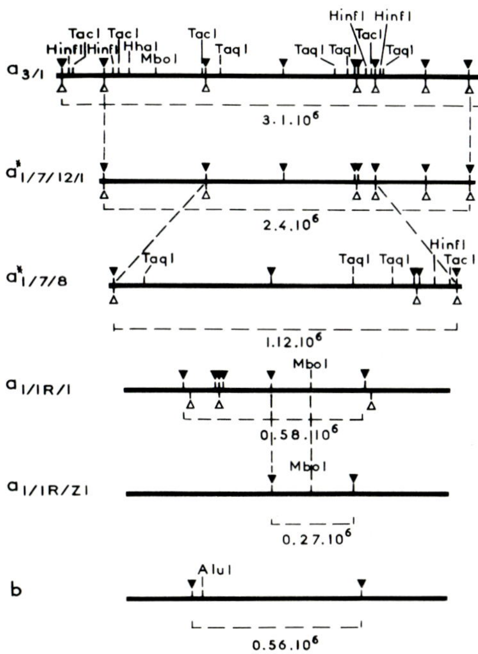

Fig. 2. Restriction enzyme maps of the repeating units of the mitochondrial genomes of six spontaneous petite mutants. The molecular weights of the repeat units are indicated, along with the positions of Hae III (Δ), Hpa II (Δ) and other restriction sites.

genomes (Hpa II and Hae III are restriction fragments splitting the sequences CCGG and GGCC, respectively); b) they hybridize on such parental fragments. Petite genomes of this class are exemplified by a*-1/7/8 and a-1/1R/1 (Fig. 2; for typographical reasons subscripts in the original code numbers of petites follow a dash here). In contrast, the repeat units of petite genomes of the second class were not excised at (CCGG, GGCC) clusters. Examples of these genomes are those of a-1/1R/Z1 and b (Fig. 2). The first of these genomes when degraded by Hpa II released a fragment corresponding in size to a Hpa II parental fragment; the petite genome did not, however, hybridize on this fragment, but did so on the two parental fragments corresponding to the two largest Hpa II fragments and to the larger Hae III fragment of a-1/1R/1 (Fig. 2). The second of these genomes, that of b, when degraded by Hpa II also released a fragment corresponding in size to a Hpa II fragment; its hybridization took place on two parental Hpa II fragments, one of which had about the same size as the petite Hpa II fragment; very recent sequences data (see below) indicate, however, that the latter finding is the result of a coincidence and that the repeat unit of b was not excised at (CCGG, GGCC) clusters nor at GC clusters containing CCGG sequence.

B. *Excision Sequences are Repeated Sequences*

So far, we have obtained definitive information on the excision sequences of two petites of the second class and preliminary results on those of one petite of the first class.

Several results strongly suggested that the repeat unit of a-1/1R/1 contained the repeat unit of a-1/1R/Z1 plus its flanking sequences (Faugeron-Fonty *et al.*, 1979): a) both petite genomes were originally present in the same spontaneous heterogeneous petite a-1/1R, from which they could be isolated by sub-cloning; b) they derived from the same region (map positions 27-46 of Sanders *et al.*, 1977) of the parental wild-type genome; c) they shared a Hpa II-Mbo I fragment (Fig. 2); d) they showed overlapping hybridization patterns on Hpa II and Hae III parental fragments (see above). All these data suggested that the genome of a-1/1R/Z1 could be a secondary excision product of a-1/1R/1; alternatively, the two petite genomes could derive from two overlapping segments of the parental wild-type genome. As expected, the 884 nucleotide sequence of the repeat unit of a-1/1R/1 (Fig. 3) was found (Gaillard *et al.*, 1980) to contain a 416 nucleotide sequence which was identical to that, previously sequenced by Gaillard and Bernardi (1979), of a-1/1R/Z1, except for two base pair changes. Very interestingly, the initial 13-nucleotide sequence of a-1/1R/Z1 was found to be repeated (with one A:T to T:A change) just after the end of the repeat unit.

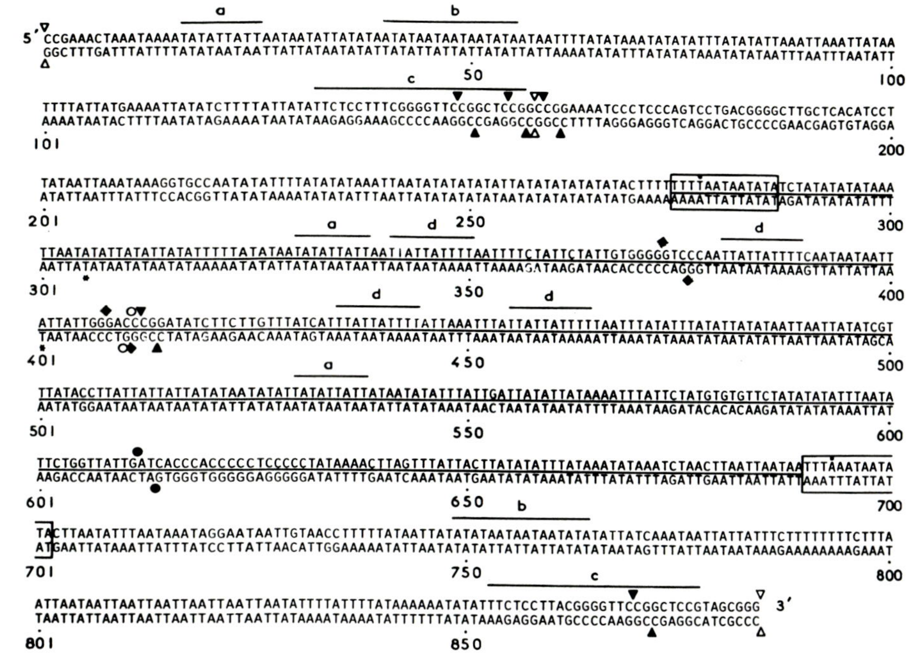

Fig. 3. Nucleotide sequence of the repeat unit of the mitochondrial genome of spontaneous petite mutant a-1/1R/1. Cutting sites of Hae III (⧸▼), Hpa II (▼), Mbo I (●), Mbo II (O) and Ava II (◆) are indicated. The sequence marked by the continous line is identical with the repeat unit of a-1/1R/Z1 (Gaillard and Bernardi, 1979), except for the asterisked base pairs. Boxed sequences are the direct repeats flanking the sequence of a-1/1R/Z1 (asterisks mark a base pair difference). Lines indicate some of the repeated sequences.

The sequence of the repeat unit of another petite, provisionally called a-1/1R/14, also related to a-1/1R/1 revealed that again an initial 13-nucleotide sequence is repeated just after the end of the repeat unit. The sequence under consideration is AATAATTATTATT at positions 394-406 and 773-785 on the a-1/1R/1 repeat (Fig. 3).

Finally, the partial primary structure of the excision sequences of a*-1/7/8 (Fig. 2) has shown that they correspond to two direct repeats of a-3/1, each one of which is characterized by the clustering of a Tac-I, a Hae III and a Hpa II site (Fig. 2).

In conclusion, the three excision sequences known so far correspond to three sets of direct repeats, two of which are in AT spacers and one in GC clusters.

C. Genomes without Genes

An inspection of the sequence of a-1/1R/1 confirms the sequence features predicted by our previous work for both AT spacers and GC clusters, but fails to reveal the presence of any gene or gene fragment, a conclusion supported by the current knowledge of codons used for the synthesis of mitochondrially encoded proteins (Macino and Tzagoloff, 1979) and of tRNA gene sequences (Martin *et al.*, 1979). In all likelihood, the only function left in this petite genome is replication. Since the excision mechanism leading to the formation of spontaneous petite genomes is extremely conservative as far as the excised sequence is concerned (in contrast to that induced by ethidium, where sequence rearrangements of different kinds are frequent; Lewin *et al.*, 1978), each repeat unit of a-1/1R/1 should contain an origin of replication. A good candidate for this role is the 80-nucleotide sequence centered around position 412, since it is characterized by features already found in replication origins of other genomes. In its left half, this sequence contains a palindrome 23 nucleotides long (378-400) and a small symmetrical sequence TTAATT (402-406), which are flanked by two inverted repeats formed by G:C base pairs only; in its right half, the sequence contains a decanucleotide formed by A:T base pairs only, which is a direct repeat of another sequence present in the left half. The central feature of the sequence is a very characteristic cluster of a Ava II, Mbo II and Hpa II cutting sites (the recognition site of Mbo II being a pentanucleotide 8 nucleotides away from the cutting site, this cluster corresponds to 13 nucleotides in a sequence of 16; see Figs. 3 and 5).

D. Origins of Replication of the Mitochondrial Genome

To test the hypothesis that the 80-nucleotide sequence just discussed corresponds to the origin of replication of a-1/1R/1 we decided to look for its most salient features, the central restriction site cluster, in the repeat units of b and a*-1/7/8, two petites whose genomes arose from two other regions (Fig. 4) of the mitochondrial genome of the parental wild-type cells. This search having proved successful, we determined the two sequences flanking the cluster (de Zamaroczy *et al.*, 1979) and found them to be identical to that of a-1/1R/1, except for a few base pair changes (Fig. 5). Incidentally, the sequence of Fig. 5 proves that the repeat unit of

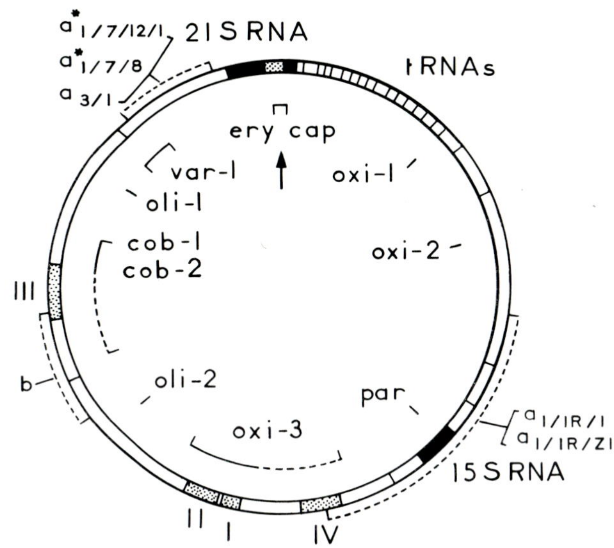

Fig. 4. Regions of excision of the petite genomes of Fig. 2 on a general map (adapted from Borst and Grivell, 1978) of the mitochondrial genome of wild-type yeast cells.

b was not excised at or around the Hpa II site, since the 80 nucleotides around it clearly were not interrupted by the excision process.

The finding of the 80-nucleotide sequence in the mitochondrial genome of a*-1/7/8 strongly suggests that this sequence is also present in the genomes of a*-1/7/12/1 and a-3/1, since their repeat units encompass that of a*-1/7/8. If such is the case, then all six spontaneous petite genomes of Fig. 2 contain the sequence under consideration, in spite of their origin from three distinct regions of the parental wildtype genome (Fig. 4). Interestingly, one of these petites, b, originated in fact from a different parental strain (strain B), compared to all the other ones, which originated from strain A; its mitochondrial DNA showed, however, the same hybridization patterns on restriction fragments from the mitochondiral DNAs of both strains A and B.

Finally, it should be mentioned that petite a-1/1R/14 lacks the 20 or so left-most nucleotides in the putative replication origin, showing that this stretch is not essential.

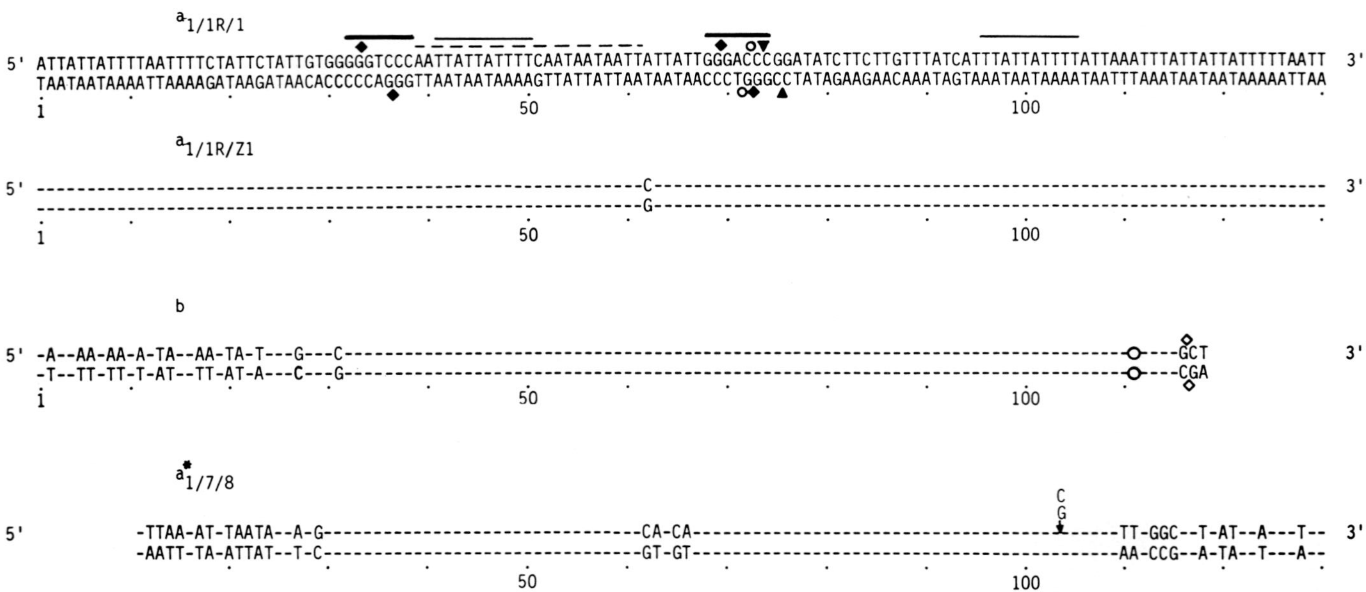

Fig. 5. *Nucleotide sequence of the repeat unit of the mitochondrial genome of a-1/1R/1 (Fig. 3) between base pairs 340 and 470, of the corresponding segment of a-1/1R/Z1 and of the region around the Hpa II, Mbo II, Ava II cluster of the repeat units of b and a*-1/7/8. Dashes indicate nucleotides identical to those found in a-1/1R/1, taken as a reference. One base pair deletion in b is indicated by a double empty circle, one base pair addition in a*-1/7/8 by an arrow. Restriction sites are indicated as in Fig. 3; Alu I by ◇. The 23-base pair palindrome is indicated by a broken line, the 7-base pair inverted repeats by heavy lines, the 10-base pair direct repeat by thin lines.*

E. Supersuppressive Petites

Even if the circumstantial evidence just presented is very strong, no direct evidence is available yet to prove that the 80-nucleotide sequences correspond to origins of replication of mitochondrial DNA. Two additional, independent lines of evidence support this view, however: a) the results presented in the previous section imply that the mitochondrial genome units of wild-type cells have at least three (actual or potential) origins of replications; in agreement with this idea is the finding that the regions from which the petite genomes of Fig. 2 arose, although in need of a better definition, overlap with the regions from which spontaneous petites arise with the highest frequency (Mathews *et al.*, 1977); b) the four petite genomes of Fig. 5 are supersuppressive (Goursot *et al.*, 1980; the evidence for a*-1/7/8 being still preliminary), namely they are selectively transmitted to the progeny from crosses with wild-type cells without any apparent sequence change; the simplest explanation for this phenomenon is that multiple replication origins are present in these petite genomes, one per repeat unit, and this allows them to compete out the mitochondrial genome of wild-type cells in crosses, and this is spite of the fact that the wild-type cells issued from the cross have a selective growth advantage over the petite mutants.

III. DISCUSSION

A. The Petite Mutation, Neutrality and Suppressivity

The results just described are directly relevant for our understanding of the mechanism of the spontaneous petite mutation as well as of the phenomena of neutrality and suppresivity. As far as the first point is concerned, while it will be of interest to know about the primary structure of more excision sequences, it is already clear that the excision mechanism experimentally found completely fulfills the hypothesis put forward ten years ago. In contrast, nothing is known about the amplification mechanism. Concerning the neutrality-suppresivity issue (crosses of wild-type cells with netural petites yield wild-type cells only; those with suppressive petites yield both wild-type cells and petite mutants in ratios essentially dependent upon the petite used in the cross), this is clarified to some extent by the results on the replication origin. Two situations appear to be quite clear. Supersuppressive petites, when crossed with wild-type cells, yield almost exclusively petite mutants, the mitochondrial genome of which seems to be identical to that of the petite entering the cross.

As already mentioned, this situation appears to stem from the high replicative efficiency of supersuppressive genomes, which are characterized by very short repeat units comprising an origin of replication, and which can therefore compete out the wild-type genomes. It is obvious that a supersuppressive genome of very low complexity and very low concentration of excision sequences can only undergo mutations affecting the replicative function; when this is lost, cells without mitochondrial genomes are formed (Nagley and Linnane, 1970; Goldring *et al.*, 1970), which will behave as neutral in crosses. The other two situations require further investigations. Some neutral petites contain a mitochondrial genome, the best documented case being that of RD1A (Moustacchi, 1972). In this case, it is known (van Kreijl and Bos, 1977) that *most* of the mitochondrial DNA is present in an extremely short (68 nucleotides) repeat unit which has nothing in common with the origins of replication sequenced here. Since this petite is the result of a drastic ethidium treatment, which is known to cause a massive breakdown of mitochondrial DNA (Goldring *et al.*, 1970) followed by genome rearrangements (Lewin *et al.*, 1978), we suggest that the lack of replicative competitivity of this genome is due to the fact that it may only carry one origin of replication translocated on the amplified repeat unit. The final case, that of suppressive petite, may be due to a number of reasons. The study of the suppressivity of a-1/1R/14, which lacks part of the origin of replication, will be of great interest in this respect. For instance, the length of the repeat unit and changes in the origin of replication may affect the replicative competitivity of a petite. Fig. 6 presents a scheme of the different situations encountered so far in petites, along with suggested abbreviations.

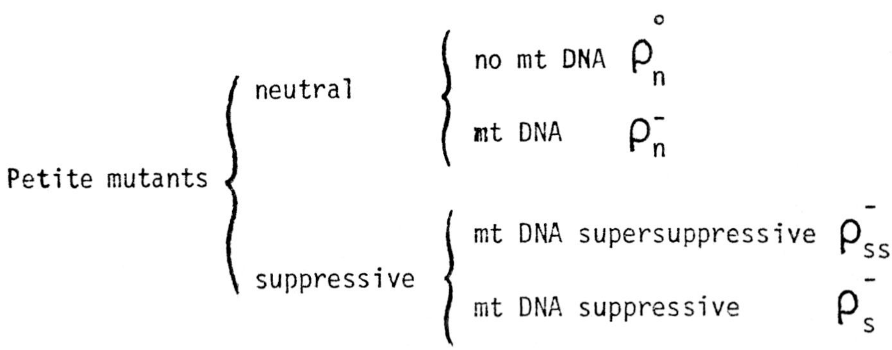

Fig. 6. Scheme of a classification of petite mutants (see text). Suggested abbreviations are indicated.

B. The General Significance of Excision Sequences

There is little doubt that the direct repeats investigated here are responsible, by a mechanism involving a site-specific, illegitimate recombination, for the excision of the repeat units of petite genomes, and therefore for the extreme instability of the mitochondrial genome of wild-type cells. It is also quite possible that mitochondrial genome segments bordered by such sequences can be transposed, by an excision-insertion mechanism, onto other mitochondrial genome units present in the same cell. This mechanism might account for three different findings: a) the different lengths of "allelic" spacers (namely of spacers contiguous to allelic genes) which were observed in the mitochondrial genomes of different yeast strains (Bernardi *et al.*, 1975; Prunell *et al.*, 1977); b) the insertions which are present in the genomes of some strains (Sanders *et al.*, 1977); c) the existence of multiple origins of replication (de Zamaroczy *et al.*, 1980).

One may wonder, however, about the "physiological" role of "excision" sequences in the mitochondrial genome of yeast. There are at least three sorts of answers to such question. Taking into account the diversity of excision sequences, these answers need not be mutually exclusive. Since excision sequences are so reminiscent of those found, for example, on both sides of insertion sequence IS1 and of transpoon Tn 9 in *E. coli* (Calos *et al.*, 1978; Grindley, 1978; Johnsrud *et al.*, 1978), the first answer may be that excision sequences play indeed the same role as those prokaryotic sequences; if this answer is correct, it implies that this type of sequence is conserved in organisms very far away from prokaryotes. The second sort of answer is that these sequences are normally used in site-specific recombination processes among genome units of the same vegetative cells or of different cells in crosses; evidences for the high frequency of both processes have been presented elsewhere (Bernardi *et al.*, 1975; Prunell *et al.*, 1977; Fonty *et al.*, 1978). The thid sort of answer has to do with regulatory functions of some kind; GC clusters, for instance, might play a primary role as sequences recognized by DNA replicase, RNA polymerase, enzymes of RNA processing (once they are transcribed), regulatory proteins (Prunell and Bernardi, 1977). Further work should clarify these issues.

REFERENCES

Bernardi, G. (1979). *Trends in Biochem. Sci.*, *4*, 197-201.

Bernardi, G., Prunell, A., and Kopecka, H. (1975). In *Molecular Biology of Nucleocytoplasmic Relationships* (Puiseux-Dao, S., ed.). pp. 85-90. Elsevier, Amsterdam.

Bernardi, G., Baldacci, G., Culard, F., Faugeron-Fonty, G., Gaillard, C., Goursot, R., Strauss, F., and de Zamaroczy, M. (1980) in FEBS Symposium on DNA (Sponar, J., and Zadrazil, S., eds.). In press, Pergamon Press, New York.

Borst, P., and Grivell, L. A. (1978). *Cell 15,* 705-723.

Calos, M. P., Johnsrud, L., and Miller, J. H. (1978). *Cell 13,* 411-418.

Faugeron-Fonty, G., Culard, F., Baldacci, G., Goursot, R., Prunell, A., and Bernardi, G. (1979). *J. Mol. Biol. 134,* 493-538.

Fonty, G., Goursot, R., Wilkie, D., and Bernardi, G. (1978). *J. Mol. Biol. 119,* 213-235.

Gaillard, C., and Bernardi, G. (1979). *Molec. Gen. Genet. 174,* 335-337.

Gaillard, C., Strauss, F., and Bernardi, G. (1980). *Nature,* in press.

Goldring, E. S., Grossman, L. I., Krupnick, D., Cryer, D. R., and Marmur, J., *J. Mol. Biol. 52,* 323-335 (1970).

Goursot, R., de Zamaroczy, M., Baldacci, G., and Bernardi, G. (1980). *Current Genetics,* in press.

Grindley, N. D. F. (1978). *Cell 13,* 419-426.

Johnsrud, L., Calos, M. P., and Miller, J. H. (1978). *Cell 15,* 1209-1219.

van Kreijl, C. F., and Bos, J. L. (1977). *Nucl. Acids. Res. 4,* 2369-2388.

Lewin, A., Morimoto, R., Rabinowitz, M., and Fukuhara, H. (1978). *Molec. Gen. Genet. 163,* 257-275.

Macino, G., and Tzagoloff, A. (1979). *Proc. Nat. Acad. Sci. USA 76,* 131-135.

Martin, N. C., Miller, D., and Donelson, J. E. (1979). *J. Supramol. Struct., Suppl. 3,* 132.

Mathews, S., Schweyen, R. J., and Kaudewitz, F. (1977). In *Mitochondria 1977,* (Bandlow, W., *et al.*, eds.). pp. 133-138, de Gruyter, Berlin.

Moustacchi, E. (1972). *Biochim. Biophys. Acta 277,* 59-60.

Nagley, P., and Linnane, A. W. (1970). *Biochem. Biophys. Res. Comm. 39,* 989-996.

Prunell, A., and Bernardi, G. (1977). *J. Mol. Biol. 110,* 53-74.

Prunell, A., Kopecka, H., Strauss,F., and Bernardi, G. (1977). *J. Mol. Biol. 110,* 17-52.

Sanders, J. P. M., Heyting, C., Verbeit, M. P., Meijlink, F. C. P. W., and Borst, P. (1977). *Molec. Gen. Genet. 157,* 239-261.

de Zamaroczy, M., Baldacci, G., and Bernardi, G. (1979). *FEBS Letters 108,* 429-432.

DISCUSSION

P. BERG: What is the state of the DNA in the yeast mitochondria in terms of the proteins with which it is associated? How does it compare to the DNA in the nucleus?

G. BERNARDI: According to a very recent paper by F. Caron, C. Jacq and J. Rouviere-Yanic (Proc. Natl. Acad. Sci. USA *76*, 4265-4269) yeast mitochondria do not contain histones but have an abundance of 20,000 dalton DNA-binding protein which introduced superhelical turns in relaxed circular DNA in the presence of a nicking-closing enzyme like histones. This protein is slightly basic and heat-stable and might play the role of histones.

P. BERG: The reason for my question is whether the kind of recombinational excision you proposed is something that could occur in genomic DNA elements with which histones and other accessory proteins are associated; or, is this phenomenon restricted to "unprotected" types of DNA genomes?

G. BERNARDI: I agree with you that one may wonder about the inhibiting role on recombination by the complex structure of chromosomes. On the other hand, it is also true that the relative amount of interspersed repetitive sequences is much higher in the mitochondrial genome of yeast compared, for instance, to its nuclear counterpart.

C. HERSCHBERGER: If you introduce suppressive petite DNA into a rho^{o} strain can you get transformation or are you dependent strictly on a genetic cross? Can you displace the wild-type mitochondrial genome in a transformation type system with the petite DNA?

G. BERNARDI: You can transfer, without any problem, any mitochondrial genome into rho^{o} cells, namely cells deprived of mitochondrial genome. In fact, this is a nice way to see the effect of different nuclear backgrounds on the mitochondrial genome function.

C. HERSCHBERGER: Are these transferred by genetic crosses?

G. BERNARDI: Yes.

C. HERSCHBERGER: Can you do a similar experiment using yeast transformation system where you introduce the purified DNA into the rho^{o} or wild-type cells?

G. BERNARDI: I suppose so, but to my knowledge nobody has managed, so far, to do it. I think some attempts have been made, and probably somebody will succeed one day or another.

A. BOLLON: Are there any common sequences at all between the mitochondrial genome and the nuclear genome?

G. BERNARDI: Some old hybridization experiments done when nick translation was not a current technique, which are against the idea of common sequences, but those experiments should be repeated because now techniques are so much more sophisticated.

CONTROL OF CELL TYPE IN YEAST BY GENETIC CASSETTES

Ira Herskowitz[1]
Jasper Rine
George Sprague, Jr.
Rob Jensen

Institute of Molecular Biology
and
Department of Biology
University of Oregon
Eugene, Oregon

The yeast genome has three loci (MAT, HML, and HMR) which can harbor either an a or an α cassette. The cassette at MAT (the mating type locus) is expressed and determines the cell type: MATa cells have 'a' mating type and MATα cells have 'α' mating type. The cassettes at HML and HMR ("storage" or "library" loci) are not expressed due to the action of Sir gene products. The information at HML or HMR becomes active by transposition to MAT, in which case the cassette previously at MAT is discarded. The transposition event is a non-reciprocal transfer (a "transmission") of information from HML or HMR to the mating locus. This process occurs at low frequency in HMLα HMRa ho strains and at high frequency in HMLα HMRa HO strains and is under a variety of controls.

I. INTRODUCTION

Haploid yeast cells undergo a remarkable switching process in which cells of one type ('α') switch efficiently to another type ('a') in a specific pattern (Strathern and

[1]Supported by a Research Career Development Award (AI 00163) and Research Grant (AI 13462) from the N.I.H., by a Molecular Biology Training Grant and a National Research Service Award in Genetics from the N.I.H., and by an American Cancer Society postdoctoral fellowship (PF 1282) to G.S.

ISBN 0-12-633360-2

Herskowitz, 1979). This change in cell type reflects a change in the alleles of the mating type locus (MAT): cells with the a allele are of 'a' mating type, and cells with the α allele are of 'α' mating type. Cells of opposite mating type mate efficiently with each other to form a/α diploids, which are unable to mate. These diploids, unlike a/a or α/α diploids (formed from a/α cells by mitotic recombination) are able to sporulate and give rise to tetrads containing four haploid meiotic products, 2 'a' and 2 'α'. This segregation pattern indicates that cell type (mating type) is determined by alleles of a single locus (MAT), which has been mapped to chromosome III (Mortimer and Hawthorne, 1969).

Because switching between mating types is very efficient (occurring in approximately 73% of cell divisions by "experienced" cells--see Strathern and Herskowitz, 1979), haploid a or α cells grow into colonies composed primarily of a/α diploid cells formed by matings between siblings of opposite mating type (Hawthorne, 1963b; Oshima and Takano, 1971; Hicks and Herskowitz, 1976). (The mating type switching system is inactive in a/α cells; hence a/α cells are stable during mitotic growth and do not give rise to a/a or α/α cells). The process by which haploids give rise to diploids is called "diploidization" (Winge and Roberts, 1949); strains with this ability are termed "homothallic".

The ability of cells to switch between a and α indicates that all yeast cells contain the genetic information to be a or α but that only one or the other is expressed. Our studies have centered on elucidating how the mating type locus determines cell type and on the mechanism of mating type interconversion. Our working hypothesis for the role of the mating type locus itself, based on a hypothesis originally proposed by MacKay and Manney (1974), is that the alleles of the mating type locus code for regulatory proteins which govern expression of genes concerned with mating and sporulation located elsewhere in the genome (see Herskowitz *et al.*, 1980). To account for mating type interconversion, we have proposed that the alleles of the mating type locus, a and α, are transposable genetic elements whose position in the genome determines whether they are expressed. According to this hypothesis (the "cassette model"), the yeast genome contains silent a and α information at "storage" loci (HML and HMR, see below); this information becomes expressed by a non-reciprocal transmission of information into the mating type locus, which acts like the playback head of a tape recorder in that it expresses whatever information is plugged into it (see model 4 below). Why a and α cassettes are silent at *HML* and *HMR* and expressed at *MAT* is discussed below. Sufficient mechanistic detail on the interconversion process is known to allow an additional useful analogy to an electronic calculator. In this analogy, *HML* and *HMR* are STORAGE registers, and the mating type locus is the DISPLAY register. Starting with α in DISPLAY (at the mating type locus), *a* information is placed in DISPLAY by replacing the *α* information with *a* information from STORAGE. In an electronic calculator as in mating type interconversion, the information previously in DISPLAY is discarded; furthermore, the information in STORAGE (at *HML* and *HMR*) remains intact.

This hypothesis has been derived from genetic and physiological analysis of natural variants and mutants of *S. cerevisiae*. In this paper, we describe the discovery of the main protagonists in mating type interconversion, the loci *HML*, *HMR*, and *HO* (Winge and Roberts, 1949; Takano and Oshima, 1967; Harashima *et al*., 1974; see also Santa Maria and Vidal, 1970, and Naumov and Tolstorukov, 1973). From these findings, a variety of models for mating type interconversion involving genetic rearrangement can be drawn in which *HML* and *HMR* play differing roles. These models are discussed in their own right and as foils to the cassette model. In particular, these models are evaluated in light of additional information on mating type interconversion and on tests of the cassette model. We conclude by discussing extensions of the cassette model to cells with multiple cassettes, in which a single cell can give rise to multiple cell types in a specific sequence.

II. IDENTIFICATION OF *HO*, *HML*, AND *HMR* LOCI

HO: Not all yeast strains switch mating types efficiently. Conventional laboratory strains interconvert mating types at a frequency of approximately 10^{-6} (Hawthorne, 1963a; Rabin, 1970; Hicks and Herskowitz, 1976). These strains (termed "heterothallic") differ from strains with high switching frequency at a single genetic locus (Winge and Roberts, 1949). *HO*, which is unlinked to the mating type locus (and is on chromosome IV; G. Kawasaki, personal communication) promotes efficient switching both from *a* to *α* and from *α* to *a*. The behavior of *HO* and *ho* can be seen clearly in a cross between *a* *HO* and α *ho* strains.

TABLE I. Segregation from an a/α HO/ho Diploid

genotype	phenotype	genotype	phenotype	genotype	phenotype
a HO	D	a ho	a	a ho	a
a HO	D	a ho	a	a HO	D
α ho	α	α HO	D	α ho	α
α ho	α	α HO	D	α HO	D
Type I Tetrad		Type II Tetrad		Type III Tetrad	

In this analysis, the a/α HO/ho diploid is induced to sporulate, and haploid segregants are allowed to grow into colonies. These spore clones are then tested both for mating ability and for ability to sporulate. ho segregants exhibit mating ability and do not sporulate, whereas HO segregants diploidize, that is, grow into a/α diploids, which sporulate but do not mate. Three types of tetrads are obtained, each of which has two mating segregants (either a or α) and two segregants which diploidize (D). In this type of segregation, unlike those described below, the number of segregants which exhibit a mating ability is equal to the number with α mating ability.

HMR: The HMR locus was identified as a natural variant allele in strains of *Saccharomyces norbensis* (Santa Maria and Vidal, 1970; Harashima *et al.*, 1974) in an analysis of the following kind. An *S. norbensis* strain was observed to have stable α mating type even though crosses to heterothallic *S. cerevisiae* strains revealed that it carried HO: diploids formed between the α *S. norbensis* strain and an *S. cerevisiae* a ho strain yielded segregants which diploidized. Why do *S. norbensis* strains maintain stable α mating type despite carrying HO? Analysis of crosses between *S. norbensis* and a HO *S. cerevisiae* strains were particularly informative. In this case, the following tetrad types were observed.

TABLE II. Segregation from a Diploid [a/α HO/HO HMR/HMR'] Formed between α S. norbensis and an a HO S. cerevisiae Strain

phenotype	inferred genotype of spore	phenotype	inferred genotype of spore	phenotype	inferred genotype of spore
D	a HMR	D	a HMR'	D	a HMR
D	a HMR	D	a HMR'	D	a HMR'
α	α HMR'	D	α HMR	D	α HMR
α	α HMR'	D	α HMR	α	α HMR'
Type I Tetrad		Type II Tetrad		Type III Tetrad	

The striking observation is that 'α' segregants were observed in approximately one quarter of the spore clones; no 'a' segregants were observed. These tetrad types (and their frequencies of occurrence, which are not discussed here) can be rationalized by proposing that *S. norbensis* contains a defect in a gene necessary for α cells to switch to a. For this discussion, the allele from *S. cerevisiae* is denoted HMR, and the allele from *S. norbensis* HMR'. This locus maps to chromosome III, distal to the *thr4* locus, and is 57 centimorgans from the mating type locus (Harashima *et al.*, 1974; Harashima and Oshima, 1976).

The presence of the HMR' allele gives rise to an additional distinctive segregation which is exhibited by diploids obtained in non-parental ditype tetrads (Table II, Type II). Since each of the spores diploidized, two of these segregants were initially MATa HO HMR' and two were initially MATα HO HMR. The diploids resulting from mating between siblings during growth of these spores would thus be MATa/MATα HO/HO HMR'/HMR' and MATa/MATα HO/HO HMR/HMR, respectively. Sporulation of these diploids showed that two produced four diploidizing segregants. The other two (MATa/MATα HO/HO HMR'/HMR') yielded a distinctive segregation, 2 'α' and 2 diploidizers in each tetrad (Harashima *et al.*, 1974).

HML: A similar analysis of a strain of *S. oviformis* revealed the existence of the HML locus, which is required for diploidization by a cells but not by α cells (Takano and Oshima, 1967; Harashima *et al.*, 1974). In this case, the *S. oviformis* strain exhibited stable a mating type, despite carrying HO, due to the presence of the HML' allele. The

HML locus is the leftmost known marker on chromosome III, 65 centimorgans from the mating type locus (Harashima and Oshima, 1976). Diploids which are MATa/MATα HO/HO HML'/HML' HMR/HMR give the distinctive segregation of 2 a and 2 diploidizers in each tetrad.

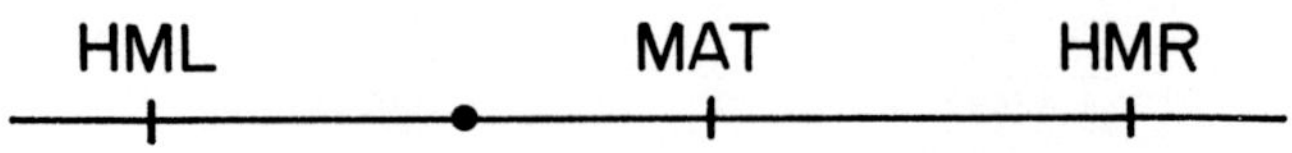

The nature of HML' and HMR': To summarize the above analysis, MATα HO HML HMR' and MATa HO HML' HMR strains are unable to diploidize, whereas MATα and MATa HO HML HMR strains are able to diploidize. In a formal sense, HML' is an inactive allele of HML (unable to promote diploidization of a spores), and HMR' is an inactive allele of HMR (unable to promote diploidization of α spores). Subsequent analysis of HML' and HMR' revealed a surprise--MATa HO HML' strains were able to diploidize if they were HMR', and MATα HO HMR' strains were able to diploidize if they were HML' (Naumov and Tolstorukov, 1973; Harashima *et al.*, 1974). MATa/MATα HO/HO HML'/HML' HMR'/HMR' diploids yielded four diploidizing segregants in each tetrad. Naumov and Tolstorukov (1973) suggested that the HML' allele was functionally equivalent to HMR and that HMR' was functionally equivalent to HML. Subsequent analysis has borne out this hypothesis and shown that HML' is not simply a recessive mutation of HML: HML and HML' are codominant alleles, as are HMR' and HMR (Klar and Fogel, 1977; Arima and Takano, 1979; Harashima and Oshima, 1979). Models presented below provide concrete roles for the HM loci.

III. MODELS FOR MATING TYPE INTERCONVERSION

A model for mating type interconversion must explain the following observations: (1) Mating type interconversion is a change of alleles of the mating type locus. (2) HO is necessary for the high frequency of mating type interconversion but is not necessary for maintenance of the new mating type. (3) HMR or HML' are necessary for switching from α to a; HML or HMR' are necessary for switching from a to α.

A. Type I models

In these models, the mating type locus contains determinants for both a and α cell types.

1. In a "flip-flop" model (model 1), the region between a and α genes is invertible by intramolecular recombination promoted by HO. This recombination event requires, in addition, either HMR (HML') for flipping from α to a or HML (HMR') for flipping from a to α. (A variation of this scheme proposed by D. Hawthorne is described by Holliday and Pugh, 1975).

Model 1

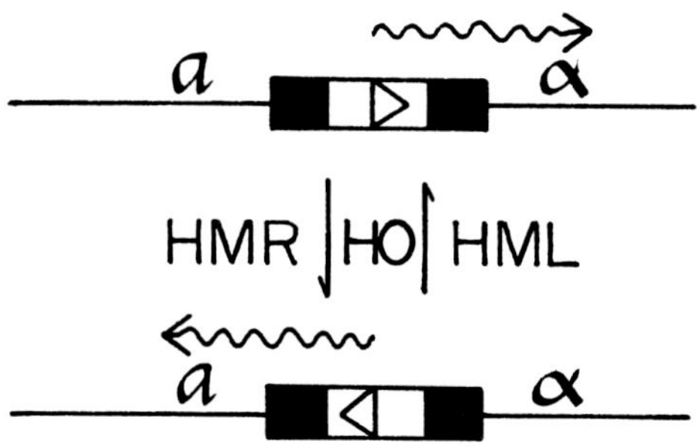

Precedent for an invertible region controlling expression of adjacent genes has been uncovered for phase variation in Salmonella by Simon and coworkers (Zieg *et al.*, 1978). Precedent for site-specific recombination enzymes with specificity cofactors comes from work on lambda site-specific recombination promoted by Int and Xis proteins: the reaction between phage and bacterial att sites requires only Int, whereas the reaction between prophage hybrid att sites requires both Int and Xis (Guarneros and Echols, 1970; Echols, 1970). HML and HMR might, therefore, be analogous to two different "Xis-like" cofactors which determine the specificity of HO.

2. Another type of inversion model (model 2), analogous to G region inversion in phage Mu, differs from the above model in that the a or α structural genes are inverted to allow expression of either a or α information. The inversion of the G region of phage Mu, which is promoted by the

Mu Gin product (Chow et al., 1977), allows expression of alternative host range determinants (Toussaint et al., 1978; Howe et al., 1979; P. van de Putte, personal communication).

Model 2

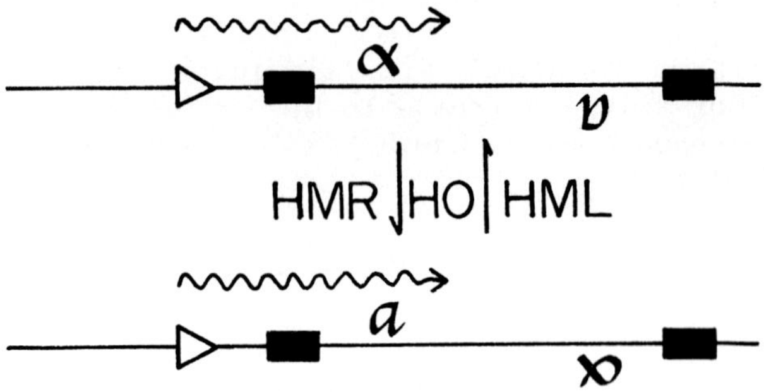

3. In a "transposable regulatory site" (TRS) model (Rine et al., 1979) expression of a or α information is governed by insertion of HML or HMR information into a site at the mating type locus (model 3). Insertion of HML (or HMR') catalyzed by HO would give rise to expression of α information, for example, due to directing transcription rightwards. Insertion of HMR (or HML') would allow expression of a information. (Another possibility is that HML and HMR contain DNA segments which can act as intervening sequences whose insertion into a or α information is required for essential RNA processing.)

Model 3

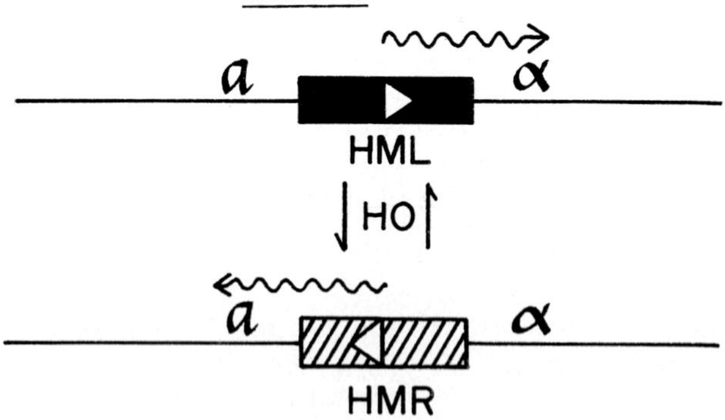

The hypothesis that HMR and HML loci code for transposable segments which associate with the mating type locus to determine cell type was originally proposed by Oshima, Takano, and coworkers (Oshima and Takano, 1971; Harashima *et al.*, 1974). Precedent for transposable regulatory sites comes from several sources, notably from studies of maize controlling elements (McClintock, 1956), in which association of a controlling element with a gene can alter expression of that gene in a variety of ways (see, for example, Dooner and Nelson, 1977). Examples from prokaryotes abound: Insertion of a lambda prophage into the *trpC* gene leads to expression of the adjacent *trpB* gene from the phage *int* promoter (Shimada and Campbell, 1974); insertion of a lambda prophage into its natural *att* site, between *gal* and *bio*, leads to phage-dependent expression of *gal* and *bio* operons from phage promoters *pL*, *pR*, and *pR'* (when regulatory proteins N and Q are supplied) (Krell *et al.*, 1972; Adhya *et al.*, 1974; Forbes, 1978); insertion of Tn3 into the sulfanilimide resistance gene leads to expression of the adjacent streptomycin resistance gene from the Tn3 transposase (*tpnA*) promoter (Heffron *et al.*, 1979).

B. Type II models

In these models, the mating type locus contains either *a* information or α information but not both.

In the "cassette model", HML and HMR harbor silent copies of α and *a* information, whose association with the mating type locus leads to the expression of whichever block is transposed (model 4). Association of HMR (or HML') with the mating type locus leads to *a* cell type; association of HML (or HMR') with the mating type locus leads to α cell type.

Model 4

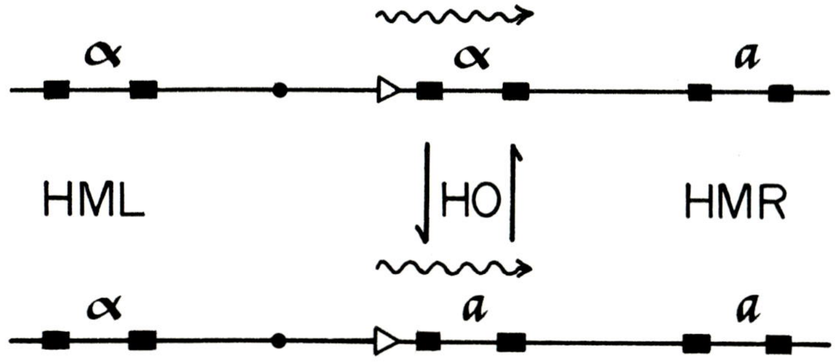

The cassette model is a specific version of the proposal that *HMR* and *HML* are "controlling elements" whose association with the mating type locus determines cell type. A distinguishing feature between the cassette and TRS models is the type of information that is transposed, structural genes in the cassette model and regulatory sites in the TRS model. Two variations can be imagined for a cassette model: the transposable unit might be an intact coding sequence, or it might become joined to other structural information at the insertion site (as occurs in immunoglobin synthesis; Bernard *et al.*, 1978).

IV. ADDITIONAL OBSERVATIONS ON MATING TYPE INTERCONVERSION AND A TEST OF THE CASSETTE MODEL

Several observations gathered over the last few years allow a distinction among the above models for mating type interconversion. These observations, discussed below, are the following: (1) Mutations of the mating type locus, including nonsense mutations, can be "healed" by mating type interconversion. (2) Two types of chromosomal rearrangements, a deletion on chromosome III and formation of a circular chromosome III, result in switching from one mating type to another, presumably by fusion of the mating type locus to *HMR*a and *HML*α, respectively. (3) The information at *HML*α and *HMR*a can be expressed *in situ* in *ho* strains carrying mutations of *SIR* genes (*SIR*, silent information regulator). These studies indicate that *HML*α is functionally equivalent to the mating type locus. (4) Mutations in *HML*α and *HMR*a give rise to mutations at the mating type locus as a result of mating type interconversion. In other words, the mating type locus is "wounded" in strains carrying *HML*α⁻ and/or *HMR*a⁻ mutations.

A. Restoration of function to mutations of the mating type locus--"healing": Mutations of both the α and *a* mating type locus alleles were originally identified by MacKay and Manney (1974) and by Kassir and Simchen (1976). Mutations of *MAT*α lead to defects in mating and define two complementation groups, α1 and α2 (Strathern, 1977; Herskowitz *et al.*, 1980). Mutations of *MAT*a (*mata** or *mata1*⁻) do not affect mating but do affect ability to promote sporulation in *mata**/*MAT*α diploids. All mutations of the mating type locus that have been tested are efficiently "healed" as a result of mating type interconversion (Hicks and Herskowitz, 1977;

D. Hawthorne, personal communication; Strathern _et al._, 1979a; Klar _et al._, 1979a). The original experiments performed with _matα1-2_ and _matα1-5_ showed that _matα1⁻ HO_ strains switched to _MATa_ and subsequently to _MATα_ within a few generations after germination. It is notable that _matα1-2_ has recently been shown to be an _amber_ mutation (Table III).

TABLE III. _matα1-2_ is an _amber_ mutation

Tetrads from _matα1-2/MATa his4-am/his4-am SUP4-3/sup4_

inferred genotype		phenotype mating	α-factor	His	number
matα1-2	_SUP4-3_	α	+	+	
matα1-2	_SUP4-3_	α	+	+	6
MATa	_sup4_	a	−	−	
MATa	_sup4_	a	−	−	
matα1-2	_sup4_	nm	−	−	
matα1-2	_sup4_	nm	−	−	9
MATa	_SUP4-3_	a	−	+	
MATa	_SUP4-3_	a	−	+	
matα1-2	_SUP4-3_	α	+	+	
matα1-2	_sup4_	nm	−	−	30
MATa	_SUP4-3_	a	−	+	
MATa	_sup4_	a	−	−	

Legend to Table III:

A diploid formed between _matα1-2 his4-am SUP4-3_ (BA8-44c) and _MATa his4-am sup4_ (BA8-20d) was sporulated and segregants tested for mating ability, α-factor production, and His by standard methods. _his4-am_ is _his4-580_; _SUP4-3_ is a temperature-sensitive _amber_ suppressor (Rasse-Messenguy and Fink, 1973); _sup4_ is the wildtype, non-suppressing allele. nm = non-mater. The above results were obtained at 23°. All segregants were His^- at 34°; all segregants which were α at 23° were non-maters and did not produce α factor at 34°. A _MATα1⁺ SUP4-3_ strain mated and produced α-factor at both temperatures.

D. Hawthorne (personal communication) has likewise shown that a *matα-ochre* mutation can be healed to *MATα*$^+$. Healing of *matα1*$^-$ and *mata1*$^-$ mutations also is observed in *ho* strains (Hicks and Herskowitz, 1976; Klar *et al*., 1979a). These results led to the proposal that the yeast genome contains cryptic copies of the mating type locus alleles which can replace defective information at the mating type locus (Hicks *et al*., 1977; Hicks and Herskowitz, 1977). The genetic analysis of *HML* and *HMR* discussed above led to the specific proposal that the *HML* and *HMR* alleles present in standard *S. cerevisiae* strains carried α and *a* information respectively. The *HML'* and *HMR'* alleles thus correspond to *a* and α information respectively. The standard alleles will be referred to hereafter in this paper as *HMLα* and *HMRa*. The ability to heal mutations of the mating type locus, in particular, nonsense mutations, is not explainable by Type I models.

B. Mating type interconversion by deletion and circle formation: Switches between *a* and α occur in *ho* strains at a frequency of approximately 10^{-6}. These rare events are easily detected by selection for matings between strains of like mating type which carry complementary auxotrophic mutations. In a rare mating between *a* strains, prototrophs are formed between cells which have switched from *a* to α and those cells of *a* mating type with complementing markers. One class of prototrophs obtained in rare matings has the properties expected of *a*/α diploids--they do not mate, but are able to sporulate and yield 2 *a* and 2 α segregants. In both *a* x *a* and α x α rare matings, a minority class of the non-mating, sporulation-proficient prototrophs are obtained which yield inviable spores due to mutations inseparable from the mating type locus. In α x α rare matings, some prototrophs yield 2 α and 2 inviable spores in each tetrad (Hawthorne, 1963a). The two inviable spores are able to mate soon after germination and exhibit *a* mating type. These *a*-lethal cells have sustained a large deletion on the right arm of chromosome III distal to the mating type locus which extends beyond *thr4*. In *a* x *a* rare matings, Mortimer and Hawthorne (personal communication) identified a prototroph which yielded 2 *a* and 2 inviable spores in each tetrad. The inviable spores exhibited α mating type. Their genetic analysis indicated also that the α-lethal spores had sustained a complex rearrangement of chromosome III. On the basis of the cassette model, it was proposed that the α-lethal mutation might result from formation of circular fragment of chromosome III. This prediction has been confirmed (Strathern *et al*., 1979b).

The ability to switch from α to a-lethal by deletion and from a to α-lethal by circle formation cannot be explained readily by models in which HML and HMR code for functions which promote interconversion (such as model I), nor can TRS models explain such events. In a cassette model, the a-lethal deletion is viewed as a fusion of an essential site at the mating type locus with information at HMRa. This explanation has been confirmed by the observation that α to a-lethal switches in strains with HMRa mutations lead to a defective a mating type locus (Kushner, 1979; A. Klar, personal communication). The α-lethal circular chromosome is viewed as a fusion of an essential site at the mating type locus with the α information of HMLα.

C. Expression of HMLα and HMRa *in situ*: Mutations in ho strains have been found which allow expression of cryptic mating type loci. The first such mutation identified [originally called ssp515 (Herskowitz *et al.*, 1977) and now sir1-1 (Rine *et al.*, 1979)] was isolated as a mutation which allowed mating by α1⁻ mutants. The striking observation was that the ability of sir1-1 to suppress the mating defect of α1⁻ mutants was dependent on HMLα: matα1⁻ HMLα HMRa ho sir1-1 strains mated as α, whereas matα1⁻ HMLa HMRa ho sir1-1 strains did not. This analysis was the first to reveal a phenotype for HMLα, originally identified by its role in mating type interconversion, in heterothallic strains. The sir1-1 mutation suppresses all mutations of the α mating type locus and of the a mating type locus. We thus have proposed that sir1-1 acts by allowing *in situ* expression of the information at HMLα (and HMRa), which supplies the functions lacking in mating type locus mutations such as α1⁻ or α2⁻. Mutations have been independently isolated by others which also appear to allow expression of HMLα and HMRa information (J. McCullough; personal communication; Klar *et al.*, 1979b; Haber and George, 1979). A systematic isolation of such mutations (Rine, 1979) reveals that there are at least four different SIR genes. The ability of HMLα to complement defects of matα1⁻ and matα2⁻ mutations provides several significant pieces of information on mating type interconversion. (1) HMLα is functionally equivalent to MATα. This finding is consistent only with the cassette model and not with models in which HML and HMR are transposable regulatory sites. (2) Since sir mutations are recessive to SIR^+, the wildtype SIR gene products appear to be essential for keeping the information at HMLα and at HMRa silent. According to this view, transposition of α and a cassettes from HML and HMR to the mating type locus leads to expression of a or α information by removing them from sites of Sir action present at HML and HMR but absent at the mating type locus (Herskowtiz et al., 1977; Rine et al., 1979). Whether Sir controls expression of HML and HMR exclusively remains to be determined (see Rine et al., 1979).

D. Wounding of the mating type locus in HMLα⁻ and HMRa⁻ mutants. The cassette and controlling element models predict that a or α strains with mutations of HML# or HMRa should give rise to strains with mutations of the mating type locus as a result of mating type interconversion. Mutants defective in HMLα and HMRa have been isolated in our lab (Kushner et al., 1979; Blair et al., 1979) and by Klar and Fogel (1979) utilizing very different isolation schemes and have been shown to wound the mating type locus as a result of mating type interconversion in both ho and HO strains. For example, MATα HO HMRa strains carrying the HMLα-66 mutation give rise to a strains and then to matα⁻ strains by mating type interconversion (Kushner et al., 1979). Klar (personal communication) has similarly shown that HMRa-nonsense mutations give rise to strains with nonsense mutations of MATa. These observations fulfill an explicit prediction of the cassette and controlling element models.

E. Structure of the mating type locus: In the above discussion of mating type interconversion, we have avoided making any assumptions about the structure of the mating type locus itself. Clearly, such information would be useful in distinguishing among the models described above. The two general types of structures described above--one in which the mating type locus contains both a and α regulatory information, the other in which it contains only one type of regulatory information--can in theory be distinguished by a simple genetic cross. If the alleles of the mating type locus contain either MATa or MATα but not both, then it should be impossible to obtain α^+ recombinants from matα⁻/MATa ho/ho diploids. If the mating type locus contains both a and α coding information, α^+ recombinants are expected to be observable. We have found that matα1⁻/MATa diploids yield MATα^+ progeny at a frequency of approximately 5×10^{-5} (G. Sprague, Jr. and J. Rine, unpublished). (Since the strains used in this cross carried HMLα, we do not know whether MATα^+ results from recombination between matα1⁻ and MATa or whether it results from mating type interconversion or by recombination with HMLα.) matα2⁻/MATa HMLa/HMLa HMRa/HMRa diploids yield MATα^{+} recombinants at a frequency of approximately 2 x 10^{-3}. For comparison, recombination between matα1 and matα2 is approximately 10^{-2} (J. Strathern and I. Herskowitz, unpublished). These results suggest that the a and α alleles are partially nonhomologous and thus are consistent with the predictions of the cassette model.

V. THE MECHANISM OF MATING TYPE INTERCONVERSION

How is the information at HML and HMR transposed to the mating type locus? Two observations provide additionl information bearing on the mechanism of mating type interconversion. (1) The cassette at the mating type locus appears to be discarded as a result of the interconversion process. The loss of the exiting cassette is indicated by studies of switching of HO MATα strains carrying only silent a information at HML and HMR. After this strain switches to a, progeny of this a cell are not found which have α mating type (in 99,000 progeny cells analyzed; Rine, 1979). Thus the α cassette originally at the mating type locus is irretrievable. (2) The information at HML and HMR remains unchanged as a result of mating type interconversion. That is, when MATα HO HMLα HMRa strains switch to MATa, they remain HMLα HMRa (Harashima et al., 1974). The transposition of information from HMRa thus is a non-reciprocal transfer of information from HMRa to the mating type locus.

Several molecular models can be proposed for the mating type interconversion transposition event. In one type of model, transposition occurs by a directed gene conversion, which might occur as follows: one DNA strand from HMRa invades the MATα locus and displaces one of the resident α strands. Degradation of the displaced and remaining α DNA, in conjunction with nicking of the invading a strand followed by its ligation, leads to formation of a chromosome with two gaps--one at MAT, the other at HMR. These gaps are filled by copying the resident a DNA strand in both cases. Other types of models involve freely diffusing intermediates. For example, by some unspecified mechanism, an a cassette is produced which either becomes aligned with the MATα cassette or which becomes inserted adjacent to it (to form a structure analogous to a tandem prophage). Replacement of α by a information would then occur by intermolecular or intramolecular recombination events respectively.

VI. EXTENSIONS AND EXTRAPOLATIONS OF THE CASSETTE MODEL

A cassette mechanism provides a means by which a cell can uniquely express one of several genes or blocks of genes. The *S. cerevisiae* genome contains two silent genes, at *HML* and *HMR*, which constitute its "library" of stored information. In this case, these genes determine the two haploid cell types. One can imagine other organisms which have a more extensive library of silent genes coding for cell type determinants or other characteristics. A cell with, for example, 26 different cassettes--*cas*A, *cas*B...*cas*Z--can uniquely express one of these simply by transposition of the appropriate cassette into a "playback" or "display" locus.

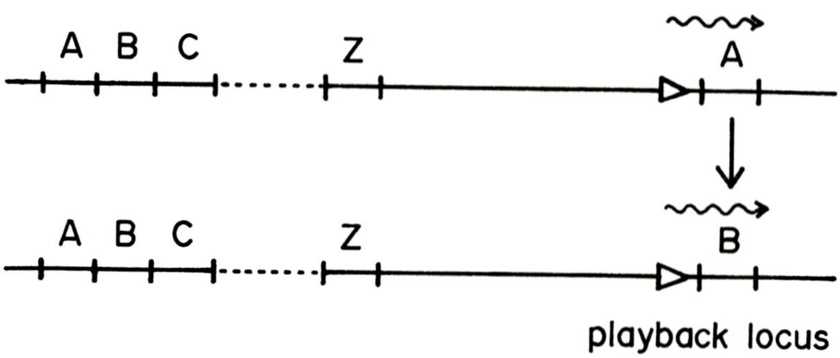

According to such a multiple cassette mechanism, a population of cells derived from a single cell can contain many different cell types. By controlling various aspects of the interconversion process, it is possible to generate clones in which cells of different types appear in a specific order. Yeast mating type interconversion exhibits directionality--cells competent to switch mating types change to the opposite mating type in more than 50% of cell divisions (Strathern and Herskowitz, 1979). Thus they do not replace the cassette at the mating type locus with a random choice of *HML*α or *HMR*a. The mechanism by which a cell senses which cassette to use remains to be determined but might be controlled by products of the active cassette or by DNA sites of the cassettes (Strathern and Herskowitz, 1979; Strathern *et al*., 1979a). For a cell with multiple cassettes, one can propose that an active A cassette programs its substitution by a B cassette, and that the active B cassette can program its substitution by a C cassette, etc. An A cell would thus give rise to B cells, B cells to C cells, etc., as a result of sequential cassette insertion. These ideas are discussed more fully elsewhere in a model to account for the ordered production of multiple cell types during development in higher eukaryotes (Strathern and Herskowitz, 1979; Herskowitz *et al.*, 1980). It should be noted that a population of stable cells, for example, J cells, can be produced if the active J cassette does not program its own removal. A cassette mechanism can also explain how pathogenic microorganisms escape from immune surveillance by changing their cell surface (Bloom, 1979). Antigenic varition seen in Trypanosomes (Vickerman, 1978) and in the agent of relapsing fever (the bacterium *Borrelia*; Felsenfeld, 1965) might occur by the activation of cassettes containing the structural genes for major cell surface proteins or for enzymes involved in construction of other antigens.

ACKNOWLEDGEMENTS

We thank Julie Dunn for preparation of the manuscript, Kerrie Rine for preparation of the figures, and Flora Banuett and our colleagues for discussion.

REFERENCES

Adhya, S., Gottesman, M., and deCrombrugghe, B. (1974). Proc. Nat. Sci. 71, 2534.

Arima, K. and Takano, I. (1979). Genetics, in press.

Bernard, O., Hozumi, N., and Tonegawa, S. (1978). Cell 15, 1133.

Blair, L.C., Kushner, P.J., and Herskowitz, I. (1979). In "ICN-UCLA Symposium on Molecular and Cellular Biology" (T. Maniatis and R. Axel, eds.), Academic Press, New York, in press.

Bloom, B.R. (1979). Nature 279, 21.

Chow, L.T., Kahmann, R., and Kamp, D. (1977). J. Mol. Biol. 113, 591.

Dooner, H.K. and Nelson, Jr., O.E. (1977). Proc. Nat. Acad. Sci. 74, 5623.

Echols, H. (1970). J. Mol. Biol. 47, 575.
Felsenfeld, O. (1965). Bacteriol. Rev. 29, 46.
Forbes, D.J. (1978). Ph.D. Thesis. University of Oregon.
Guarneros, G. and Echols, H. (1970). J. Mol. Biol. 47, 565.
Haber, J.E. and George, J.P. (1979). Genetics, in press.
Harashima, S. and Oshima, Y. (1976). Genetics 84, 4317.
Harashima, S. and Oshima, Y. (1979). Genetics, in press.
Harashima, S., Nogi, Y., and Oshima, Y. (1974). Genetics 77, 639.
Hawthorne, D.C. (1963a). Genetics 48, 1727.
Hawthorne, D.C. (1963b). Proc. 11th Intern. Congr. Genet. 1, 34 (abstr.).
Heffron, F., McCarthy, B.J., Ohtsubo, H., and Ohtsubo, E. (1979). Cell, in press.
Herskowitz, I., Strathern, J.N., Hicks, J.B., and Rine, J. (1977). In "ICN-UCLA Symposium on Molecular Approaches to Eucarotic Genetic Systems" (G. Wilcox, J. Abelson, and C.F. Fox, eds.), p. 193. Academic Press, New York.
Herskowitz, I., Blair, L., Forbes, D., Hicks, J., Kassir, Y., Kushner, P., Rine, J., Sprague, Jr., G., and Strathern, J. (1980). In "The Molecular Genetics of Development" (W. Loomis and T. Leighton, eds.) Academic Press, New York, in press.
Hicks, J.B. and Herskowitz, I. (1976). Genetics 83, 245.
Hicks, J.B. and Herskowitz, I. (1977). Genetics 85, 373.
Hicks, J.B., Strathern, J.N., and Herskowitz, I. (1977). In "DNA Insertion Elements, Plasmids, and Episomes" (A.I. Bukhari, J.A. Shapiro, and S.L. Adhya, eds.), p. 457. Cold Spring Harbor Press, Cold Spring Harbor, New York.
Holliday, R. and Pugh, J.E. (1975). Science 187, 226.
Howe, M.M., Schumm, J.W., and Taylor, A.L. (1979). Virology 92, 108.
Kassir, Y. and Simchen, G. (1976). Genetics 82, 187.
Klar, A.J.S. and Fogel, S. (1977). Genetics 85, 407-416.
Klar, A.J.S. and Fogel, S. (1979). Proc. Nat. Acad. Sci. 76, 4539.
Klar, A.J.S., Fogel, S., and Radin, D.N. (1979a). Genetics 92, 759.
Klar, A.J.S., Fogel, S., and MacLeod, K. (1979b). Genetics, in press.
Krell, K., Gottesman, M.E., Parks, J.S., and Eisenberg, M.A. (1972). J. Mol. Biol. 68, 69.
Kushner, P.J. (1979). Ph.D. Thesis. University of Oregon.
Kushner, P.J., Blair, L.C., and Herskowitz, I. (1979). Proc. Nat. Acad. Sci. 76, 5264.
MacKay, V.L. and Manney, T.R. (1974). Genetics 76, 225.

McClintock, B. (1956). Cold Spring Harbor Symp. Quant. Biol. 21, 197.

Mortimer, R.K. and Hawthorne, D.C. (1969). In "The Yeasts" (J.S. Harrison and A.D. Rose, eds.) Vol. 1, p. 385, Academic Press, New York.

Naumov, G.I. and Tolstorukov, I.I. (1973). Genetika 9, 82.

Oshima, Y. and Takano, I. (1971). Genetics 67, 327.

Rasse-Messenguy, F. and Fink, G.R. (1973). Genetics 75, 459.

Rine, J. (1979). Ph.D. thesis, University of Oregon.

Rine, J., Strathern, J.N., Hicks, J.B., and Herskowitz, I. (1979). Genetics, in press.

Rabin, M. (1970). M.S. thesis. University of Washington.

Santa Maria, J. and Vidal, D. (1970). I.N. Investigaciones Agronómicas (Madrid) 30, 1.

Shimada, K. and Campbell, A. (1974). Proc. Nat. Acad. Sci. 71, 237.

Strathern, J.N. (1977). Ph.D. thesis. University of Oregon.

Strathern, J.N. and Herskowitz, I. (1979). Cell 17, 371.

Strathern, J.N., Blair, L.C., and Herskowitz, I. (1979a). Proc. Nat. Acad. Sci. 76, 3425.

Strathern, J.N., Newlon, C.S., Herskowitz, I., and Hicks, J.B. (1979b). Cell 18, 309.

Takano, I. and Oshima, Y. (1967). Genetics 57, 875.

Toussaint, A., Lefebvre, N., Scott, J.R., Cowan, J.A., DeBruijn, F., and Bukhari, A.I. (1978). Virology 891, 146.

Vickerman, K. (1978). Nature 273, 613.

Winge, Ö. and Roberts, C. (1949). Compt. Rend. Trav. Lab. Carlsberg. Ser. Physiol. 24, 341.

Zieg, J., Hilmen, M., and Simon, M. (1978). Cell 15, 237.

DISCUSSION

A. BUKHARI: Is there, in terms of changes in chromosomes, a higher frequency of translocation of chromosome 3 in big *HO* strains?

I. HERSKOWITZ: The question is, "Does *HO* have any other phenotype besides shuffling cassettes and, in particular, does it cause an increased frequency of translocations?" This has never been seen, but actions of *HO* other than in mating type interconversion have not been looked for either carefully or exhaustively. One would like to know whether *HO* promotes a higher frequency of recombination in general; however, in order to look at recombination, you have to do it in an *a*/α diploid, and *HO* is "turned off" in these cells. So you have to be a little tricky and construct special strains to carry out these kinds of experiments, something which has not yet been done.

A. BUKHARI: I assume that the frequency of Hawthorne deletions is higher in homothallic strains than in heterothallic strains. Is that correct?

I. HERSKOWITZ: The experiment has not been done to my knowledge.

A. BUKHARI: From the structural diagrams you showed it is not entirely clear whether there is any difference between the *mat* locus and HM loci. They all appear to be totally equivalent. There must be some differences in HM loci and the *mat* locus, because you do not shuffle HM between HM and from *mat* to HM.

I. HERSKOWITZ: The question is what are the mechanisms by which an α cell switches preferentially to an *a* cell when the DNA heteroduplex studies by Nasmyth and Tatchell and by Hicks, Strathern and colleagues show that the cassettes at *HML* and *HMR* are flanked by homologous regions. If one looks closer at these structures, they do have some subtle differences, which might conceivably determine the directionality of switching. Another mechanism for determining directionality would be for the mating type locus to code for a function which directs transposition from *HML* or *HMR* to the mating type locus. Thus the mechanism for directionality may either be a structural one or a functional one - we do not yet know the answer.

W. SZYBALSKI: Ira, I have two questions. One, I remember when I was working on the yeast genetics in Carlsberg's lab with Winge in 1947-49 there were no α and a; all used cell diploidizing very well. Do you now know what was the genetic structure of those strains of *Saccharomyces* from Carlsberg's laboratory?

I. HERSKOWITZ: Waclaw notes that when he worked in Winge's laboratory the original strains of yeast all diploidized. Hence, one did not know that cells and a mating type - either *a* or α . In doing crosses between *S. chevalieri* and other yeast strains, Winge and Roberts found that there was a single determinant, called *D*, which determined the ability of cells to diploidize. Cells with *D* diploidized; those with the *d* allele exhibited a mating type, called *a* or α by Lindegren.

W. SZYBALSKI: Why do you suggest that the *SIR* genes code for a repressor?

I. HERSKOWITZ: I neglected to mention that the different mutations such as *sir* and *mar*, which allow expression of the ordinarily silent information at *HML* and *HMR*, are recessive to wildtype. Thus the SIR^+ and MAR^+ genes formally behave as if they code for inhibitors of expression of *HML* and *HMR* and therefore might be repressors.

STRUCTURE AND EVOLUTION OF GROWTH HORMONE RELATED GENES

Howard M. Goodman[1]
Frances DeNoto
John C. Fiddes[2]
Robert A. Hallewell[3]
Guy S. Page[4]
Susan Smith
Edmund Tischer

Howard Hughes Medical Institute Laboratory
Department of Biochemistry and Biophysics
University of California
San Francisco, California

I. INTRODUCTION

We have described the isolation and amplification in bacteria of the cDNA sequences for the polypeptide hormones rat insulin (rI) (1), human insulin (hI) (2), rat growth hormone (rGH) (3), human chorionic somatomammotrophin (hCS) (4), human growth hormone (hGH) (5,6), mouse lipotropin/endorphin precursor (ACTH/LPH) (7) and human chorionic gonadotropin (hCG) (8). Two of these hormones, growth hormone (GH) and chorionic somatomammotrophin (CS), along with prolactin (PL), are a set of closely related polypeptides whose genes probably arose by duplication from a common ancestral precursor (5,6,9,10). This hypothesis is consistent with the high degree of nucleotide sequence homology found when the cloned gene structures are compared (this manuscript and ref. 4, 5 and 6). The expression of these genes is also affected by other hormones. Thus, this system is interesting for the study of regulation,

[1]This work was supported by NIH grants CA14026 and AM19997, and a grant from Eli Lilly Co. H.M.G. is an Investigator of the Howard Hughes Medical Institute.

[2]Fellow of the Anna Fuller Fund

[3]Fellow of the Science Research Council of Great Britain.

[4]Supported by A.C.S. Fellowship J-398

ISBN 0-12-633360-2

structure, and expression of related genes. We summarize here our findings on the isolation of the cDNA and gene sequences coding for growth hormone and chorionic somatomammotrophin in rat and man and present a comparison of their sequences.

II. MATERIALS AND METHODS

The detailed methods involved in these studies have been reported elsewhere (1-4, 6, 8, 11, 12).

III. RESULTS

Analysis of Recombinant DNA Containing Hormone Gene Sequences

RNA containing human growth hormone (hGH) mRNA was isolated by the guanidinium thiocynate-CsCl density gradient method which we described previously (1, 11) from hGH producing tumors of pituitary origin and human chorionic somatomammotrophin (hCS) RNA from term placentae removed by Caesarian section. Double-stranded cDNA was synthesized from the MRNA by reverse transcription. 550 base pair fragments from these cDNAs were obtained by digestion with *Hae*III endonuclease. These fragments were purified to >99% as described (4, 5). Briefly, this involved polyacrylamide gel electrophoresis, internal cleavage with another endonuclease (*Pvu*II for hGH and *Hha*I for hCS), isolation and subsequent religation of the cleaved pieces, and reelectrophoresis. The self-complementary decanucleotide linkers, pC-C-A-A-G-C-T-T-G-G, containing the cleavage sites for endonuclease HindIII were ligated to the hGH *Hae*III fragment. Similarly, the *Eco*RI linker, pC-C-G-A-A-T-T-G-G, was linked to the hCS *Hae*III fragment. After digestion with *Hind*III or *Eco*RI to produce cohesive termini, the fragment was purified and ligated to the *Hind*III or *Eco*RI cut plasmid vector pBR322 (13). The EK2 host strain X1776 was transformed with this recombinant DNA and ampicillin resistant recombinants were selected. All recombinants contained the correct *Hae*III fragment from hGH or hCS cDNA (4, 5).

The inserted DNA from one recombinant in each case was removed by *Hind*III or *Eco*RI endonuclease digestion and its nucleotide sequence analyzed using the chemical degradation method of Maxam and Gilbert (14). The *Hae*III fragments correspond to amino acids 24 to 191 of the coding region and part of the 3'- untranslated region of both mRNAs (4,5).

More recently almost "full-length" hGH and hCS cDNA clones have been obtained by similar procedures, but without HaeIII endonuclease digestion (6) (with the exception that the cDNA for hCS was obtained from a first trimester placenta as described elsewhere (8)). The hGH clone has been analyzed by DNA sequence analysis using both the chemical cleavage method (14) and the chain-termination technique (15). It contains 29 nucleotides in its 5'-untranslated region, 78 nucleotides coding for the 26 amino acid presequence, 573 nucleotides coding for the 191 amino acids of growth hormone itself, and the entire 3'-untranslated region (108 nucleotides). The complete DNA sequence is presented in Figure 1 and predicts the previously unknown sequence of the signal peptide of hGH. The translation in one frame agrees with the known amino acid sequence of hGH (9, 10, 16) except at positions 29, 47, 74, 91, 107, 109, and 122 all of which involved asp<->asn or glu<->gln residues (6). An almost identical sequence has been recently published by Roskam and Rougeon (17). It differs from the sequence shown in Figure 1 as follows: it has an additional 11 bases at the 5' end; position 3 is a G in their sequence; amino acid 9, leu (CTT) in Figure 1 is pro (CCT) in their clone or leu (CTT) in another of their clones, and their sequence contains an additional C residue inserted at position 701(to the right of the asterisk in Fig. 1). The codons used in the prehGH sequence are shown in Table 1. In some cases, there is a marked selectivity in the use of one triplet over another for the same amino acid. In particular, there is a strong preference for the third position of the codon to be C or G rather than U or A. This preference is the same for all of the hormone genes we have analyzed.

The partial sequence of the hCS cDNA clone isolated from first trimester placental RNA is shown in Figure 2. The sequence has not yet been completed in the 5' and 3' untranslated regions. The translation of this sequence in one frame is identical to the reported amino acid sequence of hCS (16). However, the sequence of the preregion predicated from the cDNA clone differs from that recently reported by Sherwood et al. (18) using amino acid sequence analysis. The reported pre-sequence (18) contains only 25 amino acids and is missing the alanine at position -25 (Fig. 2). The pre-sequence of hCS containing 26 amino acids is quite similar to hGH (Fig. 1, and below). A codon usage table (Table 2) shows a distribution similar to that of hGH (Table 1).

```
                                  -26                      -20
                                  met ala thr gly ser arg thr
GGATCCTGTGGACAGCTCACCTAGCTGCA     ATG GCT ACA GGC TCC CGG ACG

                                      -10
ser leu leu leu ala phe gly leu leu cys leu pro trp leu gln
TCC CTG CTC CTG GCT TTT GGC CTG CTC TGC CTG CCC TGG CTT CAA

                1                                   10
glu gly ser ala phe pro thr ile pro leu ser arg leu phe asp
GAG GGC AGT GCC TTC CCA ACC ATT CCC TTA TCC AGG CTT TTT GAC
                                 20
asn ala met leu arg ala his arg leu his gln leu ala phe asp
AAC GCT ATG CTC CGC GCC CAT CGT CTG CAC CAG CTG GCC TTT GAC

            30                                  40
thr tyr gln glu phe glu glu ala tyr ile pro lys glu gln lys
ACC TAC CAG GAG TTT GAA GAA GCC TAT ATC CCA AAG GAA CAG AAG

                                 50
tyr ser phe leu gln asn pro gln thr ser leu cys phe ser glu
TAT TCA TTC CTG CAG AAC CCC CAG ACC TCC CTC TGT TTC TCA GAG

            60                                  70
ser ile pro thr pro ser asn arg glu glu thr gln gln lys ser
TCT ATT CCG ACA CCC TCC AAC AGG GAG GAA ACA CAA CAG AAA TCC

                                 80
asn leu glu leu leu arg ile ser leu leu leu ile gln ser trp
AAC CTA GAG CTG CTC CGC ATC TCC CTG CTG CTC ATC CAG TCG TGG

            90                                  100
leu glu pro val gln phe leu arg ser val phe ala asn ser leu
CTG GAG CCC GTG CAG TTC CTC AGG AGT GTC TTC GCC AAC AGC CTG

                                 110
val tyr gly ala ser asp ser asn val tyr asp leu leu lys asp
GTG TAC GGC GCC TCT GAC AGC AAC GTC TAT GAC CTC CTA AAG GAC

            120                                 130
leu glu glu gly ile gln thr leu met gly arg leu glu asp gly
CTA GAG GAA GGC ATC CAA ACG CTG ATG GGG AGG CTG GAA GAT GGC

                                 140
ser pro arg thr gly gln ile phe lys gln thr tyr ser lys phe
AGC CCC CGG ACT GGG CAG ATC TTC AAG CAG ACC TAC AGC AAG TTC
```

```
          150                                             160
asp thr asn ser his asn asp asp ala leu leu lys asn tyr gly
GAC ACA AAC TCA CAC AAC GAT GAC GCA CTA CTC AAG AAC TAC GGG

                                    170
leu leu tyr cys phe arg lys asp met asp lys val glu thr phe
CTG CTC TAC TGC TTC AGG AAG GAC ATG GAC AAG GTC GAG ACA TTC
            180                                     190 191
leu arg ile val gln cys arg ser val glu gly ser cys gly phe
CTG CGC ATC GTG CAG TGC CGC TCT GTG GAG GGC AGC TGT GGC TTC

AM                    *
TAG CTGCCCGGGTGGCATCCTGTGACCCCTCCCCAGTGCCTCTCCTGGCCCTGGAAGTT

GCCACTCCAGTGCCCACCAGCCTTGTCCTAATAAAATTAAGTTGCATCAAAAAAAAAAAA
```

Fig. 1. Nucleotide sequence of hGH cDNA and the amino acid sequence of human pregrowth hormone. The sequence was determined according to the procedure of Maxam and Gilbert (14) from 5' or 3'-end labeled restriction fragments of chGH800/pBR322. The sequence shown is the plus strand of DNA, i.e. the strand that has the same sequence as the mRNA. The sequence between the two internal *HaeIII* sites was taken from our previous work (5). Most of it, and all of the sequence outside these two internal *HaeIII* sites was resequenced by the chain-termination method (15) as described in detail elsewhere (6) using as single-stranded templates the chGH800 *HindIII* fragment recloned in the vector M13mp5 and as primers restriction fragments of chGH800/pBR322. The amino acid sequence has been deduced from the DNA sequence using the genetic code. The termination codon, TAG, is designated by the symbol AM for "amber". The asterisk marks the position of the extra C in the sequence of Roskam and Rougeon (17).

```
                                                              -26
                                                              met
TCTTGCCACACCTGCTTGATCCCCTTGGGTCCTGTGGACAGCTCACCTAGTGGCA ATG

                    -20
ala pro gly ser arg thr ser leu leu leu ala phe ala leu leu
GCT CCA GGC TCC CGG ACG TCC CTG CTC CTG GCT TTT GCC CTG CTC

-10                                     1
cys leu pro trp leu gln glu ala gly ala val gln thr val pro
TGC CTG CCC TGG CTT CAA GAG GCT GGT GCC GTC CAA ACC GTT CCG

                10                                      20
leu ser arg leu phe asp his ala met leu gln ala his arg ala
TTA TCC AGG CTT TTT GAC CAC GCT ATG CTC CAA GCC CAT CGC GCG

                                    30
his gln leu ala ile asp thr tyr gln glu phe glu glu thr tyr
CAC CAG CTG GCC ATT GAC ACC TAC CAG GAG TTT GAA GAA ACC TAT

                40                                      50
ile pro lys asp gln lys tyr ser phe leu his asp ser gln thr
ATC CCA AAG GAC CAG AAG TAT TCG TTC CTG CAT GAC TCC CAG ACC

                                    60
ser phe cys phe ser asp ser ile pro thr pro ser asn met glu
TCC TTC TGC TTC TCA GAC TCT ATT CCG ACA CCC TCC AAC ATG GAG

                70                                      80
glu thr gln gln lys ser asn leu glu leu leu arg ile ser leu
GAA ACG CAA CAG AAA TCC AAT CTA GAG CTG CTC CGC ATC TCC CTG

                                    90
leu leu ile glu ser trp leu glu pro val arg phe leu arg ser
CTG CTC ATC GAG TCG TGG CTG GAG CCC GTG CGG TTC CTC AGG AGT

                100                                     110
met phe ala asn asn leu val tyr asp thr ser asp ser asp asp
ATG TTC GCC AAC AAC CTG GTG TAT GAC ACC TCG GAC AGC GAT GAC

                                    120
tyr his leu leu lys asp leu glu glu gly ile gln thr leu met
TAT CAC CTC CTA AAG GAC CTA GAG GAA GGC ATC CAA ACG CTG ATG
```

```
                130                                         140
gly arg leu glu asp gly ser arg arg thr gly gln ile leu lys
GGG AGG CTG GAA GAC GGC AGC CGC CGG ACT GGG CAG ATC CTC AAG
                                    150
gln thr tyr ser lys phe asp thr asn ser his asn his asp ala
CAG ACC TAC AGC AAG TTT GAC ACA AAC TCG CAC AAC CAT GAC GCA

                160                                         170
leu leu lys asn tyr gly leu leu tyr cys phe arg lys asp met
CTG CTC AAG AAC TAC GGG CTG CTC TAC TGC TTC AGG AAG GAC ATG

                                    180
asp lys val glu thr phe leu arg met val gln cys arg ser val
GAC AAG GTC GAG ACA TTC CTG CGC ATG GTG CAG TGC CGC TCT GTC

                190 191
glu gly thr cys gly phe AM
GAG GGC ACG TGT GGC TTC TAG GTGCCCGAGTAGCATCCTGTGACCCCTCCCC

AGTGCCTCTCCTGGCC
```

Fig. 2. Partial nucleotide sequence of hCS cDNA and the amino acid sequence of human prechorionic somatomammotropin. The sequence was determined by the chain termination technique (15) using as single-stranded templates the chCS800 HindIII fragment recloned in the vector M13mp5 and as primers restriction fragments of chCS800/pBR322. The sequence between the two internal HaeIII sites was taken from our previous work (4). The pre-sequence is labeled number -26 to -1, the peptide 1-191, and the termination codon as AM for "amber".

Evolution of Hormone Genes

A comparison of the sequences of prehGH and prehCS (Fig. 3, amino acid homology; Fig. 4, nucleotide sequence homology) shows that these two hormones differ by only 33 amino acids. The amino acid homology of the prehormones is therefore 85%, while comparison of the nucleotide sequence in the same region shows that 672 out of 730 nucleotides are the same; a homology of 92%. The presequences are 85% homologous (18) (nucleotide sequence homology in the same region is 94%) and the 3'-untranslated regions are 92% homologous (only 47 bases are known for hCS (4)).

Table 1: Codon Usage in Pre hGH mRNA

4/UUU/phe	3/UCU/ser	3/UAU/tyr	2/UGU/cys
1Ø/UUC/phe	7/UCC/ser	5/UAC/tyr	3/UGC/cys
1/UUA/leu	3/UCA/ser	Ø/UAA/OC	Ø/UGA/OP
Ø/UUG/leu	1/UCG/ser	1/UAG/AM	2/UGG/trp
2/CUU/leu	Ø/CCU/pro	1/CAU/his	1/CGU/arg
1Ø/CUC/leu	6/CCC/pro	2/CAC/his	4/CGC/arg
4/CUA/leu	2/CCA/pro	3/CAA/gln	Ø/CGA/arg
16/CUG/leu	1/CCG/pro	11/CAG/gln	2/CGG/arg
2/AUU/ile	1/ACU/thr	Ø/AAU/asn	2/AGU/ser
6/AUC/ile	4/ACC/thr	9/AAC/asn	5/AGC/ser
Ø/AUA/ile	5/ACA/thr	1/AAA/lys	Ø/AGA/arg
4/AUG/met	2/ACG/thr	8/AAG/lys	5/AGG/arg
Ø/GUU/val	3/GCU/ala	2/GAU/asp	Ø/GGU/gly
3/GUC/val	6/GCC/ala	9/GAC/asp	8/GGC/gly
Ø/GUA/val	1/GCA/ala	6/GAA/glu	Ø/GGA/gly
4/GUG/val	Ø/GCG/ala	9/GAG/glu	3/GGG/gly

Table 2: Codon Usage in Pre hCS mRNA

4/UUU/phe	2/UCU/ser	4/UAU/tyr	1/UGU/cys
8/UUC/phe	8/UCC/ser	4/UAC/tyr	4/UGC/cys
1/UUA/leu	1/UCA/ser	O/UAA/OC	O/UGA/OP
O/UUG/leu	4/UCG/ser	1/UAG/AM	2/UGG/trp
2/CUU/leu	O/CCU/pro	3/CAU/his	O/CGU/arg
1O/CUC/leu	3/CCC/pro	4/CAC/his	5/CGC/arg
3/CUA/leu	2/CCA/pro	5/CAA/gln	O/CGA/arg
16/CUG/leu	2/CCG/pro	8/CAG/gln	3/CGG/arg
2/AUU/ile	1/ACU/thr	1/AAU/asn	1/AGU/ser
5/AUC/ile	6/ACC/thr	6/AAC/asn	3/ GC/ser
O/AUA/ile	3/ACA/thr	1/AAA/lys	O/AGA/arg
7/AUG/met	4/ACG/thr	8/AAG/lys	4/AGG/arg
1/GUU/val	4/GCU/ala	1/GAU/asp	1/GGU/gly
3/GUC/val	5/GCC/ala	14/GAC/asp	5/GGC/gly
O/GUA/val	1/GCA/ala	5/GAA/glu	O/GGA/gly
3/GUG/val	1/GCG/ala	9/GAG/glu	3/GGG/gly

Similar comparisons of hGH and rGH (Fig. 5, amino acid homology; Fig. 6, nucleotide sequence homology) show amino acid homologies of 58% for the presequence and 67% for the coding region. The nucleotide sequence homology is more extensive: 73% in the 5'-untranslated region, 76% in the presequence region, 76% in the coding region, and 55% in the 3'-untranslated region. PrehGH is one amino acid longer than pre-rGH. Two gaps at amino acids number 44 and 108 had to be added to the rGH sequence and one between amino acids number 183 and 184 to the hGH sequence to acheive maximum homology. Similarly, gaps of three base pairs were added to the nucleotide sequences (Fig. 6). These results confirm and extend the previous hypothesis (9,10) that the GH and CS genes probably evolved by a gene duplication mechanism, and since both species (rat and human) contain both hormones this presumably occurred independently in each species after they separated.

Extended homologies between the 3'-untranslated regions of rGH and hGH are observed if appropriate gaps are added to this part of the sequence. The homology in this region can be increased from 38% if no additions are made (not shown) to 55% (Fig. 6). This procedure also reveals a homology of 93% (27/29 nucleotides) at the ends of the molecules. This region includes the AAUAAA (AATAAA in the cDNA) sequence found in almost all eukaryotic mRNAs (19). Similar homologies exist between species when human and rabbit β-globin (19,20) or human and rat insulins (2) are compared as shown in Figure 7. There is little or no homology (except for the AAUAAA) between these different mRNAs, i.e. growth hormone, globin, and insulin, but rather only between the same mRNA in different species. This mRNA specific conservation of sequences would seem to indicate a specific functional role for these sequences perhaps as signals for polyadenylation. In addition, the last three bases (ATC) before the poly A addition site are the same in rGH and hGH and is repeated in rGH (Fig. 6).

```
-26                     -20
met ala thr gly ser arg thr ser leu leu leu ala phe gly leu
        pro                                         ala
    -1Ø                                 1
leu cys leu pro trp leu gln glu gly ser ala phe pro thr ile
                                ala gly     val gln     val

pro leu ser arg leu phe asp asn ala met leu arg ala his arg
                            his             gln

leu his gln leu ala phe asp thr tyr gln glu phe glu glu ala
ala                 ile                                 thr

tyr ile pro lys glu gln lys tyr ser phe leu gln asn pro gln
                asp                         his asp ser
 50
thr ser leu cys phe ser glu ser ile pro thr pro ser asn arg
        phe             asp                             met

glu glu thr gln gln lys ser asn leu glu leu leu arg ile ser

leu leu leu ile gln ser trp leu glu pro val gln phe leu arg
                glu                         arg
                    100
ser val phe ala asn ser leu val tyr gly ala ser asp ser asn
    met             asn             asp thr             asp

val tyr asp leu leu lys asp leu glu glu gly ile gln thr leu
asp     his

met gly arg leu glu asp gly ser pro arg thr gly gln ile phe
                                arg                     leu
                                        150
lys gln thr tyr ser lys phe asp thr asn ser his asn asp asp
                                                    his

ala leu leu lys asn tyr gly leu leu tyr cys phe arg lys asp

met asp lys val glu thr phe leu arg ile val gln cys arg ser
                                    met
                    190
val glu gly ser cys gly phe
            thr
```

<u>Fig. 3.</u> Comparison of the amino acid sequence of pre-hGH and pre-hCS. The continuous line is the amino acid sequence of pre-hGH taken from Fig. 1 and the amino acids which differ in pre-hCS are shown below (e.g. amino acid -24 is thr in prehGH and pro in pre hCS.

```
                             Met
GGATCCTGTGGACAGCTCACCTAGCTGCAATGGCTACAGGCTCCCGGACGTCCCTGCTCC
  G                     TG         C

TGGCTTTTGGCCTGCTCTGCCTGCCCTGGCTTCAAGAGGGCAGTGCCTTCCCAACCATTC
         C                             CTG     G   A    G

CCTTATCCAGGCTTTTTGACAACGCTATGCTCCGCGCCCATCGTCTGCACCAGCTGGCCT
 G                  C            AA        CGC             A

TTGACACCTACCAGGAGTTTGAAGAAGCCTATATCCCAAAGGAACAGAAGTATTCATTCC
                          A                C           G

TGCAGAACCCCCAGACCTCCCTCTGTTTCTCAGAGTCTATTCCGACACCCTCCAACAGGG
    TG  T           T    C        C                      T

AGGAAACACAACAGAAATCCAACCTAGAGCTGCTCCGCATCTCCCTGCTGCTCATCCAGT
       G              T                                 G

CGTGGCTGGAGCCCGTGCAGTTCCTCAGGAGTGTCTTCGCCAACAGCCTGGTGTACGGCG
                  G             A G          A         T A A

CCTCTGACAGCAACGTCTATGACCTCCTAAAGGACCTAGAGGAAGGCATCCAAACGCTGA
    G      G T A    C

TGGGGAGGCTGGAAGATGGCAGCCCCCGGACTGGGCAGATCTTCAAGCAGACCTACAGCA
                C       G                C

AGTTCGACACAAACTCACACAACGATGACGCACTACTCAAGAACTACGGGCTGCTCTACT
    T           G      C          G

GCTTCAGGAAGGACATGGACAAGGTCGAGACATTCCTGCGCATCGTGCAGTGCCGCTCTG
                                           G

                 Phe AM
TGGAGGGCAGCTGTGGCTTCTAGCTGCCCGGGTGGCATCCTGTGACCCCTCCCCAGTGCC
 C       CG            G      A  A

TCTCCTGGCC
```

Fig. 4. Comparison of the nucleotide sequences of prehGH and prehCS as derived from the cloned cDNA. The sequence of prehGH is shown with the changes present in prehCS below. The positions of the initiator methionine (Met, -26), the last amino acid (Phe, 191), and the termination codon (TAG, AM) are indicated.

-26						-20									
met	ala	thr	gly	ser	arg	thr	ser	leu	leu	leu	ala	phe	gly	leu	leu
		ala	asp		gln		pro	trp			thr		ser		
-10										1					
cys	leu	pro	trp	leu	gln	glu	gly	ser	ala	phe	pro	thr	ile	pro	leu
		leu		pro			ala	gly		leu		ala	met		
			10										20		
ser	arg	leu	phe	asp	asn	ala	met	leu	arg	ala	his	arg	leu	his	gln
	ser			ala			val				gln	his			
							30								
leu	ala	phe	asp	thr	tyr	gln	glu	phe	glu	glu	ala	tyr	ile	pro	lys
		ala				lys				arg					glu
	40										50				
glu	gln	lys	tyr	ser	phe	leu	gln	asn	pro	gln	thr	ser	leu	cys	phe
gly		arg			:::	ile			ala		ala	ala	phe		
					60										70
ser	glu	ser	ile	pro	thr	pro	ser	asn	arg	glu	glu	thr	gln	gln	lys
		thr			ala		thr	gly	lys			ala			arg
									80						
ser	asn	leu	glu	leu	leu	arg	ile	ser	leu	leu	leu	ile	gln	ser	trp
thr	asp	met					phe								
			90										100		
leu	glu	pro	val	gln	phe	leu	arg	ser	val	phe	ala	asn	ser	leu	val
	gly						ser	arg	ile		thr				met
							110								
tyr	gly	ala	ser	asp	ser	asn	val	tyr	asp	leu	leu	lys	asp	leu	glu
phe		thr			:::	arg			glu	lys					
	120										130				
glu	gly	ile	gln	thr	leu	met	gly	arg	leu	glu	asp	gly	ser	pro	arg
				ala			gln	glu							
					140										150
thr	gly	gln	ile	phe	lys	gln	thr	tyr	ser	lys	phe	asp	thr	asn	ser
ile				leu					asp				ala		met
									160						
his	asn	asp	asp	ala	leu	leu	lys	asn	tyr	gly	leu	leu	tyr	cys	phe
arg	ser												ser		
			170										180		
arg	lys	asp	met	asp	lys	val	glu	thr	phe	leu	arg	ile	val	gln	cys
lys			leu	his		ala			tyr			val	met	lys	
								190	191						
arg	:::	ser	val	glu	gly	ser	cys	gly	phe						
	arg	phe	ala		ser			ala							

Fig. 5. Comparison of the amino acid sequence of prehGH and pre-rGH. The continuous line is the amino acid sequence of prehGH taken from Fig. 1 and the amino acids which differ in pre-rGH are shown below. The sequence of pre-rGH is taken from Seeburg *et al.*, reference 3. The numbering of the amino acids is for hGH. Two insertions (at amino acids number 44 and 108; indicated as :::) had to be added to the rGH sequence and one (between amino acids 183 and 184) to the hGH sequence to achieve maximum homology. Human GH is 191 and rGH is 190 amino acids.

```
                      Met
GTGGACAGCTCACCTAGCTGCAATGGCTACAGGCTCCCGGACGTCCCTGCTCCTGGCTTT
        A    TG: TG  G      G   A   T A   TC  TG       A C

TGGCCTGCTCTGCCTGCCCTGGCTTCAAGAGGGCAGTGCCTTCCCAACCATTCCCTTATC
CA               TG    C        CTG    T  A  TG    G     G

CAGGCTTTTTGACAACGCTATGCTCCGCGCCCATCGTCTGCACCAGCTGGCCTTTGACAC
   T  G    C   T    G       A    G AC              TGC

CTACCAGGAGTTTGAAGAAGCCTATATCCCAAAGGAACAGAAGTATTCATTCCTGCAGAA
    A A     C  GCGT     C   T CG    G   CGC     C:::A T

CCCCCAGACCTCCCTCTGTTTCTCAGAGTCTATTCCGACACCCTCCAACAGGGAGGAAAC
TG     G TG GT     C        A C  C  AG C   A  GG  A      GG

ACAACAGAAATCCAACCTAGAGCTGCTCCGCATCTCCCTGCTGCTCATCCAGTCGTGGCT
C  G    G A TG  A G  AT    T   T    G                 A

GGAGCCCGTGCAGTTCCTCAGGAGTGTCTTCGCCAACAGCCTGGTGTACGGCGCCTCTGA
  G            T     C  GA    TA           A   TT  TA    G

CAGCAACGTCTATGACCTCCTAAAGGACCTAGAGGAAGGCATCCAAACGCTGATGGGGAG
:::CG          GAAA  G        G  A  G        GG T      CA GA

GCTGGAAGATGGCAGCCCCCGGACTGGGCAGATCTTCAAGCAGACCTACAGCAAGTTCGA
         C           T T          C       A     TGA      T

CACAAACTCACACAACGATGACGCACTACTCAAGAACTACGGGCTGCTCTACTGCTTCAG
 G C   ATG G  G         T  G     A                C        A

GAAGGACATGGACAAGGTCGAGACATTCCTGCGCATCGTGCAGTGCCGC:::TCTGTGGA
        C  C     CA      C A      GG  A  A   T    CGC T C
             Phe AM
GGGCAGCTGTGGCTTCTAGCTGCCCGGGTGGCATCCTGTGACCCCTCCCCAGTGCCTCTC
AA          CT      GCA A ACTG TGTCT   C G A TC    GT A  C C

CTGGCCCTGGAAGTTGCCACTCCAGTGCCCACCAGCCTT:GTCCTAATAAAATTAA:GTT
TGTA T    C AC       C TACA:::::::::   T                T A

GCATC
     ATATC
```

Figure 6

```
                         INSULIN

HUMAN     GCACCGAGAGAGATGGAATAAAGCCCTTGAACCAGC
RAT       C    CCTCTGC AT          T     TG

                     GROWTH HORMONE

HUMAN     CACCAGCCTT:GTCCTAATAAAATTAA:GTTGCATC
RAT       :::::::   T                T A

                        α GLOBIN

HUMAN     CCCTTCCTGGTCTTTGAATAAAGTCTGAGTGGGCGG
RABBIT      :::                            A T

                        β GLOBIN

HUMAN     :::::GGATTCTGCCTAATAAA::AAACATTTATTT
RABBIT    ::::: AC     G        GG   ::
```

Fig. 7. Homology in the 3' untranslated region of insulin, growth hormone, and globin mRNAs. The human sequences are shown with the changes found in the other species (i.e. rat for insulin or growth hormone or rabbit for globin) below. The AATAAA sequence is underlined. The insulin comparison is taken from reference 2 and the globin from 19.

Fig. 6. Comparison of the nucleotide sequences of prehGH and pre-rGH as derived from the cloned cDNA. The sequence of prehGH is shown with the changes present in pre-rGH below. The sequence for prehGH is from Fig. 1 and pre-rGH is taken from Fig. 8 of reference 3. The position of the initiator methionine (Met, -26), the last amino acid (Phe, 191) and the termination codon (TAG, AM) are indicated. Three gaps (indicated by :::) in the coding region and also four gaps (two of one bp and one of 9 bp) in the 3'-untranslated region and one gap (of one bp) in the 5' untranslated region were inserted to achieve maximum homology.

Structure of the Genes

High molecular weight DNA extracted from a human placenta obtained by Caesarian section was cleaved with EcoRI endonuclease, fractionated by electrophoresis in agarose gels and the size of the hGH and hCS containing fragments determined by hybridization to a radioactive probe synthesized from the cloned hGH or hCS cDNA. Subsequently, large amounts of DNA were digested and the resulting DNA fragments separated by preparative agarose gel electrophoresis. The fractions containing the 2.6 kb (kilobase, 1000 bp) hGH and 2.8 kb hCS fragments were identified by hybridization and ligated to the bacteriophage λ cloning vector, λgtWES.λB (21). This ligated concatameric DNA was then packaged in vitro into viable phage particles using the procedure described by Blattner et al. (22). Several hundred thousand phage plaques were screened using the method of Benton and Davis (23) and the hGH-specific cDNA probe. Phage from hybridizing plaques were plaque purified and characterized by restriction endonuclease mapping and partial DNA sequence analysis (21).

The hGH gene contains at least three intervening sequences (IVS) as follows: IVS A ~ 400-750 bp between amino acids -25 and -24; IVS B ~ 220 bp between amino acids 31 and 32; and IVS C ~ 300 bp between amino acids 71 and 72. The existence of the IVS have been confirmed by sequence analysis.

Partial DNA sequence analysis in the vicinity of the 3' junction of IVS B showed that a CCA proline codon at amino acid position 37 in the cDNA (Fig. 1) is a CTG leucine codon in the gene (21). Nucleotide sequence differences between the cDNA and the gene isolate have also been detected in the 3' untranslated region as shown in Figure 8. In particular, the HaeIII site present in the cDNA is missing in the gene sequence. The gene isolate with these differences has been designated ghGH 2.6-1.

Two additional isolates (ghGH 2.6-2 and 2.6-3) obtained subsequently from the same placental DNA but from different packaging experiments appear to be identical to each other, and differ in their HaeIII digestion pattern from 2.6-1 (see below). Nucleotide sequence analysis at the 5' junction of IVS C (Fig. 9) shows that the isolate 2.6-2 (Type A gene) has the same sequence as the cDNA, but isolate 2.6-1 (Type B gene) has two amino acid changes (number 65 and 66) as well as three third position silent changes in this region.

Thus two different types of hGH genes have been isolated, one of which appears to have a sequence which corresponds to that of the cDNA and protein while the other has sequence variations. It is not yet known whether there is a low level of expression of the variant gene.

Three forms of the hCS gene have been isolated (ghCS 2.8-1,2 and 3) each of which has a different restriction pattern. Two of these (2.8-1 and 3) differ by a BglII site: 2.8-1 does not have a BglII site while 2.8-3 is cleaved to yield a 2 kb hybridizing fragment. This is consistent with results from genomic blots where about 50% of the hybridization detected at 2.8 kb migrates at 2 kb after BglII digestion. Although the third isolate (2.8-2) has a restriction pattern quite different from the other two gene isolates and the hCS cDNA clone it hybridizes quite strongly.

To determine whether these multiple gene forms are naturally occurring, i.e. are not cloning artifacts nor confined to the one individual whose DNA was cloned, we screened a bacteriophage λ charon 4A library for hGH and hCS genes. This library was constructed by Lawn et al. (26) (obtained from Dr. T. Maniatis) by partial digestion of human fetal liver DNA with HaeIII and AluI restriction endonucleases, size fractionation, methylation with the EcoRI methylase, EcoRI oligonucleotide linker addition, and cloning in the lambda vector, Charon 4A (22). The EcoRI restriction patterns (Fig. 10A, lanes 2-5) and hybridization to the hGH cDNA probe (Fig. 10B, lanes 2-5) of four isolates is shown in Figure 10. All isolates have at least five EcoRI fragments (the two largest 19.8 and 10.9 kb are the left and right arms of Charon 4A respectively) and when hybridized to the hGH cDNA probe give the following hybridization pattern: lane 2 (λ31), 2.6 + 9.3 kb; lane 3 (λ1), 2.6 kb; lane 4 (λ10), 2.8 kb; and lane 5 (λ39), 2.6 + 2.8 kb. Isolates 19, 24, and 37 appear to be identical to isolate 1. The HaeIII restriction fragments obtained from these isolates which hybridize to the chGH probe are shown in Figure 11. This indicates that λ isolates 37, 31, 24, 19 and 1 (Fig. 11, lanes 2-6 respectively) have a HaeIII digestion pattern similar to the variant type A clone (2.6-2) from the previous packaging, while λ isolate 39 (Fig. 11, lane 1) has a type B pattern (2.6-1 type). These experiments indicate (1) that both sets (from the placental DNA and the library) of clones contain the different types of hGH and hCS genes; (2) there is a 9.3 kb hybridizing fragment previously detected in genomic blots of human DNA (21) linked to a 2.6 kb isolate both of which have a HaeIII pattern indistinguishable from the type A pattern (λ isolate 31, lane 2

Fig. 10 and lane 3 Fig 11); and (3) a hGH gene (type B) is linked to a hCS gene (type 2.8-3) in the single library isolate number 39 (lane 5 Fig. 10 and lane 1 Fig. 11). Therefore, although the data is not yet complete, there appears to be a series of different but closely related GH and CS genes some of which are fairly closely linked.

```
                              HaeIII
   TCCCCAGTGCCTCTCCTGGCCCTGGAAGTTGCCACTCCA   chGH  3' end
                        T G   :  G   T       ghGH  3' end
```

Fig. 8. Nucleotide sequence comparison of chGH and ghGH 2.6-1 in the 3'-untranslated region. The cDNA sequence (chGH) is from Figure 1. Gene isolate ghGH 2.6-1 was cleaved with endonuclease XmaI, the 5' end labeled with polynucleotide kinase and $\gamma[^{32}P]$ATP, and sequenced using the chain termination method of Maat and Smith (24) as described in Fiddes and Goodman (8). The C -> T change resulting in the loss of the HaeIII in the gene is consistent with changes in the restriction endonuclease digestion patterns.

```
                             val lys
Type B   TCC AAC AGG GTG AAA ACG CAG CAG AAA TCT  GT    IVS C

cDNA     TCC AAC AGG GAG GAA ACA CAA CAG AAA TCC AAC

         62          65                  70      72
Protein  Ser Asn Arg Glu Glu Thr Gln Gln Lys Ser Asn

Type A   TCC AAC AGG GAG GAA ACA CAA CAG AAA TCC  GT    IVS C
```

Fig. 9. Sequence comparison of hGH gene isolates. The nucleotide sequence in the vicinity of the 5' junction of IVS C of hGH gene isolates ghGH 2.6-2 (Type A) and ghGH 2.6-1 (Type B) was determined by primed synthesis using the dideoxy chain termination method (15) after transfer of the 2.6 kb EcoRI fragments in λgtWES·λB to the EcoRI site of the single-stranded phage vector M13mp2 (25).

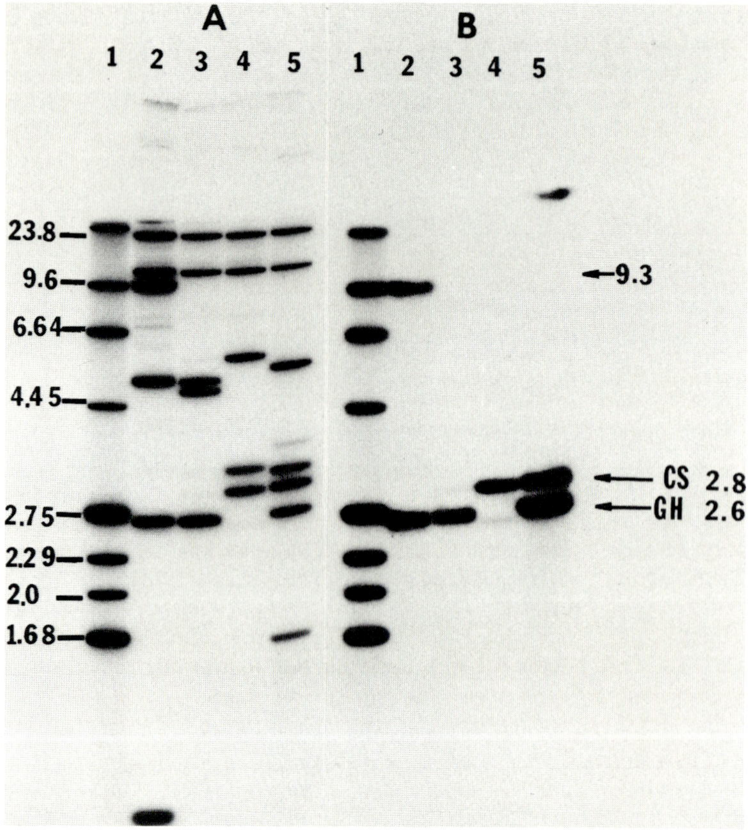

Fig. 10. Restriction and hybridization analysis of bacteriophage λ library clones. DNAs containing hGH, hCS, and related genes isolated from the λ Charon 4A library (26) were digested with EcoRI endonuclease and separated on a 1% agarose gel after labeling (Fig. 10A) with the Klenow fragment of DNA polymerase I (30). Unlabeled fragments (Fig. 10B) were run on an identical gel, blotted onto a nitrocellulose filter, and hybridized with ^{32}P-labeled chGH800 cDNA probe (21). Panel A: (1) HindIII digest of λ DNA plus MspI digest of ΦX174 R.F. DNA used as molecular weight markers; lanes 2-5; (2) λ31, (3) λ1, (4) λ10, and (5) λ39. Panel B: Lane (1) same as panel A; Lanes 2-5, blot and hybridization to ^{32}P-labeled chGH800 DNA (31, 1, 10, and 39 respectively).

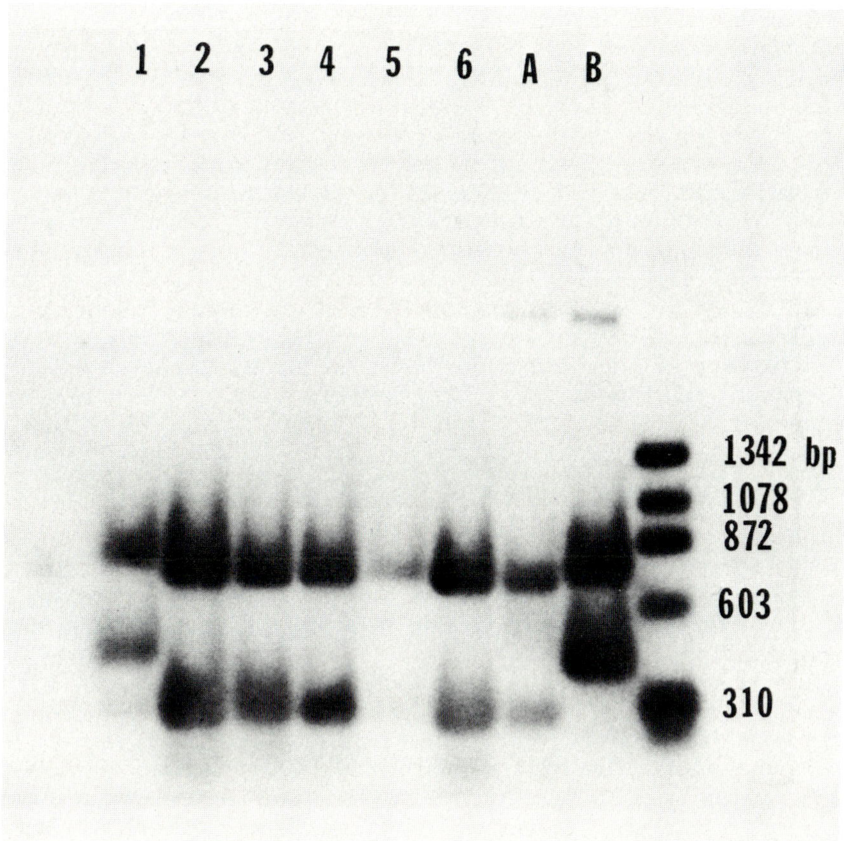

Fig. 11. Autoradiograph of a filter of recombinant λ DNAs containing hGH and related genes digested with EcoRI plus HaeIII (fractionated on a 1.5% agarose gel) and hybridized to ^{32}P-labeled chGH800 DNA. The recombinant DNAs are: Lane 1, λ39; 2, λ37; 3, λ31; 4, λ24; 5,λ19; 6, λ1; A, λ2.6-2; B, λ2.6-1. Isolates 1, 19, 24 and 37 appear to be identical.

Structure of the rGH Gene

Hooded rat DNA (27) digested with EcoRI endonuclease was fractionated by RPC-5 chromatography (28) and then by size on a sucrose gradient. An approximately 11.8 kb fragment which was partially purified by the above procedure and which hybridized to the rGH cloned cDNA probe was packaged into bacteriophage λ. A single clone was obtained from one size fraction of the sucrose gradient, while six independent but identical isolates of a second clone were obtained from the adjacent fraction. The restriction maps of these two types of clones differ radically from each other; only the former corresponds to the map expected from the known sites in the rGH cDNA clone. We suspect that the other may be rat CS or a variant of rGH, but no further evidence is currently available.

A partial sequence of the rGH gene clone (grGH) is presented in Figure 12. The positions of two intervening sequences (IVS B and C) have been identified. Intervening sequence B occurs in the same position as the corresponding IVS in the gene for hGH, i.e. between amino acids 31 and 32. Intervening sequence C, on the other hand, occurs between amino acids 70 and 71 in grGH and 71 and 72 in ghGH. However, structurally this occurs in the same place in the molecule, since a gap of one amino acid (at position 44 in Fig. 5) had to be added to the rGH sequence to achieve maximum homology with hGH. Hence, the position of this IVS also seems to have been conserved during evolution. A third intervening sequence (IVS D) has been found by sequence analysis of the rGH gene (Fig. 12). It is small (about 100 bp) and remained undetected when the hGH gene was mapped by restriction endonuclease digestion. DNA sequence analysis of an hGH gene isolate in this region indicates a corresponding IVS in the hGH gene. It probably occurs between amino acids 124 and 125 in rGH, and in exactly the same place or displaced by two base pairs in hGH, but definition of its precise boundaries require further analysis. The fourth intervening sequence (IVS A) which occurs between amino acids -25 and -24 in hGH has also been detected in rGH, however, neither its size nor location are yet accurately determined. It is also interesting to note that the 3' junctions of IVSB, C and D (Fig. 12) all have a CAG preceeded by a pyrimidine tract and hence follow the GT/AG rule described by Breathnach *et al.* (29).

```
                             -26
                             Met Ala Ala Asp Ser Gln Thr Pro Trp
GTGGACAGATCACTGAGTGGCG       ATG GCT GCA GAC TCT CAG ACT CCC TGG

Leu Leu Thr Phe Ser Leu Leu Cys Leu Leu Trp Pro Gln Glu Ala
CTC CTG AC CTTC AGC CTG CTC TGC CTG CTG TGG CCT CAA GAG GCT

Gly Ala Leu Pro Ala Met Pro Leu Ser Ser Leu Phe Ala Asn Ala
GGT GCT TTA CCT GCC ATG CCC TTG TCC AGT CTG TTT GCC AAT GCT

Val Leu Arg Ala Gln His Leu His Gln Leu Ala Ala Asp Thr Tyr
GTG CTC CGA GCC CAG CAC CTG CAC CAG CTG GCT GCT GAC ACC TAC
         31
Lys Glu Phe
AAA GAG TTC  :::AAAAGGACATGGTGGCTCACAAACCATCTGTAAAGAGATCTGA
                              IVS B
ATGCCCTCTTCTGGTGTATCTGAAGACAGCTACAGTGTACTTATATATAATAAATAAAT

AAATCTTTAAAAAAAACAAAACAAAAACAAAAACAAAACAGTAATGACAGAGAGTCACA

                                         32
                                         Glu Arg Ala Tyr Ile Pro
AGCTGGTCCCTCAGTGACTACCTTTCCTCCAG         GAG CGT GCC TAC ATT CCC

Glu Gly Gln Arg Tyr Ser Ile Gln Asn Ala Gln Ala Ala Phe Cys
GAG GGA CAG CGC TAT TCC ATT CAG AAT GCC CAG GCT GCG TTC TGC

Phe Ser Glu Thr Ile Pro Ala Pro Thr Gly Lys Glu Glu Ala Gln
TTC TCA GAG ACC ATC CCA GCC CCC ACC GGC AAG GAG GAG GCC CAG

         70                               IVS C
Gln Arg Thr
CAG AGA ACT  :::AACCTACTATCAGGCAGACAGCTCTTCCCTAGGCTAGGTAACC
                                                          71
                                                          Asp
TGGGATCCCAGGCAAGATCACTAGCTAATGCACAGCCCCTTTTTTCCCTCAG      GAC

Met Glu Leu Leu Arg Phe Ser Leu Leu Leu Ile Gln Ser Trp Leu
ATG GAA TTG CTT CGC TTC TCG CTG CTG CTC ATC CAG TCA TGG CTG

Gly Pro Val Gln Phe Leu Ser Arg Ile Phe Thr Asn Ser Leu Met
GGG CCC GTG CAG TTT CTC AGC AGG ATC TTT ACC AAC AGC CTG ATG

Phe Gly Thr Ser Asp Arg Val Tyr Glu Lys Leu Lys Asp Leu Glu
TTT GGT ACC TCG GAC CGC GTC TAT GAG AAA CTG AAG GAC CTG GAA
```

```
                                      124
Glu Gly Ile Gln Ala Leu Met Gln
GAG GGC ATC CAG GCT CTG ATG CAG  :::TTTAACCGGTCAAGCCTTAAGGT
                              IVS D
TAAT::TTCCTACTCAATTCCCTTATGGAAACTCCAGGCCCCTCTCTAAGTTCTGGAAT
                                                  125
                                                  Glu Leu Glu  sp
TGGGGAGGG:ACGGCTCTGAA:TCTTCTTTCCCACAACAG          GAG CTG GAA GAC

Gly Ser Pro Arg Ile Gly Gln Ile Leu Lys Gln Thr Tyr Asp Lys
GGC AGC CCC CGT ATT GGG CAG ATC CTC AAG CAA ACC TAT GAC AAG

Phe Asp Ala Asn Met Arg Ser Asp Asp Ala Leu Leu Lys Asn Tyr
TTT GAC GCC AAC ATG CGC AGC GAT GAC GCT CTG CTC AAA AAC TAT

Gly Leu Leu Ser Cys Phe Lys Lys Asp Leu His Lys Ala Glu Thr
GGG CTG CTC TCC TGC TTC AAG AAG GAC CTG CAC AAG GCA GAG ACC

Tyr Leu Arg Val Met Lys Cys Arg Arg Phe Ala Glu Ser Ser Cys
TAC CTG CGG GTC ATG AAG TGT CGC CGC TTT GCG GAA AGC AGC TGT

    190
Ala Phe AM
GCT TTC TAG  GCACACACTGGTGTCTCTGCGGCACTCCCCCGTTACCCCCCTGTAC

TCTGGCAACTGCCACCCCTACACTTTGTCCTAATAAAATTAATGATGCATCATATC
```

Fig. 12. Partial nucleotide sequence of the rGH gene. An approximately 1.7 kb PvuII fragment from the rGH 11.8 kb gene isolate was subcloned into M13mp5 after HindIII linker addition (27). The partial sequence was determined by primed synthesis using the dideoxy chain termination method (15). The figure is a composite of the data obtained from sequencing of the gene and the previously published cDNA sequence (3). The position of IVS A which occurs around amino acid position -25 to -24 is not shown.

IV. CONCLUSIONS

The growth hormone genes in human and rat form a set of closely related genes whose structures and relatedness provide information on how they evolved. Further, studies utilizing mutational analysis and surrogate genetic techinques should lead to a further understanding of the control of expression of these genes.

REFERENCES

1. Ullrich, A., Shine, J., Chirgwin, J., Pictet, R., Tischer, E., Rutter, W.J., and Goodman, H.M. Science 196, 1313 (1977).
2. Bell, G.I., Swain, W.F., Pictet, R., Cordell, B., Goodman, H.M., and Rutter, W.J. Nature 282, 525 (1979).
3. Seeburg, P.H., Shine, J., Martial, J.A., Baxer, J.D., and Goodman, H.M. Nature 270, 494 (1977).
4. Shine, J., Seeburg, P.H., Martial, J.A., Baxter, J.D., and Goodman, H.M. Nature 270, 494 (1977).
5. Goodman, H.M., Seeburg, P.H., Shine, J., Martial, J.A., and Baxter, J.D. in "Specific Eukaryotic Genes" (J. Engberg, H. Klenow, and V. Leick, eds.), p. 179. Alfred Benzon Symposium 13, Munksgaard, (1979).
6. Martial, J.A., Hallewell, R.A., Baxter, J.D., and Goodman, H.M. Science 205, 602 (1979).
7. Roberts, J.L., Seeburg, P.H., Shine, J., Herbert, E., and Baxter, J.D. Proc. Natl. Acad. Sci. USA 76, 2153 (1979).
8. Fiddes, J.C., and Goodman, H.M. Nature 281, 351 (1979).
9. Bewley, T.A., Dixon, J.S., and Li, C.H. Int. J. Peptide Prot. Res. 4, 281 (1972).
10. Niall, H.D., Hogan, M.L., Sauer, R., Rosenblum, I.Y., and Greenwood, R.C. Proc. Natl. Acad. Sci. USA 68, 866 (1971).
11. Seeburg, P.H., Shine, J., Martial, J.A., Ullrich, A., Baxter, J.D., and Goodman, H.M. Cell 12, 157 (1976).
12. Goodman, H.M., and MacDonald, R.J. Methods in Enzymology, in press.
13. Bolivar, F., Rodriguez, R.L., Greene, P.J., Betlach, M.C., Heyneker, H.L., Boyer, H.W., Crosa, J.H., and Falkow, S. Gene 2, 95 (1977).
14. Maxam, A.M., and Gilbert, W. Proc. Natl. Acad. Sci. USA 74, 560 (1977).
15. Sanger, F., Nicklen, S., and Coulson, A.R. Proc. Natl. Acad. Sci. USA 74, 5463 (1977).

16. Dayhoff, M.O. "Atlas of Protein Sequence and Structure", Volume 5, National Biomedical Research Foundation, Washington, D.C., pp. D-201-D-204 (1972).
17. Roskam, W.G., and Rougeon, F. Nucleic Acids Res. 7, 305 (1979).
18. Sherwood, L.M., Burstein, Y., and Schechter, I. Proc. Natl. Acad. Sci. USA 76, 3819 (1979).
19. Proudfoot, N.J., and Brownlee, G.G. Nature 263, 211 (1976).
20. Efstratiadis, A., Kafatos, F.C., and Maniatis, T. Cell 10, 571 (1977).

[[21. Fiddes, J.C., Seeburg, P.H., DeNoto, F.M., Hallewell, R.A., Baxter, J.D., and Goodman, H.M. Proc. Natl. Acad. Sci. USA 76, 4294 (1979).
22. Blattner, F.R., Blechl, A.E., Denniston-Thompson, K., Faber, H.E., Richards, J.E., Slightom, J.L., Tucker, P.W., and Smithies, O. Science 202, 1279 (1978).
23. Benton, W.D., and Davis, R.W. Science 196, 180 (1977).
24. Maat, J., and Smith, A.J.H. Nucleic Acids Res. 5, 4537 (1978).
25. Gronenborn, B., and Messing, J. Nature 272, 375 (1979).
26. Lawn, R.M., Fritsch, E.F., Parker, R.C., Blake, G., and Maniatis, T. Cell 15, 1157 (1976).
27. Cordell, B., Bell, G., Tischer, E., DeNoto, F.M., Ullrich, A., Pictet, R., Rutter, W.J., and Goodman, H.M. Cell 18, 533 (1979).
28. Tiemeier, D.C., Tilghman, S.M., and Leder, P. Gene 2, 173 (1977).
29. Breathnach, R., Benoist, C., O'Hare, K., Gannon, F., and Chambon, P. Proc. Natl. Acad. Sci. USA 75, 4853 (1978).
30. Drouin, J. J. Mol. Biol., in press (1980).

DISCUSSION

D. DRESSLER: Since the intervening sequence found in the coding region of the rat-2 insulin gene does not occur in the rat-1 insulin gene - yet both genes are expressed - can it be concluded that this insert plays no physiological role in the expression of the insulin gene?

H.M. GOODMAN: I think you will hear more definitive data on that question later from Paul Berg. In the rat, both insulin genes are expressed at approximately the same level (40%-60%), but remember they both have at least one intervening sequence so this does not really directly bear on that question.

G. KHOURY: Have you considered the possibility that the rat-1 gene perhaps lost one of its intervening sequences through a mechanism analogous to reverse transcription?

H.M. GOODMAN: This is a possibility, but we have no evidence one way or the other. The rat-1 gene for insulin does not have the insert found in the rat-2 gene but otherwise it is the same.

A.I. BUKHARI: Do you think that there was a selection for uninterrupted rat-1 gene?

H.M. GOODMAN: Both genes are expressed, probably in the same cell. The chicken gene also has two intervening sequences. Evolutionarily before and after rat, the common gene has two intervening sequences. The data are most consistent with the precise loss of one intervening sequence to generate the rat-1 gene rather than the gain of an intervening sequence to generate the rat-2 gene. Of course, I cannot rule out some kind of parallel evolution of the two types of genes.

A.I. BUKHARI: One would think that maybe the basic gene was without the insert and it came about when the gene was translocated to another place.

H.M. GOODMAN: That is possible - but then we would expect the common gene to have only one intervening sequence.

J. CHIRGWIN: Is there any structural basis that you might imagine whereby the second intervening sequence, would form a secondary structure which would be removed. Perhaps an RNA splicing enzyme makes a mistake and chops out 500 bases of single stranded DNA.

H.M. GOODMAN: There is no data at all on that at the moment.

STABLE AND UNSTABLE EXPRESSION OF GENES IN DNA TRANSFORMED CELLS

Jackson, J., Lowy, I., Ostrander, M., Pellicer, A., Roberts, J., Robins, D., Sim, G.-K., Sweet, R., Wold, B., Silverstein, S.[1] and Axel, R.[2]

College of Physicians and Surgeons
Columbia University
New York, New York

I. INTRODUCTION

Transformation provides a unique opportunity to alter the genotype of the cell by the stable introduction of new genetic information. Thus, DNA obtained either from viruses or from eukaryotic cells can be used to transfer thymidine kinase (tk) (1-4), adenine phosphoribosyl transferase (aprt) (5), and hypoxanthine guanine phosphoribosyl transferase (3,6) to mutant mouse L cells deficient in these functions.

The potential usefulness of transformation in the study of eukaryotic gene expression depends to a large extent on its generality. We have demonstrated that cells transformed with selectable genes also integrate other physically unlinked DNA fragments at high frequency (7). In this manner, we have used the tk gene to identify mammalian cells cotransformed with defined prokaryotic and eukaryotic genes. This vectoring system overcomes the limitations on the nature of donor DNA sequences and permits the introduction of any cloned gene into cultured cells. Detection of gene transfer with recessive markers, however, restricts transformation to available mutant recipients which may not provide an appropriate environment for the expression of cotransformed genes. We have therefore cotransformed cells with a dominant acting marker, the gene coding

[1] Supported by NIH grant CA-17477 and RCDA CA-00492.
[2] Supported by NIH grants CA-16346 and CA-23767.

ISBN 0-12-633360-2

for a mutant dihydrofolate reductase, which renders cells resistant to high concentrations of methotrexate (8). The use of this gene as a vector in cotransformation systems may permit the transfer of virtually any genetic element into a variety of new cellular environments.

These cotransformation procedures provide an opportunity to study the expression of cloned genes in a heterologous host. We have cotransformed four distinct cell types: murine fibroblasts, erythroleukemia cells, teratocarcinoma cells and rat hepatoma cells. The introduction of a viral tk gene into each of these cell types results in the expression of functional viral enzyme. Analysis of the expression of heterologous cotransformed genes, however, reveals transcripts that have undergone appropriate processing to remove intervening sequences but frequently display aberrant 5' termini.

Tk^+ transformants also permit a molecular analysis of mutation, since the tk gene has been cloned in bacterial plasmids and the tk^+ and tk^- phenotypes are readily selected under appropriate growth conditions. In this report, we describe a class of unstable mutants that switch phenotypes from tk^+ to tk^- to tk^+ ... at frequencies far greater than the observed rate of mutation of a hemizygous gene (9). Transformation studies in concert with restriction endonuclease mapping suggest that the determinants responsible for phenotype switching result from frequent but reversible alterations in DNA.

II. RESULTS AND DISCUSSION

A. Cotransformation of Mouse tk^- Fibroblasts with Defined Genetic Elements

We have utilized the tk gene as a vector in cotransformation experiments to construct mouse cell lines containing prokaryotic and eukaryotic genes for which no selective criteria exist (7). A representative analysis of cotransformants obtained with a variant human growth hormone gene (10) is shown in Fig. 1. DNA from five cotransformants was analyzed for the presence of growth hormone sequences by blot hybridization. The conclusions derived from these and additional experiments with several other genes permit the following conclusions: 1) The frequency of cotransformation is high; 80% of the tk^+ transformants contain sequences homologous to the nonselectable gene. 2) Most cotransformed sequences are localized in the high molecular weight DNA fraction within the cell. 3) The number of integrated copies varies from one to over 100 in independent clones. 4) The extent of the cotransformed

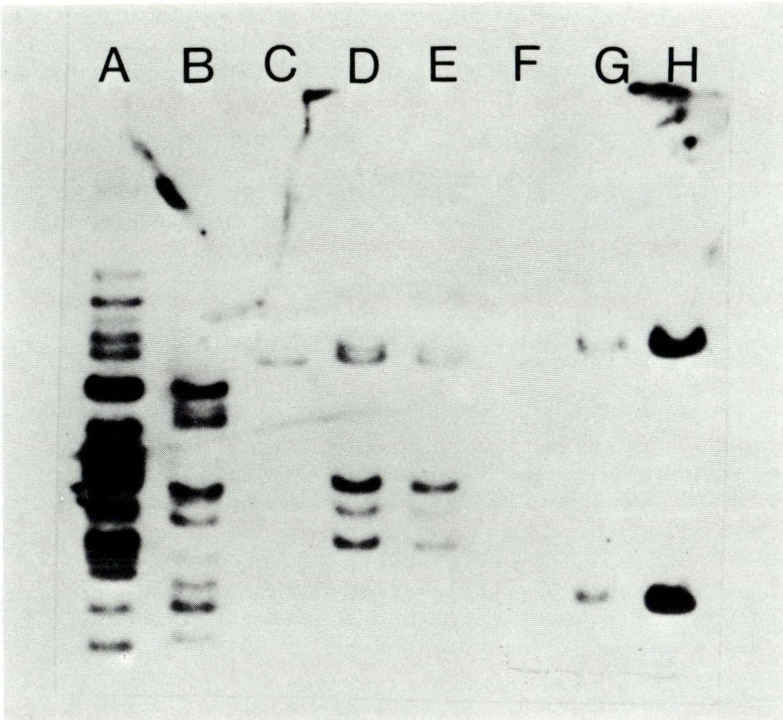

Fig. 1. Human growth hormone genes in cotransformed rat BRL cells. High molecular weight DNA from five BRL clones cotransformed with ptk and pHGH was digested with Eco RI and electrophoresed on a 0.8% agarose gel. The DNA was denatured *in situ*, transferred to nitrocellulose, and hybridized with a ^{32}P-labeled 2.6 kb Eco RI fragment containing the variant human growth hormone gene (10). Lanes A-E: DNA from five tk^+ BRL cell transformants digested with Eco RI. Lane F: BRL tk^- cell DNA. Lanes G and H: 25 pg and 250 pg respectively of pHGH digested with Eco RI (to visualize the 2.6 kb growth hormone gene insert) mixed with an equal amount of undigested pHGH (6.8 kb).

sequences maintained within a given transformant is variable; some clones integrate only 20% of the donor DNA sequences while others contain over 90% of the donor sequences. 5) The genotype of cotransformed sequences, as reflected by the blot hybridization profiles, is stable for many generations when maintained under selective pressure. 6) All of the cloned eukaryotic genes we have studied can be introduced into mammalian cells with roughly equal efficiency. The properties of cotransformants, noted above, appear to be independent of the cloned DNA sequence.

B. Expression of Cotransformed Genes

The introduction of cloned eukaryotic genes into animal cells provides an *in vivo* system to study the functional significance of various features of DNA sequence organization. Our earlier studies demonstrated that DNA from various species of mammals and birds transfer either tk or aprt activity to mutant murine cells with equivalent efficiencies. These studies suggest that the transcriptional and translational apparatus of the mouse is capable of recognizing, transcribing and translating these DNA sequences into a functional protein without regard to vertebrate species of origin.

Cotransformants containing cloned eukaryotic genes permit us to study the transcription and subsequent processing of these sequences in a heterologous host. To date, these studies have largely been restricted to transformed mouse fibroblasts. In one study, we have examined the expression of the rabbit β-globin gene in six independent transformants which contain from one to 20 copies of the cloned globin gene (11). Rabbit globin transcripts were detected in two of these transformants at steady state concentrations of five and two copies per cell. Solution hybridization experiments in concert with RNA blotting techniques indicate that in one transformed line, the rabbit globin sequences are present in the cytoplasm as a polyadenylated, 9S species. Further analysis indicates that the two intervening sequences present in the original globin transcript are correctly processed. Surprisingly, 45 nucleotides present at the 5' terminus of mature rabbit mRNA are absent from the β-globin RNA sequence detected in the cytoplasm of this transformed fibroblast. Appropriate processing of the rabbit β-globin gene has also been observed in tk^+ mouse cell transformants in which the globin and tk plasmids have been ligated prior to transformation (12). Similar results have also been obtained by using viral vectors to introduce the rabbit globin gene into monkey cells (13, 14).

Taken together, these results suggest that nonerythroid cells from heterologous species contain the enzymes necessary to correctly process intervening sequences of a rabbit gene whose expression is usually restricted to erythroid cells.

The appearance of an aberrant yet unique 5' terminus suggests the possibility of incorrect initiation of transcription in this mouse cell line. A similar analysis of cotransformants constructed by ligation of the tk and globin gene reveals transcripts with at least two discrete 5' termini: RNA with 5' termini indistinguishable from that of mature rabbit mRNA and RNA with 5' termini also lacking about 45 nucleotides (12, C. Weissman, personal communication).

Preliminary analyses of rat cells cotransformed with insulin and human growth hormone genes reveal the presence of transcripts several hundred nucleotides longer than mature insulin or growth hormone mRNA (unpublished studies from laboratory in collaboration with A. Efstratiadis and P. Seeberg). The introduction of the chick ovalbumin gene to mouse L cells results in the synthesis of immunoprecipitable ovalbumin protein (15). In a separate study, analysis of the RNA indicates that while all seven intervening sequences may be precisely removed from this avian gene by murine enzymes, the transcript is 650 nucleotides longer at the 5' terminus than mature ovalbumin (16).

These studies suggest that if the 5' terminus of mature mRNA also reflects the site of initiation of a primary transcript, murine fibroblasts do not consistently support correct initiation of heterologous genes whose expression is usually restricted to other differentiated cell types. The mere presence of a promoter sequence may not insure accurate transcription initiation in this *in vivo* system, since alternate promoters are apparently chosen. Correct initiation may be dictated by promoters acting in concert with transcription control factors which are only provided by an appropriate cellular environment. Alternatively, it is possible that correct initiation is sensitive to the relative position of the specific gene in the eukaryotic genome.

C. Cotransformation of Diverse tk^- Cell Lines

We have recently developed cotransformation systems which permit the introduction of defined genes into murine erythroleukemia (MEL) cells and teratocarcinoma (TCC) cells (17), as well as Buffalo rat liver (BRL) cells. The properties of these three recipients are unique and quite different from

TABLE I. Transformation and Cotransformation Frequency of Diverse tk^- Cell Lines

Cell line[a]	Transformation Frequency[b]	Cotransformation Frequency[c]
Mouse L	50,000	80% (>100)
BRL	5	60% (22)
TCC	0.5	27% (11)
MEL	0.5	6% (50)

[a]See text for explanation of abbreviations.
[b]Units are in colonies/μg/10^6 cells.
[c]Expressed as the frequency with which tk^+ transformants also contain at least one cotransformed element. The number in parenthesis reflects the number of tk^+ transformants analyzed.

those displayed by the mouse L cell (Table I). All three cell lines show a dramatic reduction in tk transformation efficiency when compared to mouse L cells. Transformation of L cells generates one surviving colony per 20 pg of purified gene per 5 X 10^5 cells; in MEL cells, one transformant is obtained per 3 μgs of gene. Intermediate values are obtained for TCC and BRL cell lines. While inordinately high concentrations of gene are required to effect transformation in these cells, the availability of a cloned tk gene nonetheless permits the construction and selection of numerous tk^+ transformants.

We have cotransformed these cell lines with cloned eukaryotic genes. The frequency of cotransformation in rat liver cells is high, but is reduced significantly in the murine erythroleukemia and teratocarcinoma cell lines (Table I). The number of cotransformed sequences introduced into rat liver cells ranges from one to several hundred copies per cell. However, only one to ten cotransformed sequences have been introduced into TCC or MEL cells.

We can therefore discern three classes of recipients--cells which transform and cotransform with high efficiency (L cells); cells which transform at low efficiency but cotransform at high efficiency (BRL cells); and cells which both transform and cotransform at low efficiency (TCC and MEL cells). At present, we can offer no explanation for the wide variation in transformation frequency among different cell lines, since the biological basis for competence in eukaryotic

TABLE II. Stability of the tk Phenotype

Cell line	Relative cloning efficiency in selective medium[a] HAT	BUdR
K_1	0.98	0.13
K_1B_6	0.5	0.52
$K_1B_6^N$	<0.0001	0.87
$K_1B_6H_2$	0.93	0.47
K_2	1.0	0.016
K_2B_6	0.014	0.97
K_2B_7	<0.0001	0.87
$K_2B_6H_4$	0.79	0.08

[a]100 to 1000 cells were plated in triplicate into either neutral medium (HT) or selective medium (HAT or BUdR). Cell lines are described in text.

cells is unknown. In tk^- L cells, at least 2% of the culture can be transformed to the tk^+ phenotype. Competence in this line is not a stably heritable trait and may therefore be a transient property whose expression varies greatly among different cell types.

D. Phenotypic Switching of tk Activity in Transformants

The addition of the herpes virus thymidine kinase gene to mouse Ltk^- cells results in the appearance of numerous surviving colonies expressing the viral enzyme. Transformants obtained in this way contain at least one copy of the viral tk gene integrated into high molecular weight DNA in the transformed cell. These cells have maintained the tk^+ phenotype for several hundred generations under selective conditions. The newly introduced tk gene is particularly amenable to mutational analysis, since the tk^+ phenotype can be selected in growth medium containing HAT (hypoxanthine, aminopterin and thymidine), and the tk^- phenotype can be selected in medium containing BUdR.

We have constructed two transformants, K_1 and K_2, and studied the stability of both the tk phenotype and tk genotype in several independent tk^- revertants by transformation in concert with blot hybridizations (Table II). Tk^+ transformants derived from individual surviving colonies are phenotypically stable under selective pressure and have been maintained for over 300 generations. When plated into counterselective medium (BUdR), however, surviving tk^- colonies are obtained with a frequency of 10% for K_1 and 1% for K_2 (Table II), values far greater than expected mutation rates. Several tk^- revertant clones were chosen from both K_1 and K_2 and are designated K_1B_n and K_2B_n. These revertant cell lines are stable in BUdR, but when plated into HAT, surviving colonies are again observed. The frequency with which tk^- revertants give rise to tk^+ re-revertants (designated $K_1B_nH_n$ and $K_2B_nH_n$) varies among individual revertants. Thus, K_2B_6 re-reverts to the tk^+ phenotype with a frequency of 1.0% while K_1B_6 switches phenotype with a frequency close to 50% (Table II). Tk^+ re-revertants can again be plated in BUdR, continuing the switching cycle from tk^+ to tk^- to tk^+ ... . The frequency with which transformant, revertants and re-revertants switch phenotypes varies among different clones but remains constant for each individual clone.

E. Phenotypic Switching Is Associated with Changes in DNA

The unusual pattern of phenotypic switching may be caused by self-reinforcing phenotypic changes transmissable to progeny in the absence of concomitant changes in DNA. Alternatively, this phenotypic switching may result from frequent but reversible changes in DNA. To distinguish between these alternatives, we have performed a series of transformation experiments with purified DNA from transformant, revertant, and re-revertant cell lines. If reversion to the tk^- phenotype occurs without genetic changes, DNA from these tk^- cells should transfer the tk gene to Ltk^- recipients as efficiently as DNA from tk^+ cells. On the other hand, if alterations in DNA are associated with the phenotypic switch, transfer would be far less efficient with DNA from the tk^- revertants. As expected, DNA from the tk^+ transformant or re-revertant cells is an efficient donor of the tk gene, generating from 10-30 tk^+ colonies/80 µg DNA/5 X 10^5 cells. In contrast, DNA from two independent revertants generated 50 to 100-fold fewer colonies than observed with DNA from tk^+ lines (Table III). The potential inhibition of DNA transfer by BUdR substitution in revertant DNA was avoided by growth of revertant lines in neutral medium (DME) for five generations prior to isolation.

Thus, the relative efficiency of gene transfer with tk^- DNA roughly correlates with the frequency at which the tk^-

TABLE III. Transformation Efficiency of Transformant, Revertant, and Re-revertant DNA

Cell line	tk^+ colonies[a]	Relative tk transformation efficiency[b]
K_1	75	1.0
K_1B_5	4	0.05
K_1B_6	15	0.20
$K_1B_6^N$	0	--
$K_1B_6H_1$	48	0.65
K_2	26	1.0
K_2B_6	1	0.03
$K_2B_6H_4$	42	1.6

[a]Number of surviving tk^+ colonies observed in three dishes after transformation of Ltk^- cells with 80 μg of DNA from transformant, revertant, and re-revertant cell lines.
[b]Relative transformation activity reflects the number of tk^+ colonies obtained with tk^- revertant or tk^+ re-revertant DNA compared with the number of colonies ob-obtained with the original tk^+ transformants.

revertants plate in HAT medium (Tables II and III). This observation suggests that a specific but reversible genetic alteration of the tk gene accompanies phenotypic switching.

F. Organization of the tk Gene in tk^+ and tk^- Transformants

We have analyzed the organization of the viral tk gene sequences in cell sets consisting of the transformant, revertants and re-revertants by restriction endonuclease mapping and blot hybridization. A restriction map of the 5 kb Kpn fragment of HSV DNA containing the tk gene is shown in Fig. 2. This map derives from single and double digestions of cloned fragments with eight enzymes and is in accord with previously

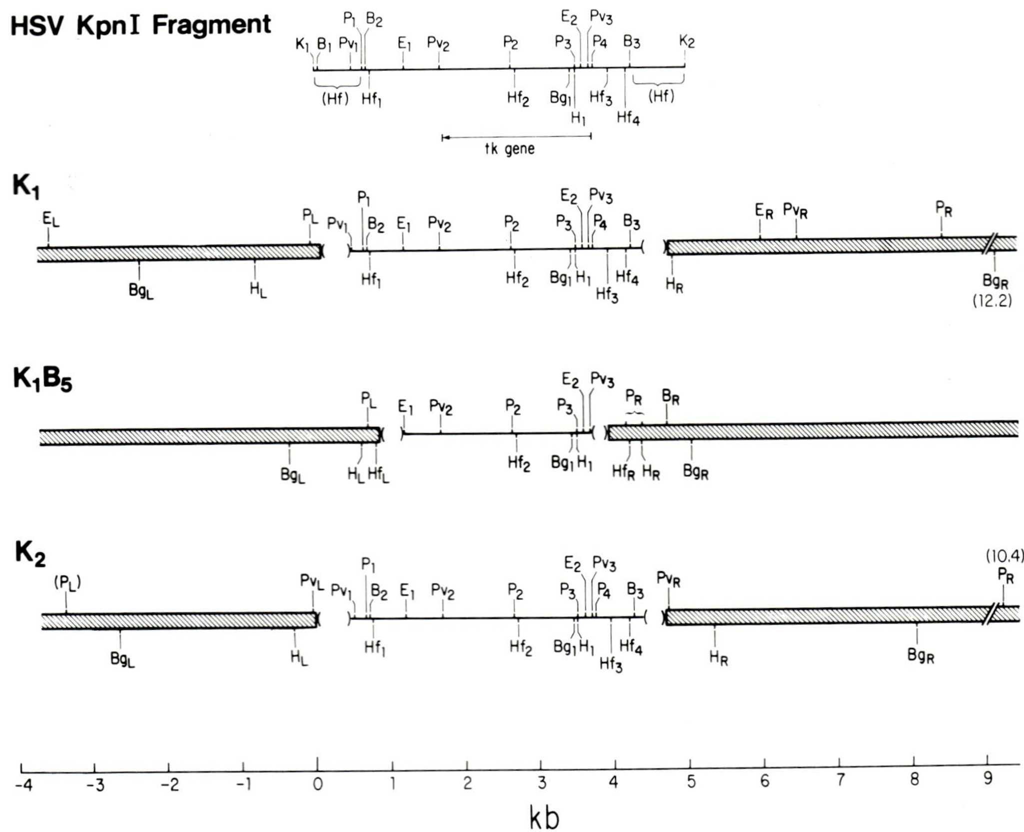

Fig. 2. Restriction endonuclease maps of the HSV tk gene region in the 5.0 kb Kpn I fragment from HSV, the transformant K_1, the revertant K_1B_5, and the transformant K_2. Viral sequences are represented by (——) and nonviral flanking DNA sequences by (▨). The region of the breakpoint between these sequences is bounded by parentheses. The functional tk gene (←—┤) resides between Pvu II sites 2 and 3 (see text). Restriction sites whose position is uncertain are shown in "()". Numbers in "()" beside restriction sites are map distances in kb. B-Bam HI, Bg-Bgl II, E-Eco RI, H-Hinc II, Hf-Hinf I, K-Kpn I, P-Pst I, Pv-Pvu II.

published data (18,19). The 2.0 kb Pvu II fragment maintains efficient transformation activity and therefore provides maximal boundaries for the functional tk gene (19).

A restriction map of the tk gene within the DNA of the transformants, K_1 and K_2, and their respective families of revertants and re-revertants was obtained by blot hybridization (Fig. 2). These results reveal three discrete classes of tk^- revertants: clones with restriction patterns identical to the original tk^+ transformants, a clone in which the tk gene has undergone a transposition, and a clone which has deleted the tk gene. All re-revertants, however, maintained restriction patterns identical to those of the tk^- clones from which they were derived.

1. <u>Tk^- Revertants with Restriction Patterns Identical to Transformants</u>. Illustrative restriction mapping data obtained with the enzyme Bgl II, which cleaves once within the tk gene, is shown in Fig. 2. K_1, two of its tk^- revertants, and numerous tk^+ re-revertants reveal identical annealing fragments, 5.8 and 8.7 kb (Fig. 3A); K_2, two of its revertants and numerous re-revertants also share identical annealing fragments (Fig. 3B). Analysis of the fragment profiles obtained after digestion with seven additional enzymes mapping 30 observable sites generate profiles which permit the following conclusions: 1) Within the limitations of this approach, four of six revertants and their corresponding re-revertants maintain a restriction pattern identical to that of the original transformant; 2) the donor tk fragment is integrated into DNA within the transformed cell; and 3) the tk gene is integrated at a single site in transformed cell DNA and therefore exists in a hemizygous state.

2. <u>Transposition of the tk Gene</u>. In one revertant, K_1B_5, changes in sequence organization have occurred at both sides of the integrated viral fragment, suggesting that K_1B_5 results from a transposition of the tk gene from one integrated location to another. The restriction map of K_1 and K_1B_5 are shown in Fig. 2, with illustrative blot hybridization data in Fig. 3A. Cleavage of DNA from K_1B_5 with Bgl II generates two annealing fragments, 4.1 and 1.6 kb in length. These fragments both differ from the fragments observed upon digestion of K_1. Since Bgl II cleaves only once within the tk gene, the molecular weight of the fragments generated reflects the distribution of Bgl II sites in DNA flanking the integrated tk gene. These data suggest that sequences flanking the tk gene on both ends differ in K_1 and K_1B_5. Significant additional mapping data (Fig. 2) support the conclusion that a transposition had occurred and further indicate that the viral

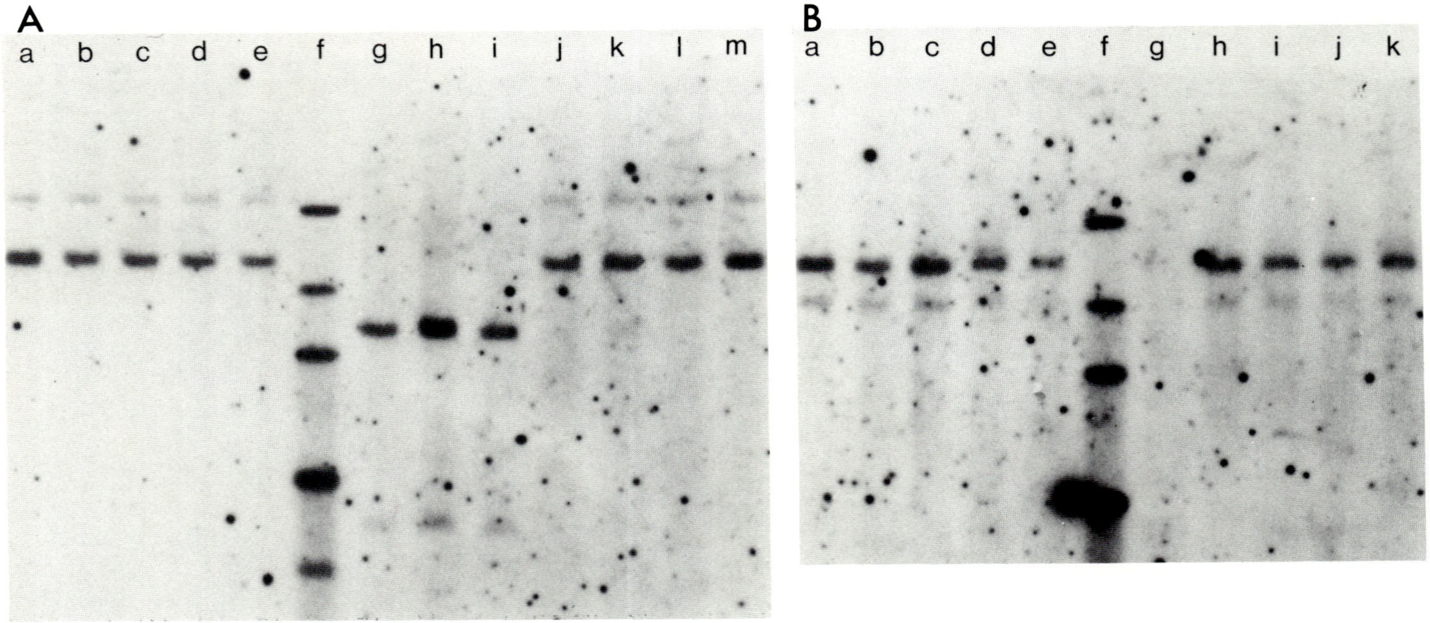

Fig. 3. Bgl II profiles of the tk fragment in sets of transformant, revertant, and re-revertant cell lines. (A) K_1 family: Lanes a-e, K_1, K_1B_4, $K_1B_4H_1$, $K_1B_4H_2$, $K_1B_4H_3$; lanes g-i, K_1B_5, $K_1B_5H_2$, $K_1B_5H_3$; lanes j-m, K_1B_6, $K_1B_6H_1$, $K_1B_6H_2$; lane f, markers (8.0, 5.0, 3.6, 2.0, 1.3 and 0.88 kb). (B) K_2 family: Lanes a-e, K_2, K_2B_6, $K_2B_6H_4$, $K_2B_6H_5$, $K_2B_6H_6$; lane g, K_2B_7; lanes h-k, K_2B_8, $K_2B_8H_4$, $K_2B_8H_5$, $K_2B_8H_6$; lane f, markers as in (A). The lower band in lane e is readily visible in the original autoradiograph.

sequences in K_1B_5 have undergone deletions at both ends of the tk-containing fragment during this rearrangement process. Tk^+ re-revertants can be selected from K_1B_5 (Table II) and these re-revertants maintain restriction profiles identical to K_1B_5. Therefore, expression of the tk gene can occur subsequent to transposition, and the initial re-arrangement may not be related to the loss of tk activity.

3. Loss of the tk Gene. Revertant K_2B_7 is unique among the tk^- lines derived in this study since it never re-reverts to the tk^+ phenotype ($<10^{-6}$). Cleavage of K_2B_7 DNA with either Bgl II or Hin II and subsequent annealing with tk probes shows no evidence of hybridizing fragments (Fig. 3B). This mutant therefore results from a deletion of most, if not all, of the integrated tk gene from the tk^+ parent, K_2.

G. Methylation Profiles of the tk Gene

Analysis of four independent revertants and eleven tk^- re-revertants at 30 restriction endonuclease cleavage sites reveals no sequence alterations associated with the reversible switching between tk^+ and tk^- phenotypes. However, short sequence alterations such as inversions or single base mutations and large rearrangements that may influence gene activity at a distance could escape detection. Alternatively, the genotypic switch may reflect DNA modification.

At present, methylation of cytosine within the DNA sequence pXpCpGpY is the predominant post-synthetic modification in mammalian DNA (20,21). Hpa II and Msp I recognize and cleave the sequence pCpCpGpG; methylation in the pCpG sequence inhibits Hpa II but not MspI (21-23). A comparison of the fragment profiles generated by this enzyme pair permits conclusions about the pattern of methylation about specific genes in cellular DNA. We therefore compared the Hpa II/Msp I restriction profiles for the tk gene in the HSV genome and in the DNA of tk^+ and tk^- transformants. Hpa II and Msp I recognize at least 26 sites within the 3.6 kb Bam segment of the tk gene (Fig. 4A). In our blots, we resolve only the largest four or five bands, 250-500 base pairs. With one exception, discussed below, viral DNA, tk^+ transformants, and tk^- mutants reveal identical fragments with Hpa II and Msp I (Fig. 4). Thus, the tk gene in the virus and in tk^+ and tk^- transformants is unmodified. Hypermethylation is therefore not associated with genetic inactivity of the tk gene.

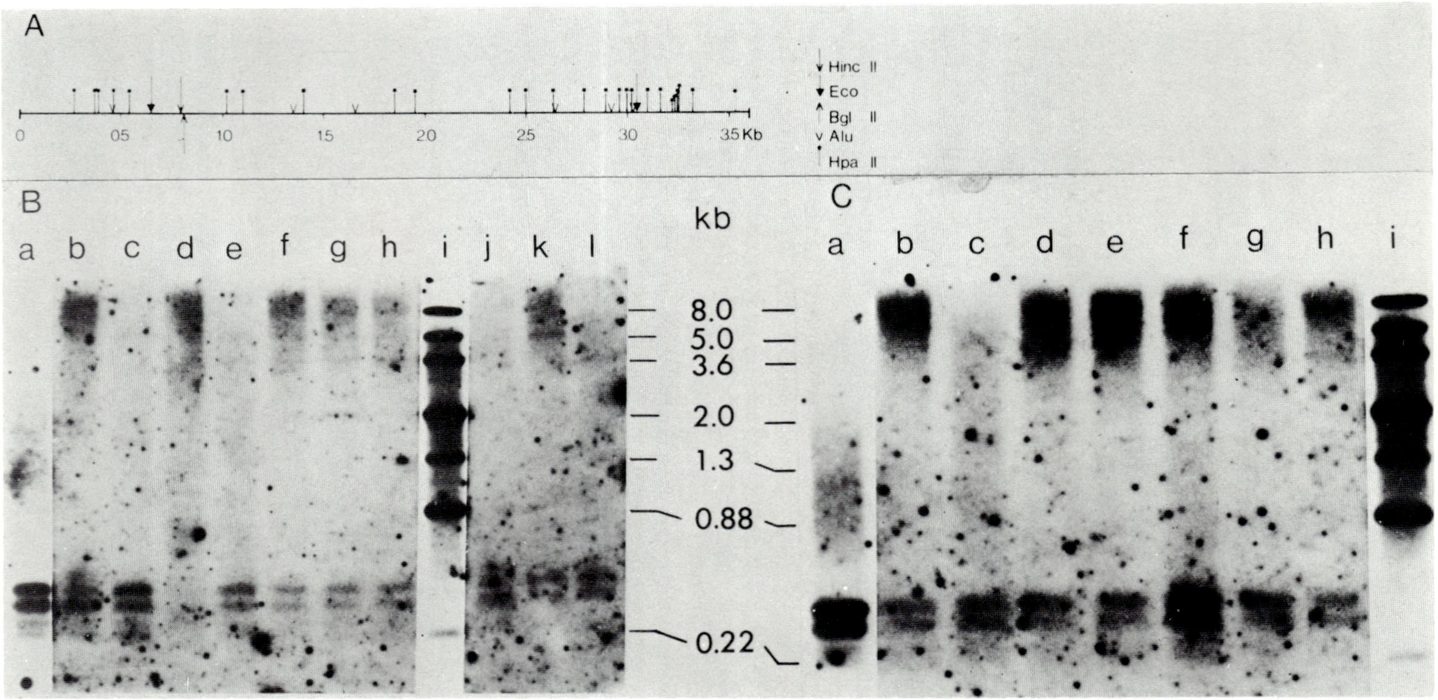

Fig. 4. Hpa I/Bam HI and Msp I/Bam HI profiles. (A) Map of the Hpa II (Msp I) sites in the cloned 3.6 kb Bam HI fragment of HSV DNA; the orientation is opposite to that in Fig. 2. (B) K_1 family: Hpa II/Bam HI digests of HSV (a), transformant K_1 (b), revertants K_1B^N (d) K_1B_6 (f), K_1B_4 (k) and K_1B_5 (l), and re-revertants $K_1B_6H_{1-2}$ (g,h); Msp I/Bam HI digests of K_1 (c) and K_1B_5 (j); markers as in Fig. 3 (i). (C) K_2 family: Hpa II/Bam digests of HSV (a), transformant K_2 (b), revertants K_2B_6 (d) and K_2B_8 (h) and re-revertants $K_2B_6H_{4-6}$ (e-g); Msp/Bam HI digests of K_2; markers as above (i). The diffuse blackening at the top of Hpa II lanes results from nonspecific binding of the labeled probe.

H. Methylation in A Non-reverting Line

Tk^- revertants maintained in culture for long periods of time can lose the ability to re-revert to tk^+. These mutants probably emerge as variants with a growth advantage in the selective medium. K_1B_6 reverts to the tk^+ phenotype with a frequency of 40%; upon long term culture in BUdR, this line lost its ability to switch to the tk^+ phenotype ($<10^{-6}$). This variant, $K_1B_6^N$ is "locked" into the tk^- phenotype. The restriction maps of the tk gene in K_1B_6 and its variant, $K_1B_6^N$, are identical. Analysis of DNA modification, however, reveals that the tk gene in $K_1B_6^N$ is extensively methylated. Digestion of $K_1B_6^N$ DNA with Msp I reveals the characteristic set of small molecular weight fragments consistent with a limit digest of the tk gene. With Hpa II, however, this set of fragments is replaced by a series of larger DNA fragments, which must result from modification at many Hpa II cleavage sites within the tk gene (Fig. 4B, lanes d, e). The number, size and intensity of these fragments suggest a cellular heterogeneity with respect to the sites of methylation. Corresponding results were obtained with two other enzymes sensitive to methylation, Hinc II and Sma I; partial digestion patterns were obtained for $K_1B_6^N$ under conditions that gave complete digestion with all other cell lines (not shown).

For comparison, we have examined the Hpa II and Msp I patterns of the β-globin genes in these same cell lines. In both K_1 and $K_1B_6^N$, Hpa II yields an identical set of fragments that differ from those obtained with Msp I. Thus, while certain Hpa II sites are modified in the β-globin sequences, the extent and position of this methylation is the same in K_1 and $K_1B_6^N$, as well as in all other tk^+ and tk^- transformants. These results suggest that specific methylation about the viral tk gene has occurred in $K_1B_6^N$, a cell line that has lost its ability to switch phenotypes.

III. CONCLUSIONS

Transformation permits the introduction of new genetic information into a cell and frequently results in a change in phenotype. We have developed cotransformation systems to study several aspects of eukaryotic gene expression: 1) as an *in vivo* system to study the functional significance of various features of DNA sequence organization; 2) as a means for gene purification where now classical routes dependent on mRNA enrichment are not feasible; 3) to analyze the molecular basis of mutations in somatic cells; and 4) to examine the fluidity of the eukaryotic chromosome.

We have studied the tk phenotype and genotype in two tk^+ transformants and several independently derived mutants. The tk^- mutants fall into two categories: a stable class which does not re-revert to tk^+ at measurable frequencies and an unstable class which undergoes phenotypic switching from tk^+ to tk^- to tk^+. One stable tk^- revertant examined in this study is typical of three additional tk^- clones previously examined in our laboratory and result from a loss of most, if not all, of the integrated viral gene. In the unstable class, our data suggest that phenotypic switching results from heritable but unstable alterations in the tk gene. In one mutant, a transposition has occurred, resulting in a change in position of the tk gene from one integrated location to another. In four cell sets, the restriction map is invariant from transformant to revertant to re-revertant. These studies are far from exhaustive, and significant rearrangements may have escaped detection. Thus, in at least two of six tk^- mutants, loss or rearrangement of large blocks of DNA, including the tk gene sequences, has occurred. We have also examined the pattern of methylation about the tk gene. Methylation in our clones is not associated, in an obligatory manner, with genetic inactivity of the tk gene. In one clone, however, extensive methylation is associated with a non-reverting tk^- phenotype. Therefore, methylation of the tk gene may prevent genetic rearrangements which result in phenotypic switching. The genome of somatic cells maybe quite fluid, and possibly, subtle alterations in DNA sequence organization may provide a common mechanism of gene regulation. Perhaps specific genes must be free of modification to undergo genetic rearrangements which lead to gene activation.

We have described the development of cotransformation systems which may now permit the transfer of virtually any cloned gene into cultured cells from diverse tissues. The introduction of wild type genes along with native and *in vitro* constructed mutants into cultured cells may provide an assay for the functional role of specific DNA sequences. Studies on the expression of cotransformed genes, restricted thus far to murine fibroblasts, reveal transcripts which are correctly processed but frequently display aberrant 5' termini. If the 5' terminus reflects the site of transcription initiation, these results suggest that multiple potential promoter sequences may exist and their choice may be dictated by the differentiated state of the host cell.

A.O. SUMMERS: Once you have transformed a cell to tk^+, what are the transformation characteristics of the cell line with respect to methotrexate resistance?

R. AXEL: Are you asking whether the competent phenotype is comparatively stable?

A.O. SUMMERS: Yes.

R. AXEL: We have not done the exact experiment you suggest, but we have done it with tk^- $hprt^-$ cell lines which we transformed, one subset to tk^+ and the second subset to $hprt^+$. Then we determined the relative frequency of transformation to $hprt^+$. It was no greater than the efficiency of the initial transformation. This suggests that competence is not stably inherited.

A.O. SUMMERS: You mentioned that different transformed clones exhibit differences in transcripts. Is it possible that by picking a clone you select for a transcript which is a minor species in the natural situation? What can you say about gene expression in your transformed cells?

R. AXEL: I doubt that we are selecting out minor species, considering the small number of clones that we have examined. Until we analyze these transcripts in somewhat greater detail by S1 mapping, I can say nothing about what they reflect.

R.A. WEINBERG: You were speaking about your collaboration with Beatrice Mintz, in which you are attempting to transfect a given gene into teratoma cells with the long term hope that these genes might be expressed when introduced into a developing mouse embryo. There is a general feeling that the expression of genes during differentiation, or more generally in mammalian cells, may be dependent on the chromosomal site into which a gene is inserted. I think there are many precedents for that. If that prejudice were correct, wouldn't it be a little unrealistic to hope that the gene introduced via the transcription process would be expressed in a meaningful and physiological way.

R. AXEL: If the prejudice were correct it would be foolish to undertake these experiments. But that remains a prejudice at the moment.

FORMATION OF β-GLOBIN FOLLOWING INFECTION WITH RECOMBINANTS CONTAINING RABBIT β-GLOBIN cDNA AT DIFFERENT LOCATIONS OF SV40's LATE REGION

Richard C. Mulligan
R. Tyler White
Paul Berg

Department of Biochemistry
Stanford University School of Medicine
Stanford, California

INTRODUCTION

Although gene cloning techniques have provided a new vision of the molecular structure of mammalian genes and their arrangement in chromosomes, such information by itself is insufficient for more than speculations about the structure-function parameters governing gene expression and regulation. Clearly, additional experimental means are needed to assay the expression of isolated genes, preferably in systems where regulated expression can be manifested. Four general approaches are being explored: (1) *In vitro* experiments with fractionated and unfractionated RNA polymerase preparations to assay the regulatory signals and factors governing transcription of isolated genes (1-3); (2) Microinjection of cloned DNAs into frog oocytes (4-6) and mammalian cells (7-8); (3) DNA-mediated transformation of mammalian cells using selective genetic markers to analyze the expression of co-transformed cloned genes (9,10); (4) Transduction of mammalian cells with SV40 recombinants containing cloned genes (11-15).

We have pursued the last mentioned approach to analyze the expression of cloned genes in mammalian cells (14). SV40 was adopted as the transducing vector because the genes responsible for almost all the viral functions have been identified and mapped (16-19), indeed, the entire nucleotide

ISBN 0-12-633360-2

sequence of the small circular genome is known (18,19), and there is extensive information available about replication, expression and regulation of the viral genome in different host cells (20).

Last year we described the construction of an SV40 recombinant genome that contains a cloned rabbit β-globin cDNA segment in place of the virus' VP1 coding sequence (14). Infection of cultured monkey cells with that particular SV40 β-globin cDNA recombinant (SVGT5-βG) resulted in the formation of cytoplasmic, polyadenylated β-globin mRNAs and authentic β-globin peptide. In the present paper we describe additional SV40 recombinants containing the rabbit β-globin cDNA segment and their ability to express β-globin mRNAs and β-globin poly peptide.

RESULTS

Construction of SVGT-β-Globin Genomes. The rabbit β-globin cDNA segment used in these experiments was obtained from T. Maniatis (21) and recloned in *E.coli* with pBR322 plasmid as previously described (14). Cleavage of the plasmid β-globin cDNA recombinant with *Hind*III and *Bgl*II endonucleases yielded the β-globin cDNA with a HindIII cohesive end 35 nucleotides 5'-proximal to the β-globin initiator AUG and a *Bgl*II cohesive end (which is also complementary to *Bam*H1 cohesive ends) just beyond the β-globin terminator UGA.

Three SVGT vectors, each lacking a different segment of SV40's late region, were constructed by partial digestion of SV40 DNA with *Hind*III endonuclease followed by complete digestion with *Bam*H1 endonuclease (See Fig. 1 and ref. 14). Following ligation of each SV40 vector to β-globin cDNA through their respective *Hind*III and *Bam*H1/*Bgl*II cohesive termini, the recombinant genomes were cloned as virions with *tsA*58 as helper (14). The β-globin cDNA in SVGT5-βG is located between SV40 map coordinates 0.945-0.145, in place of and with the same 5' to 3' orientation as the VP1 coding region (14) (See Fig. 2). SVGT7-βG and SVGT9-βG contain the β-globin cDNA segment oriented in the same 5' to 3' direction as SVGT5-βG, but located at map position 0.86-0.145 and 0.985-0.145, respectively (Fig. 1 and 2).

Noteworthy features of SVGT5-βG are the retention of the viral origin of DNA replication (ori), the putative late promoter (at or near ori), leaders, splice junctions and polyadenylation site implicated in late mRNA transcription and maturation (22-24). SVGT5-βG lacks the VP1 initiator AUG; instead, it contains the β-globin initiator AUG within 34-37 nucleotides of where the VP1 initiator AUG occurs.

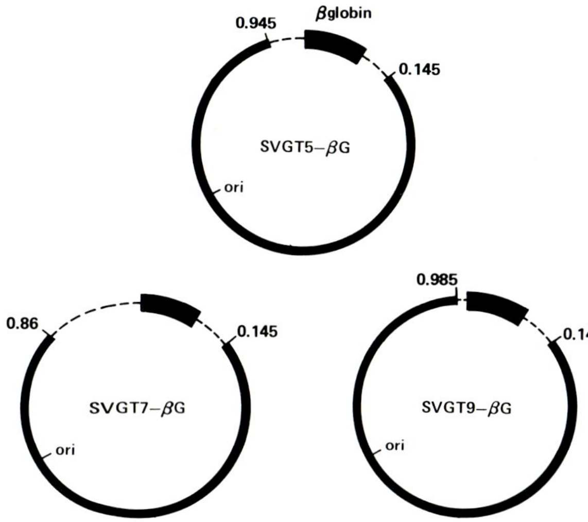

FIGURE 1. Structure of SVGT5-, SVGT7- and SVGT9 β-globin recombinants. The thick circle represents retained SV40 regions, the dotted portion spans deleted SV40 late regions and the segment replacing the deleted region is the β-globin cDNA. Ori signifies the origin of SV40 DNA replication.

SVGT9-βG retains the same regulatory and processing signals as SVGT5, but the β-globin initiator AUG is within the VP1 coding sequence, about 250 nucleotides downstream of VP1's initiator AUG. In SVGT7-βG the β-globin cDNA segment occurs between map positions 0.86-0.145; consequently, the β-globin cDNA segment lies within the coding sequence for VP2 and VP3. Particularly significant is the fact that SVGT7 lacks the splice junction implicated in splicing the 16S late mRNA (map position 0.94) but retains the leader and splice sites for 19S late mRNA (map segment 0.70-0.80) (22-24).

SVGT β-Globin Recombinants Produce β-globin mRNAs

<u>SVGT5-βG and SVGT9-βG</u>. Since SVGT5-βG and SVGT9-βG retain the DNA sequences specifying the proper 5'- and 3'-termini, leader, intervening sequences and splice junctions for 16S and 19S late mRNA synthesis, two β-globin mRNAs, each containing poly A and a leader spliced to the body of the mRNA should be formed (Fig. 2). Moreover, because the β-globin cDNA segment (0.5 kb) is smaller than the SV40 segment it replaces (1.0 kb in SVGT5 and 0.8 kb in SVGT9), the hybrid

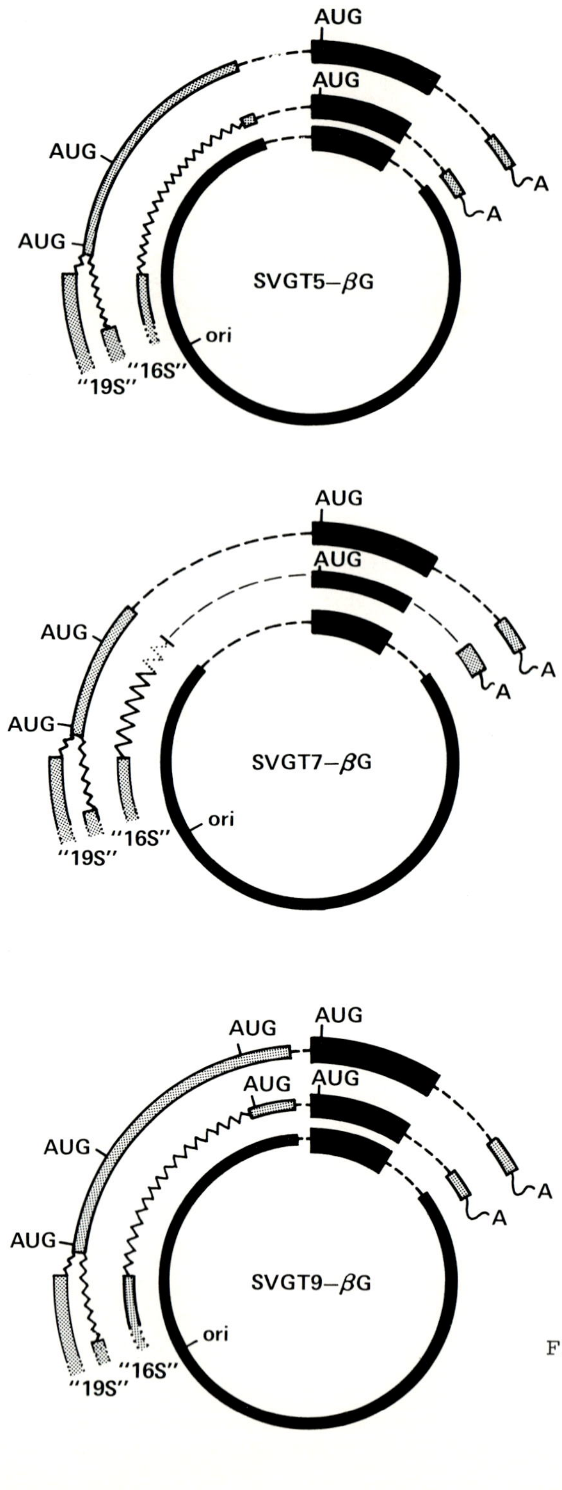
AUG
AUG
AUG
AUG
A
A
SVGT5–βG
ori
"16S"
"19S"
AUG
AUG
AUG
AUG
A
A
SVGT7–βG
ori
"16S"
"19S"
AUG
AUG
AUG
AUG
AUG
AUG
A
A
SVGT9–βG
ori
"16S"
"19S"

FIGURE 2

β-globin mRNAs will be smaller than the corresponding SV40 16S and 19S mRNAs (see Fig. 2).

To verify the structure of the mRNAs formed with SVGT5-βG and SVGT9-βG, CV1 cells were infected with virus stocks containing either one of the recombinants and tsA58 (moi 10-50 pfu/cells) and, after about 48 hours at 37°, the cytoplasmic poly A^+ RNA (freed from DNA) was isolated. Each RNA preparation was denatured with glyoxal (25), electrophoresed on 1.5% agarose gel and transferred to diazotized benzyloxymethyl paper (26). The imprints were annealed with either of two ^{32}P nick-translated probes: One was a DNA segment containing the leader sequence (map position 0.67 to 0.76) and the other was β-globin cDNA (pBR322-βG). The leader probe reveals the position of 16S and 19S mRNAs generated by the helper genome as well as any hybrid mRNAs containing leader sequences generated by the recombinants. The β-globin cDNA probe labels β-globin mRNAs (Fig. 3).

The arrows marked SVS 16S and 19S show the positions of the two SV40 (tsA58) late mRNA species. SVGT5-βG produces two β-globin mRNAs, each of which contains a leader segment and is 0.5 kb smaller than the corresponding SV40 mRNAs. (In this exposure the "19S" β-globin mRNA band in the SVGT5-βG track is barely visible, but is readily seen on longer exposures and in other experiments). Applying Berk and Sharp's S1 nuclease analysis for detecting and characterizing spliced mRNAs (27) shows that most of the β-globin mRNAs produced by SVGT5-βG and SVGT9-βG are spliced normally (data not shown), although a small amount of aberrantly spliced β-globin mRNA could escape detection.

FIGURE 2. The putative origin and structure of SVGT5-βG, SVGT7-βG and SVGT9-βG mRNAs. Two size classes of β-globin mRNAs are shown for each recombinant. Based on published data (23,24) and on the effects of mutations that delete the splice junction of the larger leader (see text), the larger β-globin mRNA species is shown as having two types of leaders. The smaller β-globin mRNAs have the leader characteristic of SV40's 16S mRNA. The precise location of the 5'-ends of each of the leader segments is deliberately unspecified because of the multiplicity of reported 5'-ends (23,24). The wavy lines represent regions that are spliced out of each mRNA. In SVGT7-βG the downstream splice junction of the small mRNA is left indeterminate (see text). AUGs in clockwise order, specify the position of the initiator codons for VP2, VP3, VP1 and β-globin proteins, respectively.

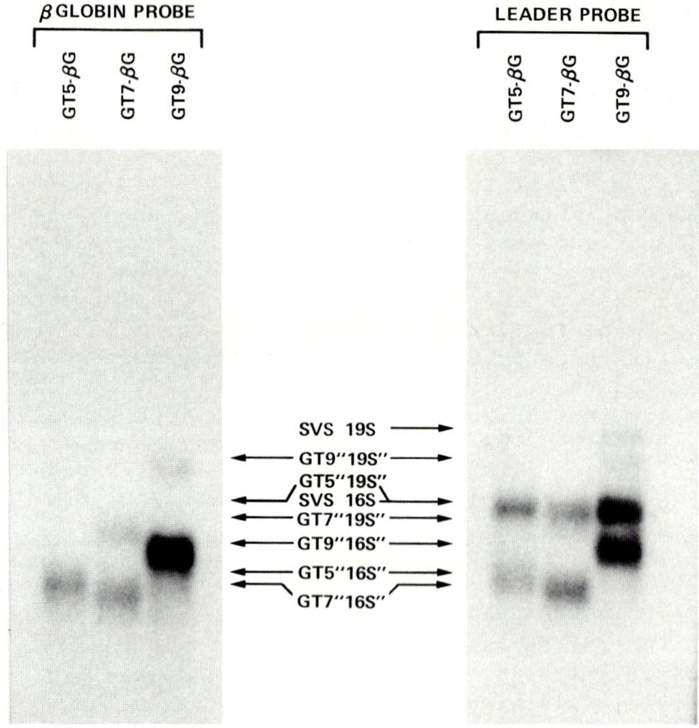

FIGURE 3. Analysis of mRNAs from infections of CV1 cells with SVGT5-, SVGT7- and SVGT9-βG recombinants. Cytoplasmic poly A containing RNA was isolated from CV1 cells 48 hours after high multiplicity infection with mixed virus stocks of *ts*A58 and either SVGT5,7, or 9-βG and analyzed as previously described (14). Tracks on the left were hybridized with ^{32}P-nick translated pBR322-β-globin cDNA ("β-globin probe") and those on the right with a ^{32}P labelled SV40 DNA fragment from 0.67 to 0.76 map units ("leader probe"). The arrows show the positions of particular SVGT-βG hybrid RNAs. Individual species of RNA are labeled '16S' or '19S' depending on whether their size is analogous to that of the 16S or 19S classes of SV40 late mRNA.

From these and other experiments we estimate that nearly equal amounts (within a factor of two) of β-globin and SV40 helper mRNAs are produced in the course of these mixed infections. This suggests that the insertion of the β-globin cDNA segment inserted into SVGT5 and SVGT9 does not alter appreciably the qualitative or quantitative parameters governing the transcription and processing of SV40 late mRNAs.

SVGT7-βG. SVGT7-βG contains the leader, intervening sequence and splice junctions used for the formation of 19S type mRNAs (23,24), but lacks the splice site at 0.94 for 16S mRNA production (22,24). Hence, we anticipated that infection with SVGT7-βG would produce β-globin mRNA with a 19S type splice only. However, following infection, isolation and analysis of the mRNAs as described above, revealed that two β-globin mRNAs were formed (Fig. 3). Since the small β-globin mRNA hybridizes with the leader probe (Fig. 3) it most likely contains leader sequences spliced to the β-globin coding segment; that likelihood was confirmed by the Berk and Sharp (27) analysis (data not presented). The large β-globin mRNA is obscured in the track hybridized with the leader probe because it is about the same size as the 16S mRNA produced by the helper.

SVGT7-βG's small β-globin mRNA is about 0.1 kb smaller than the corresponding β-globin mRNA formed from SVGT5-βG. This suggests that the small β-globin mRNA generated from SVGT7-βG is aberrantly spliced from a site within the leader segment to a sequence at or very near the upstream junction between the SV40 DNA and β-globin cDNA.

The exact structure of the novel β-globin mRNA formed by SVGT7-βG is not yet known. A clue to its structure was obtained from studies with mutants of SVGT5-βG and SVGT7-βG which contain a deletion that spans the 16S leader's splice junction at map position 0.76 (22-24). This deletion was generated in SVGT5-βG by partial digestion with Hpa I endonuclease and cloned in CV1-P cells with tsA58 helper at 41°C (see ref. 17). Nucleotide sequence analysis (28) established that SVGT5del-βG lacks 17 bp, 15 bp upstream and 2 bp downstream of the 3'-end of the 16S leader. The same mutation was introduced into SVGT7-βG by replacing the segment in SVGT7-βG containing the leader region (map position 0.72-0.82) with the corresponding segment from SVGT5del-βG.

Neither SVGT5del-βG nor SVGT7del-βG produce the small β-globin mRNA after infection, although both still produce the large β-globin mRNA (Fig. 4). The continued production of the large β-globin mRNAs is not surprising since sites upstream of the deleted nucleotides are used to splice leaders in 19S mRNAs (23,24). We interpret this finding to mean that splicing of the novel small β-globin mRNA formed by SVGT7-βG probably involves the same leader sequence used to form SV40's 16S mRNA and SVGT5-βG's small β-globin mRNA. Further studies are in progress to identify the nucleotide sequence at the splice junctions in the various hybrid β-globin mRNAs, particularly the novel site in the small β-globin mRNA produced by SVGT7-βG.

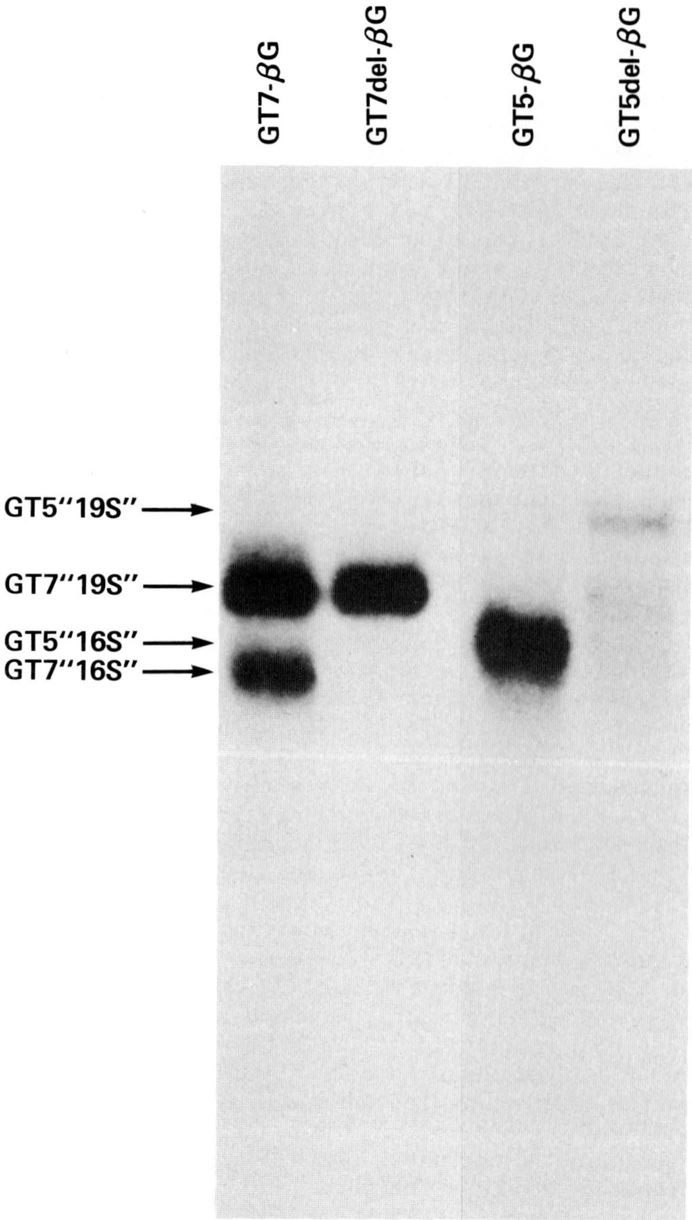

FIGURE 4. Analysis of mRNAs made in infections of CV1 cells with SVGT5del-βG and SVGT7del-βG. The figures at the left indicate the size class of the mRNAs.

Formation of β-Globin Protein

To determine if infections with SVGT β-globin recombinant genomes result in the production of β-globin peptide, CV1 cells were infected with each of the three recombinants plus tsA58. After 48 hours at 37° the cultures were labelled with ^{3}H-leucine and valine for 30 minutes; SV40 and mock-infected cultures were labeled in a comparable way. Each culture was sonicated, treated with NP40-DOC and aliquots were electrophoresed on SDS 15-20% gradient gels as previously described (14) (Fig. 5). Many labeled protein bands appear in each electrophoresed extract but the SVGT5-βG and SVGT7-βG infected cell extracts contain a protein band that migrates at the position of added authentic rabbit β-globin. However, no band of β-globin is evident in the extract from SVGT9-βG infected cells; instead, a new band with a mobility corresponding to a protein of 25,000 daltons is visible. None of these bands is discernible in either the SV40 or mock-infected extracts.

An aliquot of each extract was also immunoprecipitated with goat anti-rabbit β-globin antibody (generously provided by S.H. Boyer), adsorbed to *S. aureus* cells, the precipitated material was solubilized with SDS-urea and electrophoresed as mentioned above (Fig. 5). SVGT5-βG and SVGT7-βG infected cell extracts contain an immunoprecipitable protein with the same electrophoretic mobility as authentic rabbit β-globin. Immunoprecipitates of SV40 or mock-infected cell extract lack such a β-globin like protein. Experiments carried out with Dr. Tony Hunter of The Salk Institute showed that the putative β-globin polypeptide synthesized after infection with SVGT5-βG yields at least 8 of the 10 expected rabbit β-globin peptides after digestion with trypsin and two dimensional electrophoretic-chromatographic analysis (29). Moreover, the NH_2-terminal peptide produced from the β-globin made in SVGT5-βG infected cells co-migrates with the NH_2-terminal peptide from authentic rabbit β-globin. Hence, we conclude that the rabbit β-globin cDNA segment in SVGT5-βG and SVGT7-βG can be expressed to yield β-globin peptide.

By contrast, immunoprecipitates of SVGT9-βG infected cell extract contain little if any material corresponding to β-globin. Instead there is a faint band at the position corresponding to the 25,000 dalton protein and several proteins which migrate somewhat more slowly than authentic rabbit β-globin. The 25,000 dalton protein contains antigenic determinants from VP1 as well, since the same band is immunoprecipitated with anti-VP1 antibody (data not shown). Immunoprecipitation of SV40 infected cell extracts with the anti-VP1 antibody did not bring down any proteins with

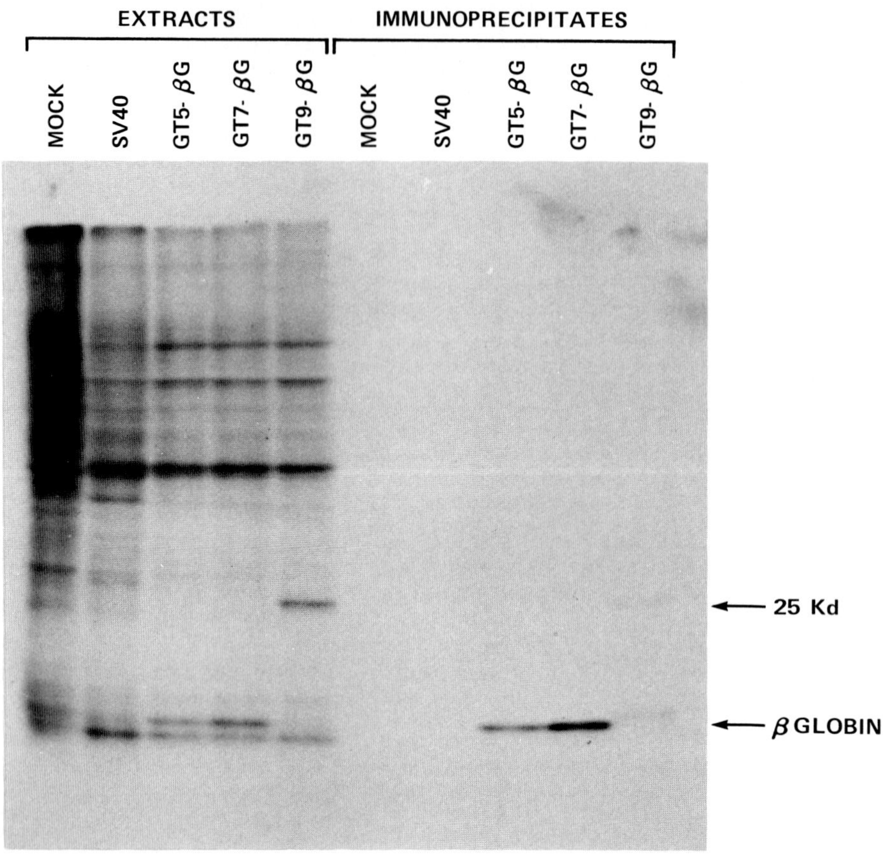

FIGURE 5. Production of β-globin in CV1 cells infected with SVGT5-, SVGT7- and SVGT9-βG recombinants. Labeled extracts and immunoprecipitates were analyzed by SDS-polyacrylamide gel electrophoresis as previously described (14). The tracks at the left show the results with unfractionated extracts and the ones at the right with immunoprecipitates obtained with goat anti-rabbit β-globin serum (14).

electrophoretic mobilities corresponding to β-globin. From the structure of SVGT9-βG and mRNAs it produces (see Fig. 2) we surmise that the 25,000 dalton peptide is a hybrid protein containing NH_2-terminal sequences of VP1 and COOH-terminal sequences of β-globin. The nucleotide sequence arrangement in SVGT9-βG is consistent with such an interpretation. Translation could initiate at the VP1 initiator AUG and proceed, first through 215 nucleotides of the VP1 coding sequence, then continue in phase, through 511 nucleotides represented in the β-globin cDNA segment. This would yield a peptide containing 72 amino acids of VP1, 12 amino acids translated from the 5'-normally untranslated region of the β-globin cDNA and 146 amino acids corresponding to β-globin. This structure would explain the hybrid protein's size and precipitability (albeit weakly) with the anti-rabbit β-globin and anti-SV40 VP1 antibodies. We suspect that the smaller β-globin like proteins arise from the 25,000 dalton protein, possibly by proteolysis at or near the VP1-β-globin amino acid junction, but further studies are needed to establish this point.

Taking account of the different number of labeled leucine and valine residues in VP1 and β-globin, the ratio of recombinant to helper genomes at the time of isotopic labeling and densitometry measurements of autoradiograms of immunoprecipitated β-globin and VP1, we estimate that in cells infected with SVGT5-βG and SVGT7-βG (each with _tsA58_), β-globin and VP1 peptides are being synthesized at nearly equal rates.

DISCUSSION

These experiments establish that an exogenous gene, in this case the coding segment for rabbit β-globin, can be recombined with SV40 DNA _in vitro_ at several positions in the late region so that the hybrid genomes can express the incorporated gene. Judging from the amount of β-globin mRNAs and β-globin protein produced during infections with these recombinants the β-globin cDNA segment appears to be transcribed, processed, transported and translated into protein about as efficiently as the SV40 segment it replaces.

Whereas the size and structure of the β-globin mRNAs formed in SVGT5-βG and SVGT9-βG infected cells were predicted on the basis of current knowledge of SV40 late mRNA formation, the result obtained with SVGT7-βG was unexpected. Although the SV40 nucleotide sequences involved in splicing 19S mRNAs are intact in SVGT7, one of the splice sites used in the formation of the small (16S) late mRNA is eliminated in this vector. Hence a large, but not a small β-globin mRNA should have been formed by SVGT7-βG. The structure of the unanticipated small β-globin mRNA is still being examined but from its size we infer that the small β-globin mRNA's leader segment is quite short and spliced very close to, or even into the 5'-untranslated region of the β-globin cDNA segment. Experiments performed by Hamer et. al. (30) are relevent to this inference. They constructed a recombinant in which the rabbit β-globin cDNA replaced an SV40 DNA segment between map coordinates 0.825 and 0.145. Therefore, in their vector, only the nucleotide sequences used for splicing the 19S mRNA species were conserved; the downstream site used for splicing the 16S mRNA was lacking. As expected, infection with this SV40-βG recombinant yielded a large, ostensibly spliced, β-globin mRNA. However, Hamer et. al. (30) also detected lesser amounts of smaller β-globin mRNAs, the origin and structure of which were not commented upon. It may be that in their, as well as our own experiments, aberrant splicing led to the production of novel β-globin mRNAs.

The failure to detect authentic β-globin peptide in cells infected with SVGT9-βG indicates that the β-globin initiator AUG codon must be the first one downstream from the leader for translation of the β-globin coding sequence to occur. This conclusion predicts that β-globin peptide is translated only from the small β-globin mRNAs of SVGT5-βG and SVGT7-βG. The finding that SVGT5del-βG, which makes the large but not the small β-globin mRNA, fails to synthesize β-globin peptide supports this prediction. In this instance the β-globin coding sequence is not translated because its initiator AUG is downstream from the VP2 and VP3 initiator (and internal) AUGs. A similar reasoning would explain why recombinants with the β-globin cDNA segment inserted at the 3'-end of the SV40's early region (map position 0.42-0.19) produce very little or no β-globin peptide even though early mRNAs containing the β-globin cDNA are formed; moreover, the coding regions for small t-antigen and truncated large T-antigen located in the 5'-proximal region of these β-globin mRNAs are translated (Southern and Berg, unpublished results).

Other genes can also be recombined with SV40 DNA so that coding sequences can be expressed as mRNAs and proteins. In experiments to be reported elsewhere an _E.coli_ gene coding for xanthine-guanine phosphoribosyl transferase (Mulligan and Berg, unpublished results) and the cDNA segment coding for mouse dihydrofolate reductase (31) have been introduced into SVGT5 and SVGT7 (Mulligan, Subramani and Berg unpublished results). In each instance the corresponding, catalytically active enzyme, XGPRTase and DHFRase, respectively, are produced during infection.

The SV40 DNA vectors developed so far offer a promising route for cloning and expressing small, well defined genes or cDNAs in certain mammalian cells. They also provide a useful approach for assessing the effects of certain mutational alterations on the clonability and/or expressability of these exogenous DNAs. For example, using expression of the β-globin gene in SVGT-βG recombinant genomes as the assay, we have explored the consequences of various deletions in the leader region on the processing and expression of SV40 late mRNAs (White and Berg, unpublished results). Such studies are readily performed because of the ability to monitor the formation and structure of β-globin containing mRNAs made in the presence of analogous SV40 mRNAs generated by the helper.

The ability to obtain expression of the β-globin cDNA sequence in, for example, SVGT5-βG also permits us to test the effect of mutational alterations in the β-globin cDNA segment itself. For example, in collaboration with H. Weber and C. Weissmann's laboratory, β-globin cDNAs with deletions of 5'-proximal untranslated nucleotides have been tested for their relative efficiencies in directing β-globin synthesis *in vivo*. These experiments show that β-globin is still synthesized when only 3 nucleotides upstream from the β-globin AUG in SVGT5-βG are present (Weber, Mulligan, Weissmann and Berg, unpublished results).

SV40-based vectors can also be used to transduce genomic DNA segments in order to assess the function of a gene's own transcription and processing signals. Such experiments are already underway in Hamer and Leder's (15), Khoury's (32) and our own laboratory.

ACKNOWLEDGMENTS

This investigation has been supported by research grants from the National Institute of General Medical Sciences (GM-13235) and The National Cancer Institute (CA-15513). R. C. Mulligan is a NSF Fellow and R. T. White is a USPHS trainee.

REFERENCES

1. Parker, C. S., Jaehning, J. A., and Roeder, R. G., *Cold Spring Harbor Symp. Quant. Biol.* *42*, 577 (1978).
2. Wu, G.-J., *Proc. Natl. Acad. Sci. USA* *75*, 2175 (1978).

3. Weil, P. A., Luse, D. S., Segall, J., and Roeder, R. G., Cell 18, 469 (1979).
4. Mertz, J. E., and Gurdon, J. B., Proc. Natl. Acad. Sci. USA 74, 1502 (1977).
5. DeRobertis, E. M., and Mertz, J. E., Cell 12, 175 (1977).
6. Sakonja, S., Bogenhagen, D. F., and Brown, D. D., Cell 19, 13 (1980).
7. Tjian, R., Fey, G., and Graessmann, A., Proc. Natl. Acad. Sci. USA 75, 1279 (1978).
8. Kriegler, M., Griffin, J., and Livingston, D. M., Cell 14, 983 (1978).
9. Wigler, M., Silverstein, S., Lee, L. S., Pellicer, A., Cheng, Y. C., and Axel, R., Cell 11, 223 (1977); Cell 16 777 (1979).
10. Wold, B., Wigler, M., Lacy, E., Maniatis, T., Silverstein, S., and Axel, R., Proc. Natl. Acad. Sci. USA 76, 5684 (1979).
11. Goff, S. P., and Berg, P., Cell 9, 695 (1976).
12. Ganem, D., Nussbaum, A. L., Davoli, D., and Fareed, G. C., Cell 7, 349 (1976).
13. Hamer, D. H., Davoli, D., Thomas, Jr., C. A., and Fareed, G. C., J. Mol. Biol. 112, 155 (1977).
14. Mulligan, R. C., Howard, B. H., and Berg, P., Nature 277, 108 (1979).
15. Hamer, D. H., and Leder, P., Cell 17, 737 (1979).
16. Lai, C.-J., and Nathans, D., Virology 60, 466 (1974).
17. Cole, C., Landers, T., Goff, S. P., Brutlag, S. M., and Berg, P., J. Virol. 24, 277 (1977).
18. Fiers, W., Contreras, R., Haegeman, G., Rogiers, R., Van de Voorde, A., Van Heuverswyn, H., Van Herreweghe, H., Volkaert, G., and Yseboert, M., Nature 273, 113 (1978).
19. Reddy, V. B., Thimmappaya, B., Dhar, R., Subramanian, K. N., Zain, B. S., Pan, J., Ghosh, P. K., Celma, M. L., and Weissman, S. M., Science 200, 494 (1978).
20. The Molecular Biology of Tumour Viruses, Cold Spring Harbor Laboratory, 1973, and 2nd edition, 1980 (J. Tooze, ed.).
21. Maniatis, T., Sim, G. K., Efstratiadis, A., and Kafatos, F. C., Cell 10, 571 (1976).
22. Ghosh, P. K., Reddy, V. B., Swinscoe, J., Choudry, P. V., Lebowitz, P., and Weissman, S. M., J. Biol. Chem. 253, 3643 (1978).
23. Ghosh, P. K., Reddy, V. B., Swinscoe, J., Lebowitz, P., and Weissman, S. M., J. Mol. Biol. 126, 813 (1978).
24. Lai, C.-J., Dhar, R., and Khoury, G., Cell 14, 971 (1978).
25. McMaster, G. K., and Carmichael, G. G., Proc. Natl. Acad. Sci. USA 74, 4835 (1977).

26. Alwine, J. C., Kemp, D. J., and Stark, G. R., Proc. Natl. Acad. Sci. USA 74, 5350 (1977).
27. Berk, A. J., and Sharp, P. A., Proc. Natl. Acad. Sci. USA 74, 1274 (1977).
28. Maxam, A. M., and Gilbert, W., Proc. Natl. Acad. Sci. USA 74, 560 (1977).
29. Gibson, W., Virology 62, 319 (1974).
30. Hamer, D. H., Smith, K. D., Boyer, S. H., and Leder, P., Cell 17, 725 (1979).
31. Chang, A. C. Y., Nunberg, J. H., Kaufman, R. J., Ehrlich, H. A., Schimke, R. T., and Cohen, S. N., Nature 275, 617 (1978).
32. Khoury, G., Alwine, J., Dhar, N. Goldman, N., Gruss, P., Jay, G., Segal, S., and Seif, I. Miami Winter Symp. 17, (1980).

DISCUSSION

C. PRIVES: What is the evidence that there are separate VP2 and VP3 messengers in SV40 infected cells?

P. BERG: The evidence is only circumstantial: Ty White has constructed a deletion that extends across the donor splice junction for the VP1 leader and just beyond the AUG for VP2. Thus this deletion takes out the acceptor splice junction for the 19S message. After infection with this mutant no 16S message can be detected, but we do find a 19S-like cytoplasmic RNA which is slightly smaller than the wild type 19S RNA. Now we have to determine whether this deletion will allow the production of VP3, (but not VP2) when it is crossed into the wild type genome.

C. PRIVES: Do you see VP3 in the cells infected with this mutant?

P. BERG: I should have said that this deletion was contructed in SVGTS-βG. This recombinant does not contain a complete VP3 coding sequence because globin sequences have removed the 3'-end of the VP3 coding region. Therefore, the deletion must be crossed back into wild type to determine if it affects VP3 production.

M. BURACZYNSKA: Are there any other animal virus vectors which would not limit the size of the inserted DNA as in the case of Ff phage?

P. BERG: In principle other viruses could also be used as vectors. Several laboratories are exploring the use of retroviruses and adenoviruses, each of which have larger genomes and would lend themselves to use as cloning vectors according to our strategy. Our approach, however, has been to eliminate the necessity of propagating the recombinant genome as a virus particle. Our aim is to propagate the recombinants as plasmids in the nuclei of the infected cell. To do that we are constructing vectors with selectable markers to ensure the maintenance of such molecules in the infected or transfected cells. The difficulty so far, however, is to insure and distinguish autonomous replicating species from integrated forms. In either case the constraint on the size of the inserted segment is eliminated. The size contraint exists only when the recombinant DNA must be package to be propagated.

REPETITIVE DNA ELEMENTS RELATED TO THE RIBOSOMAL INSERTION OF *DROSOPHILA MELANOGASTER*

Igor B. Dawid

Laboratory of Biochemistry,
National Cancer Institute
Bethesda, Maryland

Marlene Lauth
Peter K. Wellauer [1]

Swiss Institute for Experimental Cancer Research
Epalinges s/Lausanne, Switzerland

I. INTRODUCTION

The genes for ribosomal RNA (rDNA) are composed of repeated copies in all organisms (for review see Long and Dawid, 1980). These genes have been studied extensively since their products, the rRNAs, are quantitatively the most important RNAs in every cell and because the ribosomal genes are a typical example of a repeated gene family. In many eukaryotic organisms ribosomal genes occur in clusters of tandemly repeated units, each unit consisting of a region coding for the rRNA precursor molecule and a nontranscribed spacer. Different organisms may contain one or several such rDNA clusters per haploid genome.

In *Drosophila melanogaster* two rDNA clusters occur, one on the X and one on the Y chromosome. Each cluster contains 150 to 250 repeating units of rDNA. Alternation of gene and spacer regions occurs in this rDNA as it does in other cases, but a novel feature was discovered in *D. melanogaster* in the

[1] Supported by a grant from the Swiss National Science Fund.

ISBN 0-12-633360-2

form of a DNA segment that interrupts the 28S RNA coding region. This region has been called the "ribosomal insertion" (Glover and Hogness, 1977; Pellegrini, Manning and Davidson, 1977; Wellauer and Dawid, 1977; White and Hogness, 1977). Insertions occur in many but not all ribosomal repeats. They constitute a complex set of DNA regions with various sizes and sequences. DNA segments homologous to ribosomal insertions also occur in regions of the genome other than the nucleolus organizer (Dawid and Botchan, 1977).

In this paper we shall summarize the available information on structural relations between different ribosomal insertions in *D. melanogaster* and other species of *Drosophila*. We shall discuss the question whether interrupted ribosomal genes are transcribed into functional rRNA. We shall present data demonstrating that elements homologous to the insertion occur in centromeric heterochromatin of salivary gland chromosomes. And we shall summarize data on properties of centromeric insertion-like elements and their flanking sequences.

II. PROPERTIES OF RIBOSOMAL INSERTIONS

Ribosomal insertion sequences may be different in size, but the position of interruption of the 28S rRNA gene is almost always the same, with possible rare exceptions. We have cloned several repeating units of *D. melanogaster* rDNA and compared them with each other and with uncloned purified rDNA (Dawid, Wellauer and Long, 1978; Wellauer and Dawid, 1978). On the basis of these experiments we distinguish two types of insertions. Type 1 includes insertions of several size classes which are homologous to each other at their right end. This homology is the basis for considering them together. Insertions of three size classes belonging to type 1 have been isolated by cloning and studied in detail: the most common size class of 5 kb and the moderately common classes of 1 and 0.5 kb. Insertions are not internally repetitive and long insertions do not appear to have arisen by duplication of short ones.

Independently cloned molecules carrying type 1 insertions of the same size class are very similar and probably identical in sequence. Heteroduplex analysis in the electron microscope (Wellauer and Dawid, 1978) and restriction mapping of several clones obtained in our laboratory have shown this similarity. Further, the detailed restriction map of the 5 kb insertion of clone Dm103 (Glover and Hogness, 1977; Simon and Glover, personal communication) is identical to the detailed map of our cloned rDNA fragment Dmr a56 (Dawid *et al*., 1978; and unpublished experiments).

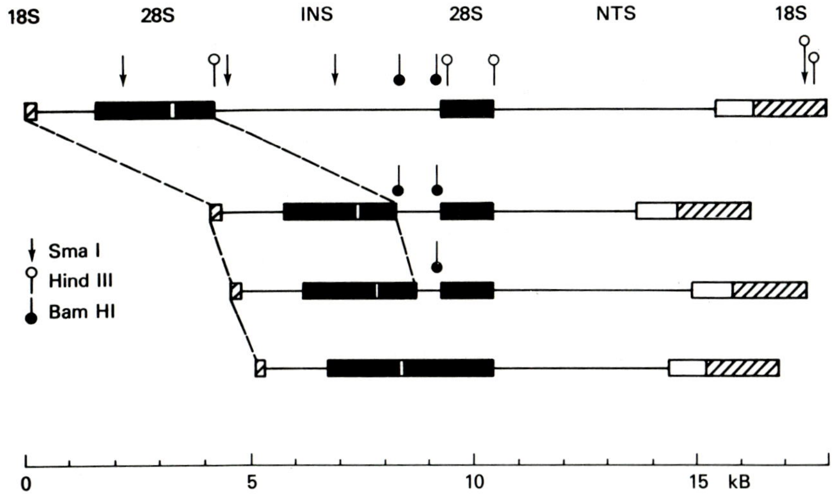

FIGURE 1. Maps of cloned EcoRI fragments of D. melanogaster rDNA. Each fragment represents a full repeating unit since a single EcoRI site occurs in the 18S RNA coding region. Maps are shown for genes carrying a 5 kb, 1 kb and 0.5 kb type 1 insertion and for an uninterrupted gene. Cleavage sites are shown for HindIII (⚲), SmaI (↓) and BamHI (●). NTS is the nontranscribed spacer. Based on data given in Dawid et al. (1978).

Type 2 ribosomal insertions are defined as DNA segments that do not share detectable homology with type 1 insertions even though they interrupt 28S rRNA genes at the same position (Dawid *et al.*, 1978; Dawid and Wellauer, 1978; Wellauer and Dawid, 1978; Wellauer, Dawid and Tartof, 1978). Type 2 insertions are also characterized by the presence of at least one EcoRI site while type 1 insertions do not carry such a site. Whether all type 2 insertions share some sequence homology is not known. It is possible that there are in fact several nonhomologous types of ribosomal insertions in *D. melanogaster*.

A. Chromosomal Distribution of Ribosomal Insertions

Two nucleolus organizers exist in *D. melanogaster* on nonhomologous chromosomes. We have studied the structure of rDNA from the two loci by comparing DNA from females where all rDNA is derived from the X chromosome with rDNA from males carrying an X chromosome with a deletion of the nucleolus organizer (the *scute* 4 - *scute* 8 inversion). In such males all rDNA is contributed by the Y chromosome. Restriction endonuclease mapping and electron microscopy showed that type 1 insertions are limited to the X chromosomal rDNA (Wellauer *et al*., 1978). About half of all rDNA repeats on this chromosome carry a type 1 insertion. In contrast, type 2 insertions are present with the similar frequency of about 15% in the rDNA of both chromosomes. It has been known for some time that the coding regions for the rRNA molecules themselves are very similar on the two chromosomes (Maden and Tartof, 1974). However, recent work by Yagura, Yagura and Muramatsu (1979) has shown that these sequences actually differ by a single transition between a C and U residue in the 18S RNA coding regions. This observation adds an interesting facet to the question of how similar but nonidentical sets of rDNA sequences can be maintained on the two nonhomologous chromosomal sites (see Wellauer *et al*., 1978).

Ribosomal gene repeats interrupted by type 1 insertions are widely interspersed with uninterrupted repeats; the pattern of this interspersion is indistinguishable from random (Wellauer and Dawid, 1977). Repeats with type 2 insertions are also interspersed with other repeats, but noticeable clustering of such repeats occurs as well (Wellauer *et al*., 1978).

B. Ribosomal Insertions in Other Species

D. virilis rDNA carries insertions in many but not all repeating units of its rDNA. This rDNA insertion has been characterized by Barnett and Rae (1979) who found that it is homologous in sequence to the type 1 insertion of *D. melanogaster*. The major *D. virilis* insertion has a length of 9.6 kb and appears to be the result of a duplication of an ancestral sequence that resembles in size the most common *melanogaster* insertion. *D. hydei* rDNA carries a 5.5 kb insertion in some repeats (Kunz and Glätzer, 1979). Other *Drosophila* species also carry sequences homologous to the *D. melanogaster* insertion that may interrupt rRNA genes (Barnett and Rae, 1979). The dipteran insects *Sciara* (Renkawitz, Gerbi and Glätzer, 1979) and *Calliphora* (Beckingham and White, 1979) also contain some rDNA repeats that are interrupted.

While the insects mentioned above have both interrupted and uninterrupted rRNA genes, all ribosomal genes are interrupted in some organisms like *Tetrahymena* (reviewed in Long and Dawid, 1980).

C. Is the Ribosomal Insertion Transcribed?

In organisms in which only a fraction of ribosomal genes is interrupted it is not clear *a priori* whether the interrupted genes are inactive or contribute to rRNA synthesis. Many genes in eukaryotes are interrupted and are expressed by transcription of the coding regions together with the introns into a large precursor RNA which is processed by splicing (see several articles in this Symposium). If such a mechanism were to operate in rRNA synthesis in *D. melanogaster* we would expect to find RNA molecules in the nucleus which contain insertion sequences. We have tested this question in the following type of experiment (Long and Dawid, 1979). RNA from embryos was separated on an agarose gel, transferred to RNA-binding paper (Alwine, Kemp and Stark, 1977) and hybridized with labeled DNA fragments derived from within the 5 kb type 1 insertion. When the gel was probed with the left portion of the insertion some heterogeneous RNA molecules between 5 and 10 kb were detected in nuclear RNA but no homologous molecules were found in cytoplasmic RNA. Using the right end of the insertion (which is homologous to small insertions, see Figure 1) we found RNA transcripts of mostly around 4 kb in the nucleus and a distinct molecule of slightly less than 1 kb in the cytoplasm. From a calibration of gels and from the kinetics of RNA excess hybridization we concluded that the RNA molecules homologous to the left half of the insertion occur at a concentration slightly less than one copy per nucleus. RNA homologous to the right end has a concentration of about 13 molecules per nucleus and 50 molecules per cell in the cytoplasm (Long and Dawid, 1979). Similar results were obtained with RNA from other stages of development. From these results we conclude that transcription of interrupted ribosomal genes followed by splicing cannot contribute significantly to rRNA production. The reasoning is based on the fact that rRNA genes are densely loaded with RNA polymerases and nascent chains (McKnight and Miller, 1976). Consequently, an embryonic nucleus contains about 4000 nascent rRNA molecules. If interrupted rRNA genes were as active as uninterrupted genes they would be equally loaded and therefore at least 50 nascent chains would cover each active 5 kb insertion. Even if splicing were to happen immediately after the polymerase had traversed the insertion, nascent chains would contribute many more insertion transcripts than we detect. Therefore we suggest that interrupted rRNA genes are usually not transcribed. The few transcripts we find may represent occasional abortive starts. The production of the 1 kb cytoplasmic insertion transcript of unknown function remains unexplained.

Electron microscopic studies of spread nucleolar chromatin (Miller spreads) have in general supported the view that interrupted rRNA genes are not transcribed. McKnight and Miller (1976) found a homogeneous matrix length interspersed with longer stretches of inactive chromatin in *D. melanogaster* embryos. Glätzer (1979) also found a homogeneous matrix length in *D. hydei* where a 5.5 kb insertion has been identified in many X chromosomal rDNA repeats (Kunz and Glätzer, 1979). The only claim for insertion transcripts has been made in a report by Chooi (1979). The distribution of matrix lengths measured by this author trails towards longer sizes. From an analysis of numbers of nascent chains attached to some of the longer transcription units Chooi concluded that such units are not due to stretching but actually represent longer transcripts. While a small number of longer transcripts may in fact occur there is no direct evidence that they include insertion sequences. It is possible that some of these matrix units represent transcripts of genes with type 2 insertions which have not yet been investigated by biochemical means. Other interpretations of Chooi's data can be made as well.

III. CHROMOCENTRIC DNA SEQUENCES HOMOLOGOUS TO RIBOSOMAL INSERTIONS

Sequences homologous to type 1 ribosomal insertions have been found in the *D. melanogaster* genome outside the nucleolus organizer (Dawid and Botchan, 1977). This set of sequences represents about as many copies as are contained within the nucleolus organizer. In this section we present evidence that most of these r-insertion-like elements are localized in the chromocenter of salivary gland chromosomes. We further summarize some properties of such r-insertion-like elements that have been isolated by cloning.

A. Hybridization of Type 1 Ribosomal Insertion to Salivary Gland Chromosomes

1. Materials and Methods Used for in situ Hybridization. The cloned rDNA repeating units and their subfragments that were used in these experiments are shown in Figure 2. These DNA fragments are described in detail by Dawid *et al.* (1978). Hybridization *in situ* to salivary gland polytene chromosomes of *D. melanogaster* (Ore-R) third instar larvae was done essentially as described by Gall and Pardue (1971) with the following modifications. The DNA of squashed chromosomes was denatured for 5 min at room temperature in absolute formamide (Merck, 2 x recrystallized) rather than in NaOH. Radioactive complementary RNA (cRNA) was used as hybridization probe. ^{3}H-cRNA (1-3 x 10^7 cpm/μg) was synthesized from agarose gel-purified DNA templates with *E. coli* RNA polymerase using 4 labeled ribonucleoside triphosphates (21-50 Ci/mmole; Amersham). Hybridization was carried out in solutions containing 0.3 M NaCl, 10 mM Tris (pH 7.0), 2-4 mg/ml unlabeled *E. coli* tRNA, and 1-2.5 x 10^7 cpm/ml ^{3}H-cRNA. Twenty μl of this solution were applied to each slide, and slides were hybridized for 15 to 18 hr at 70°C in a sealed, water saturated chamber. Slides were coated with NTB-2 autoradiographic emulsion (Kodak) and exposed in a light tight box at -20°C for between 7 and 57 days. The slides were developed, stained with Giemsa (Fluka) and mounted for light microscopy. Pictures were taken in an automatic photomicroscope (Zeiss).

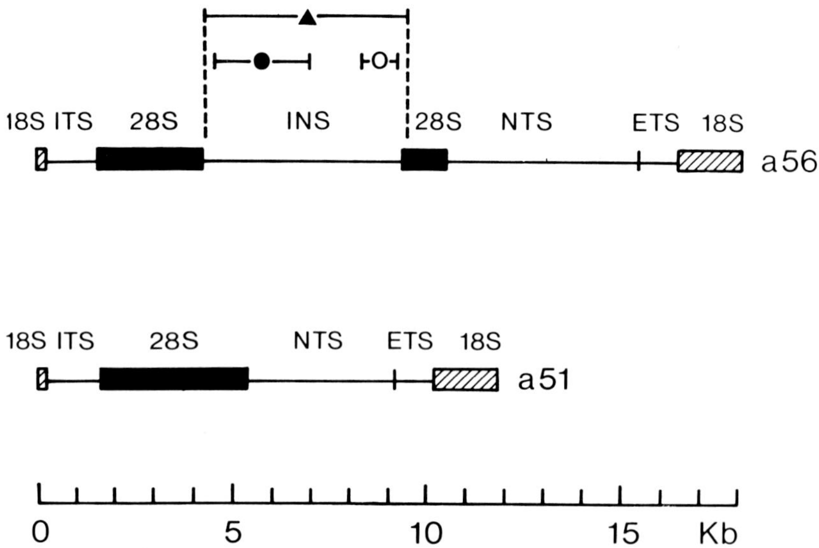

FIGURE 2. Maps of cloned rDNA repeating units of *D. melanogaster* used in chromosome hybridization experiments. a51 is a repeat which contains a contiguous 28S rRNA gene while a56 is interrupted by a 5 kb type 1 insertion sequence (Dawid *et al.*, 1978). Radioactive cRNA was synthesized against either total a56 or a51 rDNA or against restriction fragments from the a56 insertion. ——▲—— , *Hin*dIII fragment. This fragment contains mostly insertion and some 28S gene sequences at the right end; ——●—— , pure insertion fragment obtained by digestion with *Sma*I restriction enzyme; ——○——, *Bam*HI restriction fragment which contains exclusively insertion sequences. ITS, internal transcribed spacer; ETS, external transcribed spacer; NTS, non-transcribed spacer; INS, insertion.

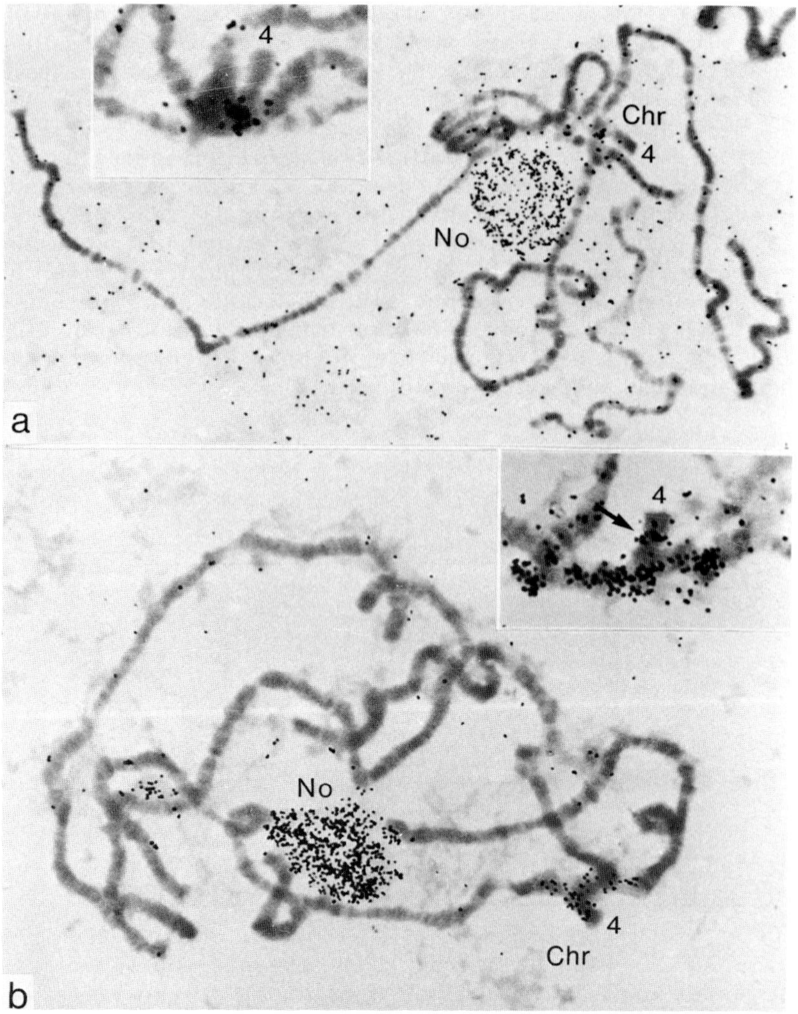

FIGURE 3. Polytene salivary gland chromosomes of *D. melanogaster* (strain Ore-R) after hybridization *in situ* with ^{3}H-labeled cRNA transcribed from cloned full-length rDNA repeating units. a) cRNA made from Dmr a51, a contiguous rDNA repeat without insertion. Note silver grains over nucleolus (No) and chromocentric heterochromatin (Chr) at the basis of chromosome 4. The inset shows the chromocenter of another chromosome set at higher magnification. Exposure time was 34 days. b) cRNA of Dmr a56, an rDNA repeat carrying a 5 kb type 1 insertion. Hybridization is over nucleolus and chromocentric heterochromatin. In addition, one band of chromosome 4 was reproducibly labeled (arrow in inset). Exposure time was 7 days.

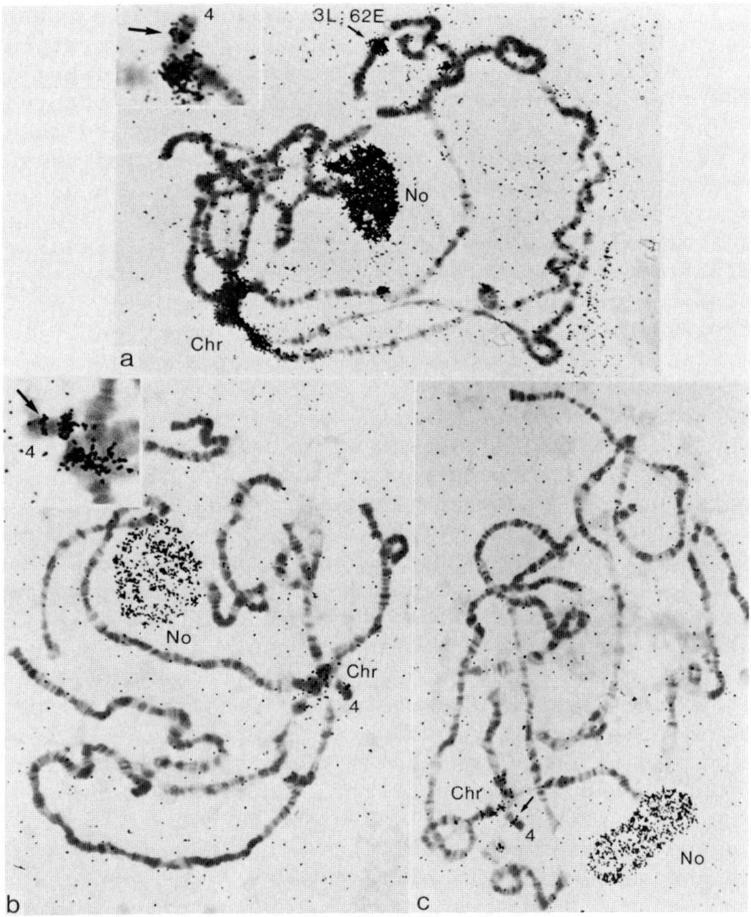

FIGURE 4. Polytene salivary gland chromosomes after *in situ* hybridization with ^{3}H-cRNA made against restriction fragments derived from the insertion of Dmr a56 (see Figure 2). a) cRNA of fragment obtained by digestion with *Hin*dIII. Silver grains are observed over nucleolus (No), chromocentric heterochromatin (Chr), and one band of chromosome 4 (arrow in inset). Hybridization to locus 62E of chromosome 3L is due to ^{3}H-cRNA against pDm4 which was included as a control in this experiment (see text). Exposure time was 57 days. b) hybridization with *Sma*I restriction fragment. Silver grains are observed over nucleolus, chromocenter and the band of chromosome 4 (arrow in inset). Exposure time was 14 days. c) hybridization with *Bam*HI restriction fragment. Nucleolus, chromocenter, and the band of chromosome 4 (arrow) display silver grains. Exposure time was 20 days.

2. Results of Chromosome Hybridization Experiments. Figure 3 shows *in situ* hybridization with complete rDNA repeating units. The uninterrupted cloned rDNA fragment Dmr a51 labels primarily the nucleolus (Figure 3a). Some label is also observed over the chromocenter (see inset). This observation may be due to the fact that the ribosomal genes are localized in the heterochromatin of the X chromosome. During polytenization the ribosomal genes expand into the nucleolus but some copies may remain with the heterochromatic region from which they derive, giving a low but distinct positive signal after hybridization with rDNA. Figure 3b shows the result of hybridization with the cloned fragment Dmr a56 which carries a 5 kb insertion. The chromocenter is now more strongly labeled and we also detect a labeled site on the 4th chromosome (inset).

Figure 4 shows hybridization with purified insertion fragments. The *Hind*III fragment which includes the entire insertion and a very short portion of the 28S RNA gene labels the nucleolus, the chromocenter and the site on the 4th chromosome that we mentioned before. To test whether a single copy of the insertion in any other euchromatic band would have been detected under these conditions we did the following control. Plasmid pDm4 contains a single copy sequence from *D. melanogaster* known to be located at position 62E (Wensink *et al.*, 1974). cRNA copied from pDm4 was mixed with cRNA copied from the ribosomal insertion and the mixture was used in the hybridization experiment. As shown in Figure 4a the site of origin of pDm4 (3L:62E) is strongly labeled. No other euchromatic band showed any reproducible significant label. Figure 4b and c shows that the left and right halves of the insertion (see legend and Figure 2) hybridize with the chromocenter and the 4th chromosome in very similar ways.

We conclude from these results that sequences homologous to type 1 ribosomal insertions are located at the chromocenter region of salivary gland chromosomes. Quantitation of hybridization to chromosomes is not precise but it is clear that multiple copies of such sequences occur at the chromocenter. Comparison of the chromosome hybridization data with biochemical data that will be discussed below suggest a possible quantitative discrepancy. Biochemical analysis showed that embryo or whole adult DNA contains at least as many r-insertion-like sequences not linked to rDNA as there are within rDNA. The *in situ* hybridization data illustrated in Figures 3 and 4 suggest that the chromocentric sequences are less abundant than insertions in the nucleolus by at least a factor of 3. This situation need not represent a discrepancy since it could be due to underreplication of chromocentric heterochromatin during polytenization in salivary glands. In addition to multiple copies of r-insertion-like sequences at the chromocenter we find one to a few copies located on the 4th chromosome. No other site of homology was observed. On the basis of the control experiment with pDm4 we conclude that no r-insertion-like sequences of a length more than a few hundred nucleotides occur anywhere else in the *D. melanogaster* genome. This conclusion contrasts with a reference to unpublished work by D. J. Finnegan made in the paper by White and Hogness (1977), where the presence of r-insertion-like elements on many euchromatic sites was suggested. We are unable to confirm that suggestion. Elements homologous to the ribosomal insertion are predominantly chromocentric with some representation on the 4th chromosome.

B. Isolation and Characterization of Chromocentric Ribosomal-Insertion-Like Elements

The first evidence for existence of r-insertion-like elements outside the nucleolus organizer was obtained from density gradient centrifugation experiments (Dawid and Botchan, 1977). We showed that a class of DNA molecules which contained segments homologous to type 1 insertions could be separated from rDNA itself. From these gradient data and from gel transfer experiments we concluded that the r-insertion-like elements were interspersed with other kinds of DNA (Dawid and Botchan, 1977; Dawid and Wellauer, 1978). This conclusion is supported by experiments summarized below.

1. *Isolation of Chromocentric R-Insertion-Like DNA by Cloning*. DNA enriched for r-insertion-like elements by gradient centrifugation was inserted into the vector pMB9 with poly(dA):poly(dT) tails. Recombinant clones were isolated that showed homology to the insertion but not to uninterrupted rDNA sequences. They were named pDmI and numbered from 101. About 40 independent clones were picked and digested with three restriction endonucleases. The digests were separated on gels and tested (Southern, 1975) for homology to ribosomal insertion. In this way we searched for cloned DNA molecules that contained an r-insertion-like element as well as a DNA segment not homologous to the insertion. Such nonhomologous regions we shall call flanking sequences. Sixteen cloned DNA molecules that contain flanking sequences were obtained. This finding confirms our earlier suggestion that r-insertion-like elements are usually interspersed with other DNA (flanking sequences).

2. Flanking Sequences are Heterogeneous. We wondered whether flanking sequences were all the same or whether they constituted a heterogeneous group. To test this question flanking sequences were isolated from restriction digests of two cloned DNA molecules, labeled, and hybridized to the set of pDmI clones. Figure 5 shows part of such an experiment. The flanking fragment from pDmI 158 hybridized with 4 other clones in addition to itself. In contrast, flanking fragments from pDmI 137 hybridized only to the parental clone in our collection of recombinant molecules (see also Dawid and Long, 1979). Thus, flanking sequences may recur in the group of 40 DNA molecules that were tested but not all flanking sequences are homologous to one another. In fact, from our preliminary data it is clear that several more nonhomologous flanking sequences must occur.

3. Restriction Endonuclease Analysis of R-Insertion-Like Elements in Different Strains of *D. melanogaster*. Earlier work has shown that multiple bands are obtained when total *Drosophila* DNA is digested with a restriction endonuclease and the fragments are probed with labeled ribosomal insertion (Dawid and Botchan, 1977). This observation was interpreted in the following way: multiple copies of r-insertion-like elements occur. These elements are all similar but not necessarily identical in sequence so that variable restriction fragments can arise from within the elements. Even more importantly, the elements are interspersed with various flanking sequences which yield different restriction patterns. Consequently sequences homologous to the insertion occur in many different restriction fragments. This interpretation is supported by the finding of different flanking regions in cloned molecules containing r-insertion-like elements, as described above. More recently we have compared the DNA from two other strains of *D. melanogaster* with the original Oregon R strain that we had used heretofore. These two strains, G-1 and G-X11 were chosen because they are particularly well suited for the analysis of chromosome hybridization patterns. The question behind this experiment concerns the possibility that r-insertion-like elements might rearrange relatively rapidly during evolution so that they might find themselves in a different environment in the three strains. Great strain differences in the location of certain repeated DNA sequences have been reported by Strobel, Dunsmuir and Rubin (1979). Figure 6 shows the results of such a comparison. Total DNA from the 3 strains was digested with 3 restriction endonucleases, the fragments were separated on an agarose gel and transferred to nitrocellulose filters (Southern, 1975). The filters were then hybridized with labeled DNA fragments

corresponding to the entire 5 kb type 1 insertion. The autoradiogram shows a complex pattern of bands, as reported previously (Dawid and Botchan, 1977). A comparison of the different strains reveals that most bands are at the same position, suggesting that the r-insertion-like elements find themselves in the same environment in these strains. Some differences, however, are apparent in these gels. In earlier work different developmental stages of animals from the same strain had been compared and found to be indistinguishable by

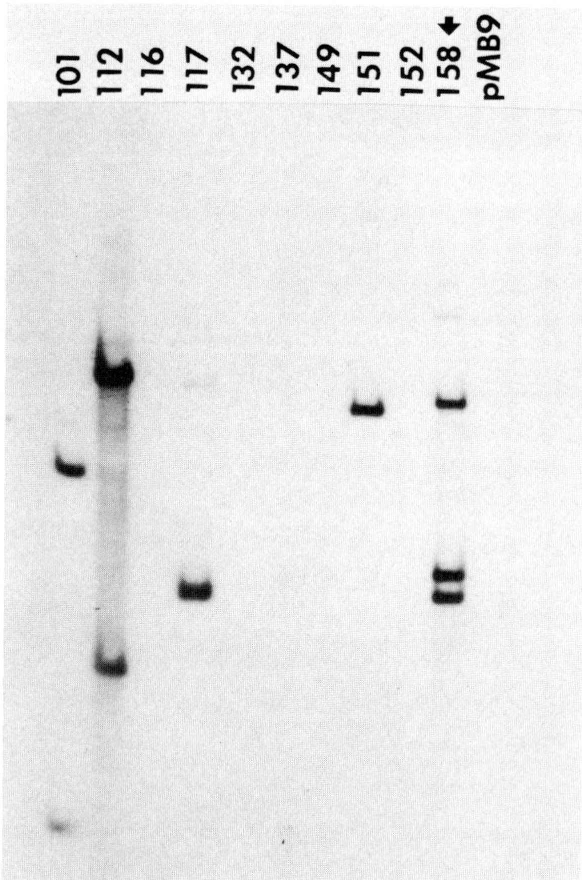

FIGURE 5. Hybridization of flanking sequences to pDmI clones. Cloned DNAs containing chromocentric r-insertion-like elements were digested with HindIII, separated on an agarose gel and transferred to a filter. The filter was hybridized with labeled flanking sequences from clone pDmI 158. Only a portion of the pDmI clones tested is shown here including all clones that gave a positive signal with pDmI 158.

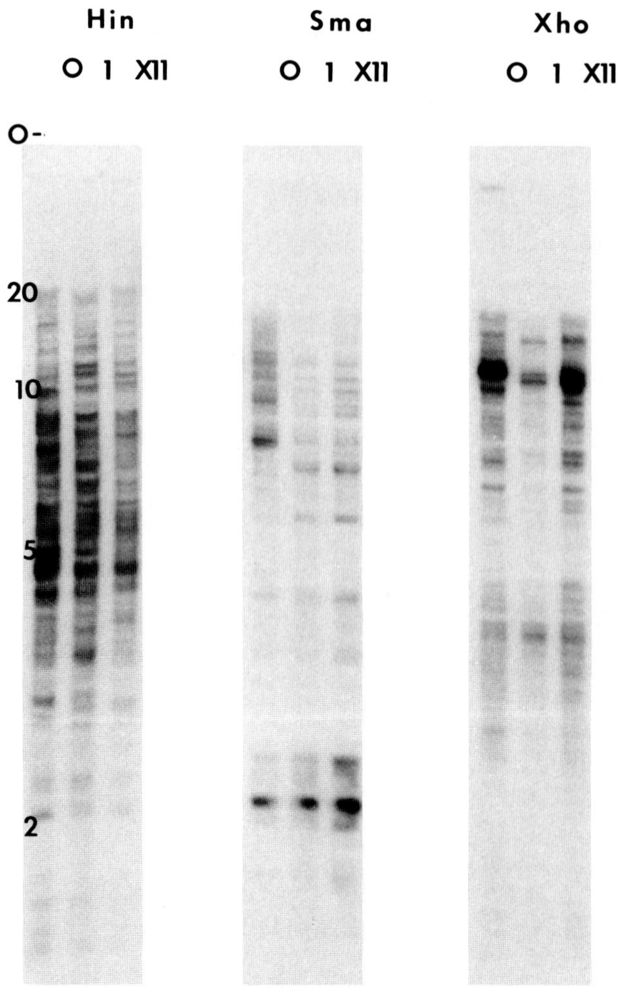

FIGURE 6. Samples of total DNA from D. melanogaster strain Oregon R (O), strain G-1 (1) or strain G-X11 (X11) were digested with HindIII, SmaI or XhoI. The digests were separated on agarose gels, transferred to nitrocellulose filters (Southern, 1975) and hybridized with labeled restriction fragments covering essentially the entire 5 kb insertion (from the HindIII site at the left to the rightmost BamHI site, see Figure 2). Sizes of bands in kb are shown at the left.

this kind of analysis (Dawid and Botchan, 1977; Dawid and Wellauer, 1978). We conclude that r-insertion-like elements are quite stable during the life cycle of the individual and show only a modest degree of rearrangement in the time that separates different strains of the same *Drosophila* species.

IV. CONCLUSIONS

The data reviewed and presented in this article show that the ribosomal insertions in *D. melanogaster* are a complex set of repetitive sequences. At least two types of nonhomologous insertions occur within the ribosomal genes. Type 1 insertions which are abundant on the X chromosome but absent from Y chromosomal rDNA, are homologous to a set of sequences located in the chromocenter. Chromocentric r-insertion-like elements are homologous to but not identical with ribosomal insertions, and are interspersed with other DNA called flanking sequences. Several and perhaps many nonhomologous flanking sequences exist in the chromocenter of *D. melanogaster*. R-insertion-like elements do not rearrange detectably during development. Some sequence rearrangements involving these elements appear to have occurred since the separation of three *D. melanogaster* strains.

The observations summarized above suggest that ribosomal insertions are not in the group of highly mobile DNA sequences exemplified by the copia element (Potter *et al.*, 1979; Strobel *et al.*, 1979). High mobility might have suggested a basis for the presence of insertions in a fraction of the ribosomal genes of *Drosophila*. One might have argued that a highly mobile sequence became inserted in the recent past into some ribosomal genes and that there has not been enough time to eliminate this sequence. The fact that r-insertion-like elements are reasonably stable in their arrangement argues against such an interpretation. Even more relevant is the point that insertions in ribosomal genes occur in at least several and possibly all *Drosophila* species (Barnett and Rae, 1979; Kunz and Glätzer, 1979), and that these insertions retain a substantial level of sequence homology (Barnett and Rae, 1979). Nucleolus organizers containing both interrupted and uninterrupted rRNA genes are therefore quite ancient among dipteran insects. It is still quite possible that this pattern arose by the insertion of members of a chromocentric family of repeated DNA sequences into the ribosomal genes. Once this event took place it appears to have been stabilized and retained for a considerable evolutionary time.

ACKNOWLEDGMENTS

We thank Drs. Maxine Singer and Eric Long for suggestions on the manuscript, Dr. Pieter Wensink for providing pDm4, and Martha Rebbert for excellent assistance.

REFERENCES

Alwine, J.C., Kemp, D.J., and Stark, G.R. (1977). Proc. Natl. Acad. Sci. U.S.A. 74, 5350.
Barnett, T., and Rae, P.M.M. (1979) Cell 16, 763.
Beckingham, K., and White, R.L. (1979). J. Supramol. Structure Suppl. 3, 55.
Chooi, W.Y. (1979). Chromosoma 74, 57.
Dawid, I.B., and Botchan, P. (1977). Proc. Natl. Acad. Sci. U.S.A. 74, 4233.
Dawid, I.B., and Long, E.O. (1979). In "Specific Eukaryotic Genes" (J. Engberg, H. Klenow, and V. Leick, eds), p. 103. Munksgaard, Copenhagen.
Dawid, I.B., and Wellauer, P.K. (1978). Cold Spring Harbor Symp. Quant. Biol. 42, 1185.
Dawid, I.B., Wellauer, P.K., and Long, E.O. (1978). J. Mol. Biol. 126, 749.
Gall, J.G., and Pardue, M.L. (1971). In "Methods in Enzymology" (L. Grossman, and Moldave, eds), Vol. 21, p. 470, Academic Press, New York.
Glätzer, K.H. (1979). Chromosoma, 75, 161.
Glover, D.M., and Hogness, D.S. (1977). Cell 10, 167.
Kunz, W., and Glätzer, K.H. (1979). Hoppe-Seiler's Z. Physiol. Chem. 360, 313.
Long, E.O., and Dawid, I.B. (1979). Cell 18, 1185.
Long, E.O., and Dawid, I.B. (1980) Ann. Rev. Biochem., in press.
Maden, B.E.H., and Tartof, K.D. (1974). J. Mol. Biol. 90, 51.
McKnight, S.L., and Miller, O.L., Jr. (1976). Cell 8, 305.
Pellegrini, M., Manning, J., Davidson, N. (1977) Cell 10, 213.
Potter, S.S., Brorein, W.J., Jr., Dunsmuir, P., and Rubin, G.M. (1979). Cell 17, 415.
Renkawitz, R., Gerbi, S.A., and Glätzer, K.H. (1979). Molec. Gen. Genet. 173, 1.
Southern, E.M. (1975). J. Mol. Biol. 98, 503.
Strobel, E., Dunsmuir, P., and Rubin, G.M. (1979). Cell 17, 429.
Wellauer, P.K., and Dawid, I.B. (1977). Cell 10, 193,

Wellauer, P.K., and Dawid, I.B. (1978). *J. Mol. Biol.* *126*, 769.

Wellauer, P.K., Dawid, I.B., and Tartof, K.D. (1978). *Cell* *14*, 269.

Wensink, P.C., Finnegan, D.J., Donelson, J.E., and Hogness, D.S. (1974). *Cell* *3*, 315.

White, R.L., and Hogness, D.S. (1977). *Cell* *10*, 177.

Yagura, T., Yagura, M., and Muramatsu, M. (1979). *J. Mol. Biol.* *133*, 533.

DISCUSSION

J.H. ROGERS: Have you considered the possibility that the limited changes you do see in the content of the 5 kb sequences may be due to unequal crossing over within the ribosomal RNA locus and not to any specialized mobility of those sequences?

I.B. DAWID: Yes, this is certainly possible. I cannot say yes or no to that. One thing is possibly relevant - if you look within the ribosomal genes themselves, the insertions are a complex set of sequences, but insertions of one size class, e.g. 5 kb insertions, all seem to be the same. "All" may be saying too much, but three separate cloned examples that have been analyzed in detail seem to be identical, so I do not think that there is a lot of variability within the ribosomal locus itself. It is certainly possible that recombination explains the changes we see, and I hope I made it clear that I do not take our data as evidence for transposons in any regular sense.

P. SZOKA: I was wondering if you looked at any of the *bobbed* mutants to see if you get the effect of losing preferentially the inserted genes?

I.B. DAWID: Right. We have done some experiments on *bobbed* mutants but they are not complete. There seem to be more transcripts of the insertion in some *bobbed* mutants but nevertheless these transcripts are still rare and they are too short to be precursors to ribosomal RNA. Therefore, we still get the impression, though it is preliminary, that even in *bobbed* mutants which have lost a good portion of their ribosomal genes, the interrupted genes are not productive in rRNA synthesis.

P. SZOKA: Isn't there also a situation where using the mutants, you can increase the amount of rRNA in the fly? Have you looked at that?

I.B. DAWID: No. That is the phenomenon of magnification of ribosomal DNA. We have not studied that yet.

T.J.M. SCHOPF: It may be the case that these genes that are inactive only create a problem if one assumes that one is looking at some kind of a "final solution". On the other hand, if they represent a dynamic equilibrium which is continually turning over, then there is no reason to suspect that every gene that you find should be "active". In other words, the genome may be turning over far more rapidly than one's preconceived notion assumes it to be.

I.B. DAWID: I agree. It is certainly a possible interpretation that we are looking at a relatively recent insertion into rDNA. There is a problem with this interpretation which I did not get into. Other species of *Drosophila* have been looked at and some of their ribosomal genes are interrupted and these insertions share sequence homology. This includes *Drosophila* species that are quite distant from each other. So, unless that same kind of insertion happens again and again at the same locus, there is a common ancestor which already had interrupted ribosomal genes and then this arrangement would be fairly ancient.

TRANSPOSABLE ELEMENTS IN THE *DROSOPHILA* GENOME[1]

Gerald M. Rubin
William J. Brorein, Jr.[2]
Pamela Dunsmuir
Robert Levis[3]
S. Steven Potter[4]
Edward Strobel[3]
Elihu Young[5]

Sidney Farber Cancer Institute and
Department of Biological Chemistry
Harvard Medical School
Boston, Massachusetts

I. "*COPIA*-LIKE" ELEMENTS CONSTITUTE A DISTINCT CLASS OF MODERATELY REPETITIVE DNA SEQUENCES IN THE *DROSOPHILA* GENOME

A class of repeated genes of *Drosophila melanogaster* has been described in which the elements are widely dispersed throughout the genome, rather than tandemly arranged (Rubin, Finnegan and Hogness, 1976). Two such dispersed repeated gene families, *412* and *copia*, have been studied in detail and share several characteristic features although they show no sequence homology (Finnegan *et al.*, 1978). *412* and *copia* elements are each found at approximately thirty widely

[1] Supported by NIH Grant GM23614.
[2] Supported by an NIH Predoctoral Training Grant.
[3] Supported by NIH Postdoctoral Fellowship Awards.
[4] Present address: Biology Department, Wesleyan University, Middletown, Connecticut.
[5] Supported by fellowship from the Damon Runyon-Walter Winchell Cancer Fund.

ISBN 0-12-633360-2

scattered locations in the chromosomes of *D. melanogaster* and both code for abundant poly(A)-containing cytoplasmic RNAs. The *412* and *copia* elements are 7 and 5 kb, respectively; their sequences are closely conserved at each chromosomal site, and it is in this respect that these families differ markedly from certain other dispersed, moderately repetitive DNA sequences found in the Drosophila genome (Wensink, Tabata and Pachl, 1979). *412* and *copia* elements are terminally redundant: *412* elements carry direct repeats of 0.5 kb and *copia* elements have direct repeats of 0.3 kb. Moreover, as described below, *412* and *copia* elements are transposable. We will refer to the members of those families of dispersed repeated DNA sequences in which the elements are closely conserved, terminally redundant, transcribed, and transposable as "*copia*-like" elements.

412 and *copia* are not the only examples of "*copia*-like" elements in *D. melanogaster*. We have demonstrated the existence of a third family of "*copia*-like" elements, *297*, which shares the features outlined above for *412* and *copia* (unpublished results). Furthermore, results obtained in this and other laboratories (Ilyin *et al.*, 1978; Tchurikov *et al.*, 1978; W. Bender, P. Spierer, M. Goldberg, R. Lifton, R. Karp and D.S. Hogness, unpublished results) suggest the existence of many additional such families. Together these putative "*copia*-like" elements may account for nearly 5% of the *Drosophila melanogaster* genome.

II. THE CHROMOSOMAL LOCATIONS OF "*COPIA*-LIKE" ELEMENTS ARE HIGHLY POLYMORPHIC

We have compared the locations of *412*, *copia*, and *297* elements in the genomes of four different strains of *D. melanogaster*, using filter hybridization to restriction endonuclease digests of embryo DNA and *in situ* hybridization to salivary gland polytene chromosomes (Strobel, Dunsmuir and Rubin, 1979). The Oregon-R, Canton-S, Seto, and Swedish-C strains used in these experiments are derived from widely separated wild populations of *D. melanogaster*; they have been maintained in the laboratory for at least thirty years (Bridges and Brehme, 1944). The four strains are morphologically indistinguishable, fully reproductively compatible isolates of the same species; their polytene chromosome banding patterns are homosequential. Our *in situ* hybridization studies demonstrated that elements of the *412*, *copia*, and *297* families have different distributions in the salivary gland chromosomes of the four *D. melanogaster* strains. The number and location of chromosomal sites observed for each dispersed repeated gene family differ dramatically among these strains; few sites are common to all four strains. Analysis of restriction enzyme cleavage sites within genomic sequences homologous to these sequences indicates that most, if not all, of these sequences occur as intact elements in the genomes of each strain. Analysis of the restriction enzyme cleavage sites in DNA flanking these elements confirms that the local genomic environments in which these elements occur differ among the strains. These results indicate that *412, copia,* and *297* elements undergo evolutionarily rapid transposition within the *D. melanogaster* genome.

Individuals from within a strain also exhibit differences in the arrangement of *412, copia,* and *297* elements; homologous chromosomes within individual flies are often polymorphic in the organization of these elements. Our results and similar studies carried out by Ilyin *et al.* (1978) document the most dramatic genetic differences detected to date among *D. melanogaster* strains, and as such represent a previously undetected type of genetic variability in Drosophila.

III. THE GENOMIC LOCATIONS OF "*COPIA*-LIKE" ELEMENTS ARE PARTICULARLY UNSTABLE IN DROSOPHILA CELL CULTURES

We have examined the stability of elements of three different dispersed repeated gene families in the genome of Drosophila tissue culture cells (Potter, Brorein, Dunsmuir and Rubin, 1979). Different amounts of sequences homologous to elements of the *412, copia,* and *297* dispersed repeated gene families are found in the genomes of *D. melanogaster* embryonic and tissue culture cells. In general the amount of these sequences is increased in the cell lines; in some cases by more than five-fold. Analysis of restriction enzyme cleavage sites within genomic sequences homologous to these elements indicates that most, if not all, of these sequences occur as intact elements in the cell culture genome, indistinguishable from those found in the embryo. Analysis of restriction enzyme cleavage sites in DNA flanking these elements indicates that the local genomic environments in which the elements occur differ not only among embryos and cell cultures, but also among individual cells in the same cell culture population. These differences were shown to be the result of transpositions of *412, copia,* and *297* elements. It appears that these elements can insert at many alternative sites.

IV. INSERTION OF *COPIA* CREATES A FIVE BASE PAIR DUPLICATION

We are examining in detail the structure of the "*copia*-like" elements and their sites of insertion. To do this we have isolated pairs of DNA segments derived from the same genomic location in two different cells, one with and one without an inserted *copia* or *297* element at that site. To date we have isolated four such pairs of DNA segments for *copia* (Potter *et al.*, 1979; Dunsmuir, unpublished) and one for *297* (Young, unpublished). Restriction enzyme mapping and electron microscopic heteroduplex analysis suggest that the two DNA segments of each pair differ only by the simple insertion of the element. We are extending this analysis to the nucleotide sequence level and data for two cases of *copia* insertion (Dunsmuir, unpublished) have been obtained. In both cases a five base pair sequence, which was present once in the target DNA before insertion of the *copia* element, is found after insertion on both sides of the *copia* element. While the sequence of the duplicated 5 base pairs is different in the two cases, the ends of the two *copia* elements are the same and correspond precisely to the ends of the terminal direct repeats (see below). Thus the creation of a small duplication upon insertion appears to be a universal feature of both the prokaryotic and eukaryotic transposable elements (reviewed in Reed *et al.*, 1979; Fink *et al.*, this volume).

V. THE ENDS OF *COPIA* ARE IMPERFECT INVERTED REPEATS

The nucleotide sequence of the terminal direct repeats of the *copia* element have been determined (Levis, unpublished). The repeat is 276 nucleotides long and has a high adenine and thymine content (200 out of 276 residues). The most striking feature of the sequence, however, is that the ends of the direct repeat are imperfect inverted repeats; of residues 1-17, 13 bases are complementary to the corresponding residues in 276-259. Since the repeats themselves are present in direct orientation at the ends of the element, the termini of the *copia* element itself are inverted repeats with respect to each other. The presence of terminal imperfect inverted repeats is a feature common to all prokaryotic IS elements and those transposons that carry terminal IS elements in direct orientation (reviewed in Reed *et al.*, 1979). While their significance is unclear, the structural analogies between the "*copia*-like" elements of Drosophila and certain prokaryotic transposons is indeed striking. It is likely that these Drosophila elements produce effects on gene structure and expression as profound as those produced by prokaryotic transposable elements.

ACKNOWLEDGMENTS

We are indebted to Brian Backner and Meredith Simon for their excellent technical assistance.

REFERENCES

Bridges, C.B., and Brehme, K.F. (1944). Carnegie Inst. of Washington Publication 552.

Finnegan, D.J., Rubin, G.M., Young, M.W., and Hogness, D.S. (1978). Cold Spring Harb. Symp. Quant. Biol. *42*, 1053.

Ilyin, Y.V., Tchurikov, N.A., Ananiev, E.V., Ryskov, A.P., Yenikolopov, G.N., Limborska, S.A., Maleeva, N.E., Gvozdev, V.A., and Georgiev, G.P. (1978). Cold Spring Harb. Symp. Quant. Biol. *42*, 959.

Potter, S.S., Brorein, W.J., Jr., Dunsmuir, P., and Rubin, G.M. (1979). Cell *17*, 415.

Reed, R.R., Young, R.A., Steitz, J.A., Grindley, N.D.F., and Guyer, M.S. (1979). Proc. Nat. Acad. Sci. *76*, 4882.

Rubin, G.M., Finnegan, D.J., and Hogness, D.S. (1976). Prog. Nucl. Acid Res. Mol. Biol. *19*, 221.

Strobel, E., Dunsmuir, P., and Rubin, G.M. (1979). Cell *17*, 429.

Tchurikov, N.A., Ilyin, Y.V., Ananiev, E.V., and Georgiev, G.P. (1978). Nucl. Acids Res. *5*, 2169.

Wensink, P.C., Tabata, S., and Pachl, C. (1979). Cell *18*, 1231.

DISCUSSION

K. LUEDERS: You said that these sequences were transcribed into RNA. I was wondering if you had tested these RNA's for translation in a cell free system and, if they do make proteins, is the total capacity of the coding sequences used?

G.M. RUBIN: We have not done any experiments of that type. Two other laboratories have looked for the presence of *copia* transcripts on polysomes in cultured cells. These two laboratories got different results. One laboratory says they are on polysomes, the other says they are not on polysomes. I think it is still very much an open question.

A. SKALKA: In your talk you made a very interesting analogy with bacterial transposable elements, but it seems to me that an equally interesting analogy with the aukaryotic retroviruses could be made. Would you comment on that?

G.M. RUBIN: The structure that we find for the "*copia*-like" elements is indeed also analogous to that of retroviruses. Elements with this structure occur throughout evolution; thus the retroviruses may be related in at least an evolutionary sense both to the "*copia*-like" elements of the Drosophila and to the TY1 element of yeast. At the moment it appears that elements which have this structure but which do not occur as viruses are very much more wide-spread in evolution than the retroviruses. One possible explanation is that retroviruses evolved from these elements by developing the ability to package their transcripts into virions. Alternatively these elements may have been introduced into yeast and *Drosophila* by viral infection. There is, however, no evidence that any of these elements presently occur as viruses either in yeast or *Drosophila*; although there are many known insect viruses, none of them are retroviruses. While there is no evidence that any of the "copia-like" elements occur as viruses, it is not possible to say for certain that there is not some condition, some helper virus, which could cause their transcripts to be packaged.

M. GUYER: You made the comment that these kinds of sequences comprised a strikingly high fraction of the *Drosophila* genome. I would just like to comment that a similar calculation made for *E. coli* shows that something like 1% of the *E. coli* genome is devoted to the known insertion sequences.

B. LANSMAN: How rigorously have people looked for low molecular weight copies of these sequences in cultured cells?

G.M. RUBIN: No rigorous analysis has been done. Heterogeneous circular DNA's have been found in *Drosophila* cells, but there is no indication that these are homologous to the "*copia*-like" elements.

SITE-SPECIFIC RECOMBINATION IN BACTERIOPHAGE λ: REQUIREMENTS FOR THE PHAGE *att* SITE AND ORGANIZATION OF THE RECOMBINATION FUNCTIONS

Pei-Ling Hsu
Wilma Ross
Ronald H. Hoess[1]
Monika Buraczynska
Arthur Landy

Division of Biology and Medicine
Brown University
Providence, Rhode Island

I. INTRODUCTION

Bacteriophage λ integrates into its host, *E. coli*, via a reciprocal recombination between the phage *att* site (*att*P or POP') and the bacterial *att* site (*att*B or BOB'). The reverse reaction, excision, occurs between the left (*att*L or BOP') and the right (*att*R or POB') prophage *att* sites (for reviews see Gottesman and Weisberg, and Nash (1,2)). All four *att* sites have in common a sequence which is 15 base pairs in length. The crossover event presumably takes place within the limits, or at the boundaries, of this "common core" (3). Both the integration and excision reactions require the phage-encoded protein Int as well as the products of host *E. coli* genes (4). An additional requirement for the excision reaction is the phage-encoded protein Xis.

The phage *att* site, *int* gene and *xis* gene map adjacent to one another and form a compact 1.5 kb cluster of functions devoted to site-specific recombination (5,6). Control of site-specific recombination and its direction are dependent upon the regulatory mechanisms governing the production of Int and Xis proteins. During integration, Int is translated from a mRNA transcript initiated at promoter P_I (7), located within or adjacent to the *xis* gene. Synthesis of this transcript is under positive control by the phage cII and cIII gene products (8-10). For excision, both Int and Xis are

[1]Present address: Frederick Cancer Research Center, Frederick, Maryland.

ISBN 0-12-633360-2

made from a single transcript initiated at P_L, approximately 14% λ units upstream of these genes (11) and this transcript is independent of cII/cIII control.

We have undertaken studies to characterize the overall organization of the *att-int-xis* region. This paper provides an overview of several experiments in which we determined the functional limits of the phage *att* site, the Int binding properties of the *att* site DNA, and the DNA structure of the *int-xis* region.

II. RESULTS AND DISCUSSION

A. Functional Limits of the Phage *att* Site

The phage *att* site is located within a 492 bp *Hin*dIII-*Bam*HI restriction fragment extending from position -251 to +242 of the *att*P region. (For numbering the DNA sequence in the *att* region, the center of the common core is taken as 0 with positive numbers extending to the right, through the *int* and *xis* genes and negative numbers extending to the left.) A supercoiled plasmid carrying this restriction fragment (pWR1) is capable of recombining with a linear *att*B fragment under *in vitro* conditions (see below) where similarly cloned *att*L or *att*R fail to recombine with the same *att*B (12). This result suggests that the *Hin*dIII-*Bam*HI fragment from the *att*P region contains all the information necessary to specify a functional phage *att* site. In other words, the functional limits of *att*P must lie within this 492 bp fragment.

Our approach to defining the functional limits of the phage *att* site has been to construct a series of plasmids carrying resected *att* sites, each having varying lengths of DNA removed from the end of the P arm or P' arm. It was thus possible to ascertain how much DNA could be removed from around the phage *att* site without impairing its function (12). In order to preserve the orientation of *att*P on each plasmid and to facilitate the subsequent sequence determination, all plasmids were constructed so that a *Hin*dIII site is always on the P arm and a *Bam*HI site is always on the P' arm of each resected *att* site (13).

1. Minimal P Arm of the Phage *att* Site. A list of P arm resected plasmids is shown in Table I. The end point of each plasmid on the P arm was determined either by restriction fragment analysis or by DNA sequencing. When tested *in vitro* for their phage *att* site functions, pWR1, pPH52, pPH53 and

TABLE I. Integrative Recombination of Plasmids with Resected P Arms*

Plasmids	End points		% Relative recombination	
	P arm	P' arm	Exp. 1	Exp. 2
pWR1	-251	+241	100	100
pPH52	-174	+241	100	48
pPH53	-160	+241	75	95
pPH54	-160	+241	106	99
pPH55	-115	+241	<0.5	<0.5
pPH56	- 90	+241	<0.5	<0.5
pPH57	- 70	+241	<0.5	<0.5
pPH58	- 50	+241	<0.5	<0.5
pPH59	- 30	+241	<0.5	<0.5
pPH60	- 6	+241	<0.5	<0.5

*Each reaction mixture contained approximately 1μg of supercoiled plasmid DNA and 1μg of ^{32}P-labeled linear DNA with a bacterial *att* site (*Eco*RI-*Bam*HI fragment; 1600 bp) at a molar ratio of 1 to 3. The recombination was carried out in a 20μl mixture containing 50mM Tris-HCl (pH7.4), 5mM EDTA, 40mM KCl, 12.5mM spermidine, 4μl of sonicated bacterial extract and 1.5μl of Int fraction I (15). After incubation at 25°C for 40 minutes, 30μl of a buffer containing 0.3M NaCl, 10mM Tris-HCl (pH7.4) was added and the samples were deproteinized and electrophoresed on a 1% agarose gel at 35 volts for 16 hrs. in E buffer (40mM Tris, 20mM Na-Acetate, 2mM EDTA, 18mM NaCl, adjusted to pH7.9 with acetic acid). The gel was stained with EtBr (1μg/ml) and portions of the gels corresponding to the unreacted linear DNA and the recombination product were excised and counted individually for ^{32}P radioactivities (12). The percentage of supercoiled phage *att* site molecules that recombined with labeled bacterial *att* site DNA (present in 3-fold molar excess) was calculated as follows:

$$\text{\% phage } att \text{ sites recombined} = \frac{^{32}\text{P radioactivity in the product band}}{\text{total } ^{32}\text{P radioactivity}} \times 3$$

The results of quantitation are shown here normalized to 100% for pWR1. (The absolute values for pWR1 are 30% and 41% in Expts. 1 and 2 respectively.) The P arm end point of each plasmid was determined by DNA sequencing (pWR1, pPH53,54,55,56,60) or by restriction enzyme analysis (pPH52,57,58,59).

TABLE II. Integrative Recombination of P' Arm Resected Plasmids*

Plasmids	End points		% Relative recombination	
	P arm	P' arm	Exp. 1	Exp. 2
pWR1	-251	+241	100	100
pPH303	-251	+124	80	113
pPH304	-251	+100	75	95
pPH307	-251	+ 82	90	69
pPH308	-251	+ 64	<0.5	<0.5
pPH310	-251	+ 46	<0.5	<0.5

*The *in vitro* recombination reactions, gel electrophoresis, and quantitation were carried out as described in Table I. The results of quantitation are normalized to 100% for pWR1. (The absolute values for pWR1 are 20% and 32% in Expts. 1 and 2 respectively.) The P' arm end point of each plasmid was determined by DNA sequencing (pWR1, pPH303,307,308,310) or by restriction enzyme analysis (pPH304).

pPH54 recombined with *att*B with approximately equal efficiency, whereas all other plasmids failed to recombine significantly under the same experimental conditions. A comparison of the end point of the P arm with the recombinational activity of each plasmid indicates that plasmids carrying a P arm extending to or beyond position -160 retain normal phage *att* site functions, whereas plasmids carrying a P arm ending at, or before, position -115 do not contain competent phage *att* sites. Therefore it is concluded that some DNA sequence which lies between positions -160 and -115 is essential for the *att*P functions. This segment of DNA contains an Int protein binding site as defined by the DNAase footprinting assay (see below). For convenience, we shall refer to the DNA between -160 and the common core as the "minimal P arm" of the phage *att* site.

2. Minimal P' Arm of the Phage *att* Site. A list of P' arm resected plasmids, along with their end points and the results of *in vitro* recombination with *att*B is shown in Table II. The data indicate that pWR1. pPH303, pPH304 and

pPH307 recombine with *att*B with approximately equal efficiency, whereas pPH308 and pPH310 do not recombine with *att*B to a detectable extent in this assay. Since the shortest functioning phage *att* site contains DNA sequence from −251 to +82 and the longest non-functioning phage *att* site contains DNA sequence from −251 to +64, it is concluded that the DNA region between +64 and +82 is crucial for the phage *att* site functions. This segment of DNA coincides with the right half of the previously identified P' arm Int protein binding site (14), thus implicating the essential role of the "intact" P' arm binding site in recombination. Hereafter, we shall refer to the DNA region between the common core and +82 as the "minimal P' arm of the phage *att* site.

3. Functional Limits of the Phage *att* Site. That the minimal P arm and minimal P' arm together constitute a functional phage *att* site was confirmed by testing the recombinational competency of a plasmid (pPH507) carrying *att*P sequence from −160 to +82. This plasmid was constructed by cloning a *Hin*dIII-*Hin*fI fragment of pPH54 (−160 to −115) and a *Hin*fI-*Bam*HI fragment of pPH307 (−115 to +82) into the

TABLE III. Integrative Recombination of Plasmids Carrying Phage *att* Sites with Resected P Arms and Resected P' Arms*

Plasmids	End points		% Relative recombination	
	P arm	P' arm	Exp. 1	Exp. 2
pWR1	−251	+241	100	100
pPH507	−160	+ 82	100	64
pPH510	−160	+ 46	<0.5	<0.5
pPH7	−115	+ 82	<0.5	<0.5
pPH10	−115	+ 46	<0.5	<0.5

*The *in vitro* recombination, gel electrophoresis and quantitation were carried out as described in Table I. The results of quantitation are normalized to 100% for pWR1. (The absolute values for pWR1 are 38% and 31% in Expt. 1 and Expt. 2 respectively.) The structure and boundaries of the cloned fragments were confirmed by restriction fragment analysis and by DNA sequencing.

*Hin*dIII-*Bam*HI site of pBR322. The resultant plasmid (pPH507) recombines with linear *att*B with efficiency comparable to that of pWR1 (see Table III). In contrast, three other plasmids which lack the DNA sequences either from -160 to -115 (pPH7), or from +46 to +82 (pPH510), or both (pPH10) all fail to recombine with *att*B to a detectable extent in this assay. Therefore the sequences required for a functional phage *att* site are contained within the 243 bp DNA segment that extends from -160 to +82 around the common core and the DNA sequence between -160 and -115 and that between +46 and +82 must be present simultaneously for the full expression of the phage *att* site functions. In a series of experiments similar to ours, it was noted that a phage *att* site carrying a bacterial DNA substitution to the left of position -152 retains the *att* site functions (K. Mizuuchi, personal communication). Therefore, it is likely that the DNA sequence from -152 to +82 would constitute a functional phage *att* site.

B. Interaction of Int Protein with the Phage *att* Site and its Relationship to Recombination

We have previously reported the identification by DNAase footprinting methods of two sites of interaction of purified Int protein with the *att*P region (14). At one of these sites, Int protects a 30 bp segment which includes the common core sequence, and therefore the site of the recombination crossover event. Int binding at this site is not unexpected and is consistent with the observed cutting and religating activity of this protein (15). The second Int-protected sequence is approximately 35 bp in length (+50 to +86), and extends 3-4 bases beyond the functional limit of the P' arm, as defined by the results described above. The requirement for Int interaction at this P' site as part of the phage *att* site function is clearly demonstrated by the observation that clones lacking either all, or the right half, of this site fail to function as competent phage *att* sites in recombination (see Tables II and III).

The interaction of Int with the P arm was analyzed using a *Hin*dIII-*Bam*HI fragment containing DNA sequence from -251 to +46 (from pPH310)(Fig. 1), as well as with other P arm-containing fragments (12). As shown in Figure 1, bound Int protein specifically protects two regions of the P arm from neocarzinostatin (NCS) digestion. In one of these regions, P arm site 1, sequence from -143 to -129 is protected, while in the other, P arm site 2, sequence from -114 to -101 is

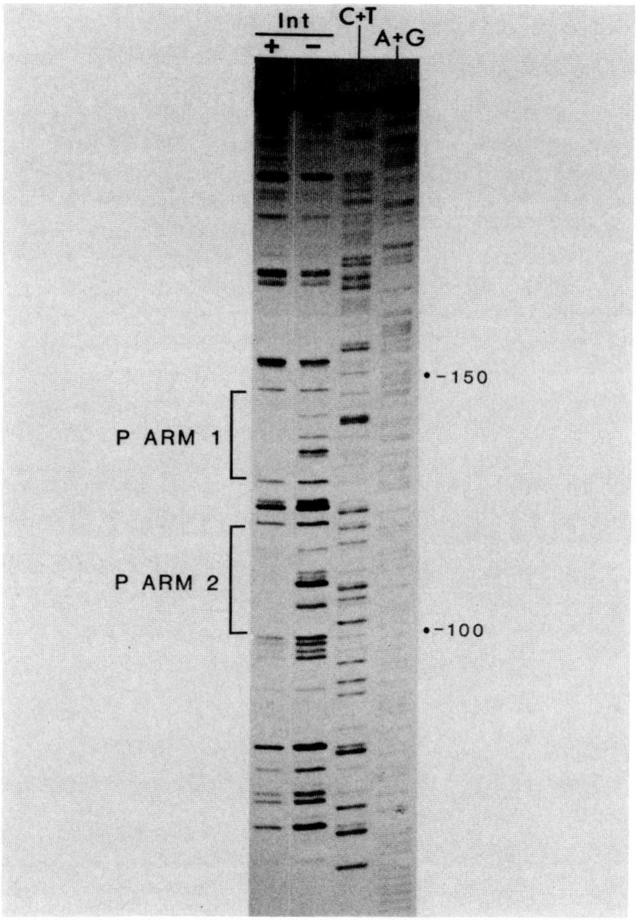

Fig. 1. Neocarzinostatin (NCS) footprint of Int interacting with a fragment containing P arm sequence. The *Hind*III-*Bam*HI fragment from pPH310, containing *att* region sequence from −251 to +46, was 5'-labelled at the *Bam*HI linker adjacent to position +46. The fragment was incubated with purified Int (gift of Y. Kikuchi and H. Nash) at 2.5μg/ml (lane 1) or without Int (lane 2), and subsequently digested with the AT-specific reagent NCS as described previously (14). Sequence markers (A+G and C+T) and conditions of electrophoresis have been described (14). Brackets identify 2 sequences, P arm 1 and P arm 2, that are specifically protected from NCS digestion by bound Int protein (see text and ref. 12).

protected. Both of these regions are contained within the functional limits of the phage *att* site. The involvement of P arm site 1 in site-specific recombination is inferred from the observation that a clone lacking this site fails to function as a phage *att* site in the *in vitro* assay. However, the present data does not allow us to distinguish whether P arm site 2 is needed for recombination; we know only that its presence without that of site 1 is not sufficient. Construction of additional plasmids which carry point mutations or deletions within P arm site 2 will be needed to ascertain the functional significance of Int binding at this site.

The locations of the four Int binding sites in the *att*P region, along with the functional limits of phage *att* site DNA, is shown in Figure 2. The obligatory presence of multiple widely separated Int binding sites for the phage *att* site function is unexpected since the cutting and rejoining event of recombination occurs within the 15 bp common core sequence. The role played by Int binding at the distant sites is not clear at this time and further experiments will be needed to explore this question as well as to clarify the requirement for a supercoiled *att*P substrate (16) and the role played by the host factor proteins.

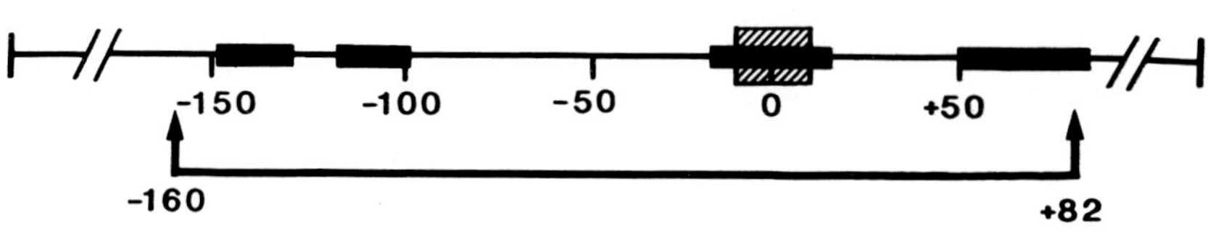

Fig. 2. Binding sites for Int protein in the *att*P region and functional limits of the phage *att* site. Binding sites for Int protein (▬) and the 15 bp common core sequence (▨) are indicated on the linear map of the *att* site region. The boundaries (-160 to +82) of the smallest functional *att*P region found in our experiments are also indicated (↑___↑).

C. A "Donor-Recipient" Paradigm for Site-Specific Recombination

The bacterial *att* site (*att*B) differs from the phage *att* site (*att*P) in many respects, such as its behavior in pairwise crosses with other *att* sites (17-19) and in its Int binding characteristics (14). In order to determine the possible features of DNA that confer the recombinational character of the bacterial *att* site (or its equivalent), we have examined the role of Int binding sites in the behavior of a given *att* site. Linear plasmids carrying all the elements of a phage *att* site (pPH507), or lacking one or more of the Int binding sites (pPH510,7 and 10; see Table III and Fig. 2) were tested *in vitro* for their competence in recombining with a supercoiled phage *att* site. As seen in Figure 3, the recombination between *att*P and pPH10 or pPH510 is almost as efficient as that between *att*P and a linear plasmid carrying the bacterial *att* site. On the other hand, the recombination between *att*P and pPH507 or pPH7 is not detected. The only difference between two *att* sites which recombine efficiently with supercoiled *att*P and two which fail to recombine significantly in this assay is the presence of the P' arm Int binding site. Thus, it appears that the presence of the P' arm Int binding site on the linear recombination partner interferes with its ability to recombine with a supercoiled phage *att* site. This interfering effect is not due to the titration of functional Int protein by the P' arm site since higher concentrations of Int do not abolish the interference (data not shown).

The results discussed above lead us to suggest a model that involves two unequal partners for integrative recombination. The phage *att* site, which provides most of the structural information and specificity for the reaction, is referred to as the "donor" molecule. A supercoiled "donor" *att* site recombines efficiently with an appropriate "recipient" *att* site, such as the bacterial *att* site or its analog. In this work we have defined the minimal requirement for a donor site, which involves multiple interactions with Int protein. Although the minimal requirements for "recipient" *att* sites have not been determined, we know that Int binding to the bacterial *att* site is much less complex and results in the protection of a short (15 bp) DNA sequence (14). The present results establish that (in the *in vitro* conditions used here) a perfectly functional "recipient" *att* site can be generated by removal of the P' arm Int binding site from a phage *att* site. Thus, whereas the presence of the P' arm Int binding site is necessary for a "donor" *att*

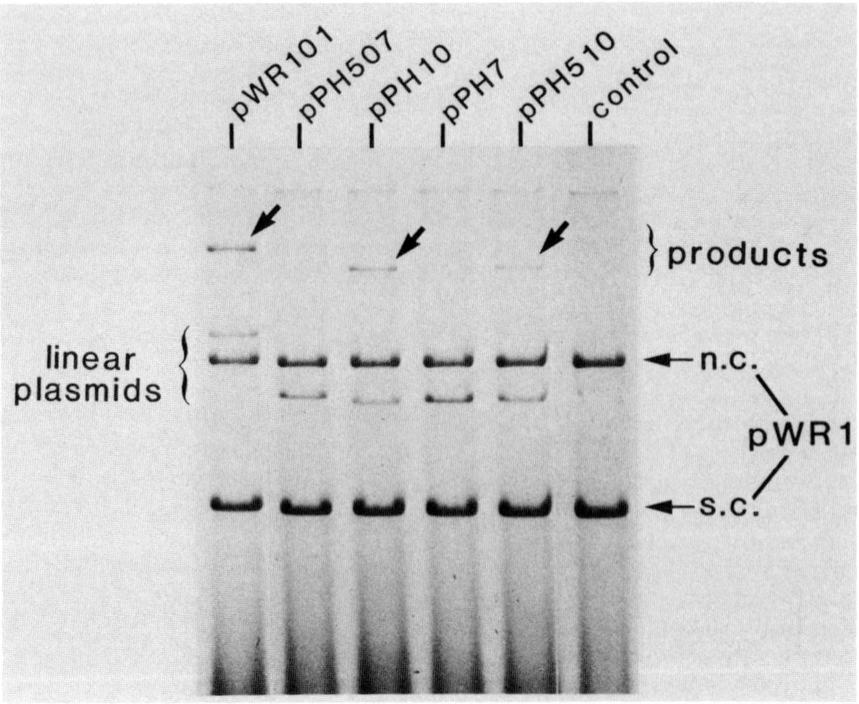

Fig. 3. "Recipient" function of the resected phage *att* sites. The *in vitro* recombination reactions and gel electrophoresis were performed as described in Table I. Each reaction mixture contained 1.5μg of supercoiled pWR1 DNA and a 3-fold excess of linear recipient plasmid DNA. The linear plasmids were generated by *Pst*I digestion of pPH7,10,507,510 or by *Eco*RI digestion of pWR101, a plasmid carrying the *Eco*RI-*Bam*HI-1600 bp fragment of the bacterial *att* site. The origin and boundaries of the restriction fragments carried by each plasmid are described in the text and in Table III. Positions are indicated on the electropherogram for DNA containing different "recipient" *att* sites (linear plasmid), the "donor" circular plasmid DNA (pWR1) containing the phage *att* site and migrating as supercoils (s.c.) or nicked circles (n.c.) and the linear product of recombination (➘).

site, its absence is necessary for a good "recipient" *att* site. Further experiments will of course be necessary to confirm, and possibly extend, the applicability of the "donor-recipient" paradigm for integrative recombination.

D. Structure of the *xis* and *int* genes

The structure of the *int* and *xis* genes and the upstream regulatory region was determined by DNA sequence analysis (20). The region corresponding to the amino termini of these two proteins and the upstream regulatory sequence is shown in Figure 4. Note that transcription of *int* and *xis* is opposite to the direction of numbering.

1. The *xis* gene. Identification of the *xis* gene and determination of its reading frame was based on the sequence analysis of the wild type DNA, the *xis*6 amber mutant and the λb538 deletion. The *xis*6 mutation occurs at position +1332 and represents a CG to TA transition, resulting in the amber codon TAG. The coding portion of *xis* extends from the ATG initiation codon at position +1347 to the TGA chain termination codon at position +1130. From the DNA sequence, Xis consists of 72 amino acids and has a molecular weight of 8,630 daltons. The very high proportion of basic amino acids (25% lysine and arginine residues) may reflect a capacity for interacting with DNA.

2. The *int* gene. Identification of the *int* gene was primarily based on the analysis of the three possible reading frames in the region sequenced. The longest, uninterrupted reading frame is 1068 bp in length, with an ATG initiation codon at position +1151 and a TAA termination codon at position +83. This 1068 bp sequence yields a protein of 356 amino acids with a calculated molecular weight of 40,330 daltons, which is in close agreement with the size estimate made from genetic analysis and protein chemistry (6,21). The next two longest open reading frames in this region are 57 and 50 codons in length and therefore do not code for a protein large enough to be Int. Int is a basic protein with 69 amino acids being basic. Furthermore, in the NH_2-terminal portion of Int, 35% of the first 20 amino acids are basic. This could reflect a common feature with other DNA binding proteins, such as the *lac* and λ repressors, which interact with DNA through their NH_2-termini (22,23).

E. Features of DNA in the *xis-int* Region

1. Overlapping Structural Genes. As shown in Figure 4, the *int* and *xis* genes contain a 20 bp overlap (positions +1132 to +1151) in their coding regions. This overlap is potentially an additional means for controlling the amounts of Int and Xis made from transcripts initiated at P_L. For instance, the potential ribosomal binding site for the *int* gene occurs at +1161 to +1158, which lies within the coding region for the Xis protein. Ribosomes engaged in the translation of the *xis* region could stearically hinder the binding of additional ribosomes to the initiation site of the *int* gene and thereby reduce the efficiency of the translation of the *int* region. Enquist *et al.* (24) have shown that the levels of Int required for excision are lower than those required for integration. Thus, reducing the level of Int production from transcription initiated at P_L might help to insure against the reintegration of phage during excision. The observations that, in phage infected cells, the amount of Int made from P_L is 3 to 4 fold higher in *xis* amber mutants (S. Adhya, personal communication; K. Abremski, personal communication) would also support the above model. However, there are other studies which yield somewhat different results (M. Gottesman, personal communication).

2. P_I Promoter and the *int*C226 mutation. During integration, the transcription of the *int* gene is originated at promoter P_I and is under the positive control of cII and cIII gene products (7-10). The *int*C mutations cause constitutive production of Int protein independent of cII and cIII control. We have sequenced the *int*C226 mutation and found that it is a single base transition (G to A) at position +1345. Several other constitutive *int* point mutants have also been sequenced by Abraham *et al.* (J. Abraham, R. Fischer and H. Echols, personal communication) and these have been found to have the same base changes as the *int*C226. The same group has also obtained DNA sequence for the region upstream of *xis*.

In Figure 4 it is seen that the *int*C226 shows a number of matches with a promoter consensus sequence (25). More important, the base change in *int*C226 from a G to an A occurs in the second position of the proposed Pribnow sequence. This position is highly conserved for either an A or T in all but one known promoter sequence, and *int*C226 matches the idealized Pribnow sequence in 5 out of 7 positions. In contrast to this comparison, the DNA sequence in the immediate vicinity of the *int*C226 mutation does not share features with

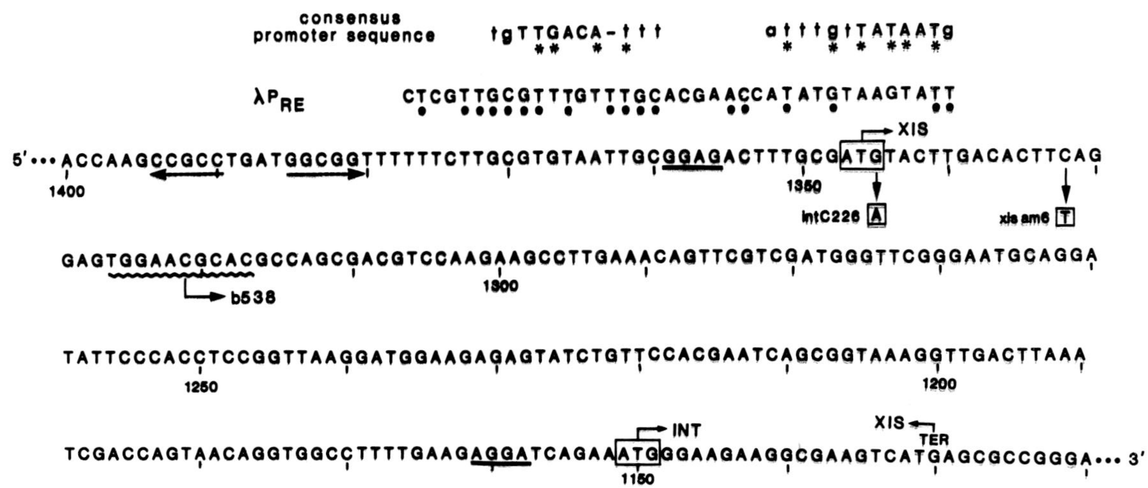

Fig. 4. Nucleotide sequence of the regulatory region for the *xis-int* genes and its comparison with a consensus promoter sequence and the λP_{RE} promoter. Transcription is from left to right, i.e. towards the common core. Numbers signify the distance in base pairs from the center of the common core. Initiation codons for *xis* (1347) and *int* (1151),(□); potential ribosome binding sites (28),(——); termination codon for *xis*, (TER); base changes for *int*C226 and *xis*6amber (□); and the approximate endpoint for the b538 deletion, (〰) are indicated. In the consensus sequence of 46 promoters recognized by *E. coli* polymerase (25) the most highly conserved and less well conserved positions are upper and lower case respectively; homologies with the putative P_I promoter are indicated (*). Homologies between λP_{RE} and the putative P_I are (●). A potential RNA termination site upstream of *xis* is identified by an inverted repeat (→←) which might form a stem and loop structure.

known transcription termination signals (25). These observations lead to the inference that the *int*C226 phenotype is not due to disruption of a termination signal as suggested in some models for cII/cIII action, but is rather the result of a promoter mutation for *int* transcription. This is consistent with the *in vitro* finding that the *int*C226 mutant enhances binding of RNA polymerase to this region of DNA and also provides a start point for RNA transcription (26).

3. Site of cII/cIII Action. When the region of DNA upstream of the *int*C mutation site was compared to the DNA sequence around the λP_{RE} promoter, which is also under positive control by cII and cIII, a strong homology (11 out of 14 bases) is found around positions +1360 to +1373 of the *att-int-xis* sequence. This region of homology does not display the features of a consensus promoter sequence. Further, λP_{RE} mutations occurring in this region appear to alter the cII and cIII regulation (25). It is very likely that this region of DNA comprises the site of cII/cIII action. Recent results with fusion deletions in the region upstream of *xis* also identify this region as the target site of cII/cIII transcriptional activation (D. Masearenhas, M. Benedik and A. Campbell, personal communication).

ACKNOWLEDGMENTS

We thank Carl Foeller and Karen Bidwell for sequence work and restriction fragment mapping. We also thank Julie Smith for invaluable help in preparing the manuscript and Eric Johnson for excellent technical assistance. This work was supported by grant no. AI13544 from the National Institutes of Health and grant no. 1-543 from the National Foundation-March of Dimes. P.-L.H. and R.H. were recipients of National Institutes of Health Postdoctoral Research Fellowships GM07046 and GM05269 respectively. W.R. is a recipient of a Brown University Graduate Fellowship. A.L. is a Faculty Research Associate of the American Cancer Society.

REFERENCES

1. Gottesman, M.E. and Weisberg, R.A., *in* "The Bacteriophage Lambda" (A.D. Hershey, ed.), p. 113. Cold Spring Harbor Laboratory, Cold Spring Harbor, N.Y. (1971).
2. Nash, H.A., *Current Top. Microbiol. Immunol. 78*, 171 (1978).
3. Landy, A. and Ross, W., *Science 197*, 1147 (1977).
4. Miller, H.I., Kikuchi, A., Nash, H.A., Weisberg, R.A. and Friedman, D.I., *Cold Spring Harbor Symp. Quant. Biol. 43*, 1121 (1979).
5. Davidson, N. and Szybalski, W., *in* "The Bacteriophage Lambda" (A.D. Hershey, ed.), p. 45. Cold Spring Harbor Laboratory, Cold Spring Harbor, N.Y. (1971).
6. Enquist, L.W. and Weisberg, R.A., *J. Mol. Biol. 111*, 97 (1977).
7. Shimada, K. and Campbell, A., *Proc. Natl. Acad. Sci. USA 71*, 237 (1974).
8. Katzir, N., Oppenheim, A., Belfort, M. and Oppenheim, A.B., *Virology 74*, 324 (1976).
9. Chung, S. and Echols, H., *Virology 79*, 312 (1977).
10. Court, D., Adhya, S., Nash, H. and Enquist, L.W., *in* "DNA Insertion Elements, Plasmids and Episomes" (A.I. Bukhari, J. Shapiro and S. Adhya, eds.) p. 389. Cold Spring Harbor Laboratory, Cold Spring Harbor, N.Y. (1977).
11. Gottesman, M.E. and Weisberg, R.A., *in* "The Bacteriophage Lambda" (A.D. Hershey, ed.), p. 113. Cold Spring Harbor Laboratory, Cold Spring Harbor, N.Y. (1971).
12. Hsu, P.-L., Ross. W. and Landy, A., submitted.
13. Hsu, P.-L. and Landy, A., in preparation.
14. Ross, W., Landy, A., Kikuchi, Y. and Nash, H.A., *Cell 18*, 297 (1979).
15. Kikuchi, A. and Nash, H.A., *Proc. Natl. Acad. Sci. USA 76*, 3760 (1979).
16. Mizuuchi, K. and Mizuuchi, M., *Cold Spring Harbor Symp. Quant. Biol. 43*, 1111 (1979).
17. Guerrini, F., *J. Mol. Biol. 46*, 523 (1969).
18. Echols, H., *J. Mol. Biol. 47*, 575 (1970).
19. Guarneros, G. and Echols, H., *Virology 52*, 30 (1973).
20. Hoess, R.H., Foeller, C., Bidwell, K. and Landy, A., *Proc. Natl. Acad. Sci. USA*, in press.
21. Kikuchi, Y. and Nash, H.A., *J. Biol. Chem. 253*, 7149 (1978).
22. Ogata, R.T. and Gilbert, W., *Proc. Natl. Acad. Sci. USA 75*, 5851 (1978).

23. Pabo, C.O., Sauer, R.T., Sturtevant, M. and Ptashne, M., *Proc. Natl. Acad. Sci. USA 76*, 1608 (1979).
24. Enquist, L.W., Kikuchi, A. and Weisberg, R.A., *Cold Spring Harbor Symp. Quant. Biol. 43*, 1115 (1979).
25. Rosenberg, M. and Court, D., *Adv. in Genetics*, in press.
26. Fischer, R.L., Takeda, Y. and Echols, H., *J. Mol. Biol. 129*, 509 (1979).
27. Honigman, A., Hu, S.-L. and Szybalski, W., *Virology 92*, 542 (1979).
28. Shine, J. and Dalgarno, L., *Proc. Natl. Acad. Sci. USA 71*, 1342 (1974).

DISCUSSION

A. BUKHARI: You said that the int gene overlaps the att site to some extent. The att sites differ from place to place; the bacterial site is different from the phage site. Now, if you cross in the bacterial att sequences into the phage, how will that affect the int activity?

A. LANDY: The carboxyl terminus of the int gene overlaps with P' binding site of the phage att site. Therefore, you might image that the P' binding site has something to do with regulation of int gene expression. All the genetic evidence thus far suggests this is not the case; the expression of downstream genes appears to be independent of the amount of int present.

If we take away the P' arm, it destroys the int gene so we have to provide Int in trans whenever we do that.

J.S. KNEPPER: Is the relative efficiency of insertion into the secondary attachment sites at all related to the extent of homology with the "common core" sequence?

A. LANDY: That is a good question. The reason we do not know the answer to it is because the efficiencies of these secondary att sites are so low that they cannot be meausred. In order to isoalte them in the first place it was necessary to use selection protocols. We are now attempting to generate a set of secondary attachment sites in which we can correlate sequence and function.

W. SZYBALSKI: Was the "footprinting" done with purified Int protein or was it done in the crude system? We both know there are several host proteins involved, and I am curious whether there are any additional binding sites for the host proteins?

A. LANDY: As I mentioned, the purified Int protein was made by Yoshiko Kikuchi and Howard Nash. It is a highly purified preparation that appears homogeneous in SDS gel electrophoresis.

W. SZYBALSKI: Did you do any analogous experiments with Xis protein?

A. LANDY: No we did not do anything with Xis protein.

THE GENESIS OF AVIAN RETROVIRUS ONCOGENES

J. Michael Bishop[1], Thomas Gonda, Stephen H. Hughes, Diana K. Sheiness, Elton Stubblefield*, Bjorn Vennstrom and Harold E. Varmus

Department of Microbiology and Immunology
University of California
San Francisco, Califronia

*Department of Cell Biology
University of Texas Cancer Center
Houston, Texas

A. The Argument

A number of retroviruses possess genetic loci whose expression transforms susceptible host cells to a neoplastic phenotype (1). These loci (or "oncogenes") serve no apparent purpose in the replication of retroviruses. Why then do they persist in the viral genome, and from whence did they originally arise? The first of these question remains an enigma: we have not identified the selective pressures responsible for the maintenance of oncogenes in their varied forms. By contrast, an answer to the second question appears to be at hand: retrovirus oncogenes probably originated from the genomes of uninfected vertebrate cells. We will defend this conclusion by reviewing our evidence that the oncogenes of several avian retroviruses are derived from physiologically active loci in the genome of normal chickens. We suggest that the evolutionary lineage of oncogenes presages their mechanisms of action and the restriction of their effects to specific tissues. Retrovirus oncogenes may provide valuable experimental analogues for genetic loci that regulate the normal growth and development of eukaryotic cells.

[1]Supported by grants from the National Cancer Institute and the American Cancer Society.

ISBN 0-12-633360-2

B. Strategy

1. Prototypes of Avian Retrovirus Oncogenes. Table 1 describes the suite of avian retroviruses that we have under study. Each member of the suite transforms a specific type (or types) of cell - both in culture and in infected animals; in each instance, transformation has been attributed to a locus ("oncogene", or *onc*) in the viral genome. The best characterized of these oncogenes is the gene *src*, which is responsible for neoplastic transformation of fibroblasts by avian sarcoma virus (ASV). The protein encoded by this gene has been identified (2,3), progress has been made in the study of its mechanism of action (4,5), and its role in transformation has been rigorously established by genetic analysis (6). The remainder of the loci (onc_{MCV}, onc_{AEV} and onc_{AMV}) have been mapped to approximate locations in the viral genomes (Figure 1), but we presently have only circumstantial evidence that these loci mediate neoplastic transformation, and we know nothing of their function.

TABLE 1. Some Avian Retroviruses.

Virus	Pathogenicity	Target in Culture	Oncogene
Avian sarcoma virus (ASV)	sarcoma	fibroblast	yes
Myelocytomatosis virus (MCV)	carcinoma myeloid leukemia sarcoma	epithelial cell macrophage fibroblast	probable
Avian erythroblastosis virus (AEV)	erythroleukemia sarcoma	erythroblast fibroblast	probable
Avian myeloblastosis virus (AMV)	myeloblastic leukemia	myeloblast	probable
Avian leukosis virus (ALV)	B cell lymphoma	?	uncertain

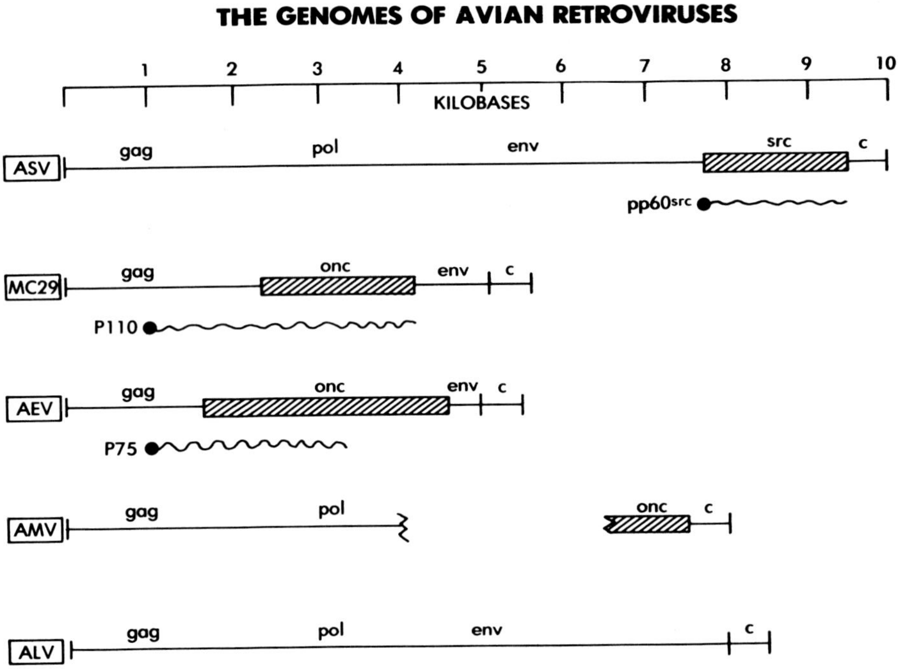

FIGURE 1. The genomes of some avian retroviruses.

Virus strains are denoted by conventional abbreviations: ASV, avian sarcoma virus; MCV, avian myelocytomatosis virus (#29); AEV, avian erythroblastosis virus; AMV, avian myeloblastosis virus; and ALV, avian leukosis virus. Replicative genes are defined as follows: gag, structural proteins of the virion interior; pol, reverse transcriptase; env, glycoprotein of the viral envelope. The region denoted C is an untranslated , domain shared by many avian retroviruses. Cross-hatched bars denote the position and extent of alleged viral oncogenes; ALV is not known to have an oncogenic locus and is so represented. As illustrated, the mapping of AMV is presently incomplete; in particular, the full extent of onc has not been defined.

The advent of molecular cloning has greatly enhanced our ability to identify and characterize viral oncogenes. For example, we have isolated the genomes of ASV, MCV and AEV as chimeric DNAs in prokaryotic vectors. The cloned viral DNAs are biologically active and can therefore be used for site-specific mutagenesis in efforts to locate nucleotide sequences responsible for specific forms of oncogenesis. Proper identification of $\underline{onc}_{MCV}$ and $\underline{onc}_{AEV}$, and exploration of domains of function within these loci, are in the offing. In addition, molecular cloning has facilitated the preparation of specific reagents for use in molecular hybridization (see below).

ASV is genetically self-sufficient: the three viral genes (gag, pol and env) required for replication are intact and the virus therefore propagates without the assistance of a helper virus. By contrast, the other viral genomes illustrated in Figure 1 have all suffered deletions in one or more of the replicative genes and thus require complementing functions from another, related retrovirus for growth (7). The helper viruses that provide the essential functions characteristically contain all of the replicative genes but have no oncogene - an important consideration in the work that follows.

2. Molecular Probes of Genetic Composition. In former times, the origins of animal virus genes received only occasional and generally gratuitous consideration. Work on retroviruses has changed all that (8): their replicative genes have been found in the germ lines of numerous species, where they can evolve in concert with the remainder of the host genome; they are reputed to have moved from the germ line of one species to that of another during the course of evolution;

and their genetic transmission from one generation to the next within a species can establish neoplasia as an inheritable disease. These (and other) circumstances prompted two conjectures: i) that all retrovirus genomes are genetic pastiches assembled by permutation and rearrangement of cellular genes (9); and ii) that uninfected cells harbor viral oncogenes whose induction provides a final common pathway for various forms of oncogenesis (10). The first of these conjectures is not easily tested and remains moot. On the other hand, we found the second postulate ripe for test by the techniques of molecular hybridiztion. To that end, we prepared radioactive DNAs (cDNA) complementary to identified viral oncogenes and asked whether uninfected cells contain nucleic acid homologous to these cDNAs.

We evolved two strategies for preparing oncogene-specific cDNA. In the first, we exploited the existance of large deletions in *src*; selective hybridizations with RNA from these mutants permitted the purification of $cDNA_{sarc}$, which reacted only with viral genomes bearing the *src* gene of ASV (11). For want of suitable deletions affecting other retrovirus oncogenes, we turned to a second strategy for preparation of cDNAs representing the *onc* loci of MCV, AEV and AMV. We assumed that the genomes of these viruses were homologous with the genomes of their respective helper viruses in all regards except for the presence or absence of *onc*. The assumption proved correct; selective hybridization with the RNAs of the helper viruses permitted the isolation of cDNAs specific for the various forms on *onc* (12,13). These strategies served us well, but they have now been superseded by our ability to clone specific regions of any of the viral genomes in prokaryotic vectors. The purity and specificity of cDNAs prepared from these clones is irrefutable and has enhanced the rigor of our analyses.

C. Experimental Findings and Conclusions

1. *Detection of Oncogene Progenitors in Vertebrate DNA.* Using specific cDNAs, we first found that uninfected vertebrate cells contain DNA homologous to *src* of ASV (14,15); similar findings were later reported by ourselves and others for the *onc* loci of MCV, AEV and AMV (12,13). As summarized in Table 2, each form of *onc* was most closely related to its homologous DNA in chickens; the related DNA in other species diverged in accord with phylogenetic distance from chickens. We drew three conclusions from these data. First, avian and mammalian genomes apparently contain loci homologous to all identified forms of avian retrovirus oncogenes. Second, these

TABLE 2. Homologues of Avian Retrovirus Oncogenes in Vertebrate Genomes.

Species	Myrs from chicken	Cellular homologue of oncogene				Divergence of homologue from onc	
		ASV	MCV	AEV	AMV	ASV	MCV
Human	325	+	+	+	+	13%	nt
Mouse	250	+	+	+	+	13%	nt
Cow	240	+	+	+	+	13%	nt
Duck	80	+	+	+	+	10%	4%
Chicken	0	+	+	+	+	4%	0%
Pheasant	10	+	+	+	+	7%	nt
Quail	20	+	+	+	+	nt	nt
Emu	80	+	nt	nt	nt	11%	nt
Rhea	100	nt	+	nt	nt	nt	5%
Salmon	225	+	+	+	+	15%	nt
Sea Urchin	450	-	nt	nt	nt	nt	nt
Drosophila	>600	-	-	nt	nt	-	-
E. coli	>600	-	-	-	-	-	-

The table summarizes results from the authors and from Roussel *et al.* (13). nt denotes "not tested".

loci probably took form with the emergence of vertebrates and then evolved during the course of speciation. Third, the loci have been conserved during the course of evolution, as if they might serve a vital function. These conclusions in turn raised several issues. First, what is the evolutionary origin of the cellular homologues of viral oncogenes? Are they genes of endogenous retroviruses, or are they normal loci in the vertebrate genome? Second, do these homologues function in the normal cell, and, if so, how? Third, do the homologues of different oncogenes belong to an interrelated battery of cellular genes, and if so, is this interrelationship manifested by either genetic linkage or coordinate function? Fourth, did the cellular homologues give rise to viral oncogenes, and if so, what bearing does this have on the mechanism(s) of viral oncogenesis.

2. Vertebrate Homologues of Retrovirus Oncogenes: Cellular Genes or Viral Genes? Our first clue came from studies with a cDNA specific for the env gene of ASV (16). We found appreciable homology with env only in chicken DNA (Table 3), which has long been known to harbor replicative genes of one or more avian retroviruses (17). It appeared that endogenous retroviruses probably entered the germ line of chickens subsequent to speciation, a conclusion that has since been confirmed by persuasive further findings (18,19). More importantly, we were able to deduce that the replicative genes of endogenous retroviruses in chickens and the vertebrate homologues of viral oncogenes probably arose and evolved independently.

How might we distinguish viral from cellular loci in the chicken genome? Three criteria have emerged from recent work. First, retrovirus genes have inevitably been found in DNA proviruses whose distinctive structure serves to signal the viral nature of the genetic unit (20). To the best of our knowledge, cellular homologues of oncogenes are not contained within such units. Second, restriction endonucleases can be used to enumerate and characterize genetic loci in eukaryotic DNA. With this approach, we and others found that both the number and the apparent chromosomal location of the proviruses of endogenous retroviruses varied from one chicken to another, even within partially inbred lines (18). By contrast, loci harboring cellular genes such as globin, ovomucoid, ovalbumin and transferrin, and the homologues for the various retrovirus oncogenes, were situated at constant sites within the chicken genome; all of the variation that we observed could be attributed to genetic polymorphism in and adjacent to the loci (18,22). Third, we were able to

TABLE 3. Phylogenetic Distribution of an Oncogene Homologue and an Endogenous Retrovirus Gene.

DNA	src	env	Phylogenetic distance (years x 10^{-6})
Chicken	+	+	0
Ring necked pheasant	+	-	20
Japanese quail	+	-	20-40
Duck	+	-	80

The table summarizes data reported elsewhere (15,16).

test for physical linkage of viral and cellular genes by analysing the DNA of chicken chromosomes that had been fractionated by rate-zonal centrifugation. By this means, we found that the homologues for the oncogenes of ASV, MCV and AEV (AMV has yet to be tested) were located on separate chromosomes, and that none of these loci were linked to replicative genes of endogenous retroviruses (21,22). Moreover, all of the homologues (ASV, MCV, AEV and AMV) are present in the DNA of chickens whose genomes contain no identifiable endogenous retrovirus (unpublished data of authors).

We conclude that the homologues of avian retrovirus oncogenes are disposed on the chicken genome in the manner of cellular rather than endogenous viral genes. Several of the oncogene homologues have now been isolated from chicken DNA by molecular cloning in prokaryotic vectors; analysis of these isolates should properly test the view that the homologues are cellular, not viral, genes.

3. Expression of Oncogene Homologues in Uninfected Cells. The cellular homologues for avian retrovirus oncogenes have been conserved throughout vertebrate evolution, as if each might serve an essential purpose. In accord with this suggestion, we have found that all of the homologues are apparently expressed in uninfected cells. The evidence is most advanced for the homologue of *src*; cells from numerous vertebrate species contain a 60,000 dalton phosphoprotein ($pp60^{proto\text{-}src}$) whose structure and apparent function are very similar to those of the viral gene product, $pp60^{src}$ (23). For want of suitable immunological reagents, we have been unable to search for normal cellular proteins related to the transforming proteins of other avian retroviruses. We have demonstrated, however, that uninfected cells contain cytoplasmic RNA transcribed from the oncogene homologues of MCV and AEV (Table 3). The homologue of onc_{AEV} gives rise to multiple forms of RNA whose distribution varies among different tissues (Table 4), whereas the forms of transcripts derived from the homologues of *src* and onc_{MCV} are alike in all cells (and all species) examined to date.

We have found no apparent variation in the expression of oncogene homologues from one chicken to another; by contrast, expression of endogenous viral genes is covariant with the genotype of individual chickens (17,24). These findings conform to our previous conclusion that oncogene homologues and endogenous retrovirus genes are independent and unrelated genetic loci. We know much less of possible functional relationships among the oncogene homologues themselves. Although they are not physically linked in the chicken genome (see above), it remains possible that they are members of a

TABLE 4. Transcription from Vertebrate Homologues of Avian Retrovirus Oncogenes.

	RNA size in kilobases						
	ASV	MCV	AEV				
Chick embryos	4.5	2.8	12	8.7	6.0	4.8	3.4
Chick fibroblasts	4.5	2.8	12	8.7	6.0		
Chick bone marrow	4.5	2.8				4.8	3.4
Chick yolk sac	4.5	2.8				4.8	3.4
Duck embryo	4.5	2.8	11.5			5.0	3.2
Quail embryo	4.5	2.8	11.5			4.3	
Mouse fibroblasts	4.5	2.8					2.7
Rat fibroblasts	4.5	2.8					2.6

Polyadenylated RNA was fractionated by electrophoresis under denaturing conditions, immobilized on chemically-activated paper, and then analysed by molecular hybridization (29). The table summarizes the sizes (in kilobases) of RNAs detected in each sample with cDNAs representing the oncogenes of ASV, MCV and AEV.

battery of genes whose expression and function are interrelated. For example, we have shown previously that related genes - such as α- and β-globin, and genes whose functions are coordinately induced by estrogen - are situated on different chromosomes in the chicken (21).

D. The Genesis of Retrovirus Oncogenes

Our search for viral oncogenes in normal cells has identified, instead, a set of normal cellular genes that emerged with early vertebrates, evolved during subsequent speciation, and eventually gave rise to the oncogenes of avian retroviruses. Several facets of this argument deserve individual attention.

i) Why do we identify the cellular loci as progenitors of the viral genes? The kinship between cellular and viral loci seems irrefutable. But which are parents, which are progeny? We take our answer from the phylogenetic distribution of the contenders: the cellular loci (and presumed progenitors) are found throughout vertebrates, the viral loci (and presumed progeny) only in chicken retroviruses; moreover, each of the viral oncogenes is most closely related to its homologue in the genome of the chicken (Table 3) - the species from which the viruses apparently originated.

ii) By what means were cellular loci assimilated into viral genes? We know little of the mechanisms by which this might have occurred, although it is generally assumed that retroviruses acquired the cellular genes by recombination. This assumption has been given substance by recent experiments in which recombination between a deletion mutant of ASV and the src homologue in the chicken genome reconstituted a viral oncogene virtually indistinguishable from wild type src (25). Recombination among retroviruses is frequent; occurs principally, if not exclusively, between homologous elements; and may employ the RNA form of the viral genome (26). Together, these considerations confound all present efforts to explain the capture of an oncogene homologue by a retrovirus genome; in particular, the homologues share no apparent homology with the helper virus genome into which they are alleged to enter (see Table 5), nor are they linked to DNA with suitable homology.

TABLE 5. RNA Carrying the Vertebrate Homologue of src is not Homologous with Other Portions of Avian Retrovirus Genomes.

Proto-src: ca. 1.6 kb
Remainder: ca. 3.0 kb

Hybridization with	
$cDNA_{sarc}$:	+
$cDNA_{3'}$:	-
$cDNA_{gp}$:	-
$cDNA_{gag/pol}$:	-
$cDNA_{RAV-0}$:	-

Uninfected avian cells contain a 4.5 kilobase RNA that carries the transcript from the vertebrate homologue of ASV src. The table summarizes efforts to hybridize that RNA to cDNAs representing other portions of the ASV genome; $cDNA_{3'}$ and $cDNA_{5'}$ represent conserved domains of 500 and 100 nucleotides, respectively, at the ends of the viral RNA; $cDNA_{RAV-0}$ represents the genome of an endogenous retrovirus of chickens; the remaining cDNAs represent the indicated genes.

iii) Are parent and progeny alike? Viral oncogenes are pathogenic in the extreme. Might they nevertheless be identical with their cellular progenitors, or have they been altered as a consequence of their entry into viral genomes? We have few clues. The proteins encoded by ASV src and its cellular homologue are extraordinarily similar (23), but a comprehensive study of their respective functions has not been accomplished. Otherwise, we know nothing of the comparative structure and function of avian retrovirus transforming proteins and their presumed cellular homologues.

E. Implications

Expression of viral oncogenes perturbs the control of cellular growth and/or development. We suggest that the cellular progenitors of these genes are regulatory loci active in specific tissues - accounting on the one hand for the effect of viral oncogenes on cellular phenotype, and on the other hand for the tissue-specificity of viral oncogenesis. This suggestion may have implications for the mechanisms of oncogenesis by both viruses themselves and other, possibly more prevalent means of tumorigenesis.

i) If viral oncogenes and their cellular progenitors have identical functions, transformation by retroviruses is likely to be a consequence of dosage: perhaps the viruses merely overload cells with normal gene products. Alternatively, viral oncogenes and their progenitors may differ in subtle but important ways that render the activity of the viral gene product anomalous. In either event, it appears likely that oncogenic retroviruses have excerpted viral regulatory functions from the vast pool of cellular genes, making these elements readily accessible for study.

ii) The cellular progenitors of viral oncogenes may be equivalent to "cancer genes" - either the genetic loci at fault in human pedigrees afflicted with a high genetic risk of cancer, or their alleged counterparts, the genes whose mutation in somatic cells leads to spontaneous or chemically induced neoplasia (27). Both versions of "cancer genes" probably number between a few dozen and a few hundred (27, 28). Distinct forms of retrovirus oncogenes could well be as numerous; perhaps their cellular progenitors are "cancer genes" in another guise (1).

REFERENCES

1. Bishop, J.M., in "The Molecular Basis of Microbial Pathogenicity" (H. Smith, J.J. Skehel and M.J. Turner, eds.) Dahlem Konferenzen. Weinheim/Deerfield Beach, FL/Basel: Verlag Chemie, in press (1980).
2. Brugge, J.S., and Erikson, R.L., Nature 269, 346 (1977).
3. Purchio, A.F., Erikson, E., Brugge, J., and Erikson, R., Proc. Nat. Acad. Sci. USA 75, 1567 (1978).
4. Collett, M.S., and Erikson, R.L., Proc. Nat. Acad. Sci. USA 75, 2021 (1978).
5. Levinson, A.D., Oppermann, H., Levintow, L., Varmus, H.E., and Bishop, J.M., Cell 15, 561 (1978).
6. Vogt, P.K., in "Comprehensive Virology" (H. Fraenkel-Conrat and R. Wagner, eds.) Vol. 9, p. 341. Plenum Press, (1977).
7. Graf, T., and Beug, H., Biochim. Biophys. Acta 516, 269 (1978).
8. Bishop, J.M., Ann. Rev. Biochem. 47, 35 (1978).
9. Temin, H.M., Ann. Rev. Genetics 8, 155 (1974).
10. Todaro, G.J., and Huebner, R.J., Proc. Nat. Acad. Sci. USA 69, 1009 (1972).
11. Stehelin, D., Guntaka, R.V., Varmus, H.E., and Bishop, J.M. , J. Mol. Biol. 101, 349 (1976).
12. Sheiness, D., and Bishop, J.M., J. Virol. 31, 514 (1979).
13. Roussel, M., Saule, S., Lagron, C., Rommens, C., Beug, H., Graf, T., and Stehelin, D., Nature 281, 452 (1979).
14. Stehelin, D., Varmus, H.E., Bishop, J.M., and Vogt, P.K., Nature 260, 170 (1976).
15. Spector, D., Varmus, H.E., and Bishop, J.M., Proc. Nat. Acad. Sci. USA 75, 4102 (1978).
16. Fujita, D., Tal, J., Varmus, H.E., and Bishop, J.M., J. Virol. 27, 465 (1978).
17. Robinson, H.L., Current Topics Microbiol. and Immunol. 83, 1 (1978).
18. Hughes, S.H., Payvar, F., Spector, D., Schimke, R.T., Robinson, H.L., Bishop, J.M., and Varmus, H.E., Cell 18, 347 (1979).
19. Astrin, S.M., Buss, E.G., and Hayward, W.S., Nature 282, 339 (1979).
20. Taylor, J.M., Current Topics Microbiol. and Immunol. 87, 23 (1979).
21. Hughes, S.H., Stubblefield, E., Payvar, F., Engel, J.D., Dodgson, J.B., Spector, D., Cordell, B., Schimke, R.T., and Varmus, H.E., Proc. Nat. Acad. Sci. USA 76, 1348 (1979).

22. Sheiness, D.K., Hughes, S.H., Varmus, H.E., Stubblefield E., and Bishop, J.M., submitted to Virology (1980).
23. Oppermann, H., Levinson, A.D., Varmus, H.E., Levintow, L., Bishop, J.M., Proc. Nat. Acad. Sci. USA 76, 1804 (1979).
24. Hayward, W.S., Braverman, S.B., and Astrin, S.M., Cold Spring Harbor Symp. Quant. Biol. 44, in press (1979).
25. Karess, R.E., Hayward, W.S., and Hanafusa, H., Proc. Nat. Acad. Sci. USA 76, 3154 (1979).
26. Hunter, E., Current Topics Microbiol. and Immunol. 79, 295 (1978).
27. Knudson, A.G., N. Engl. J. Med. 301, 606 (1979).
28. Waldren, C., Jones, C., and Puck, T.T., Proc. Nat. Acad. Sci. USA 76, 1358 (1979).
29. Alwine, J.C., Kemp, D.J., and Stark, G.R., Proc. Nat. Acad. Sci. USA 74, 5350 (1977).

DISCUSSION

G. BERNARDI: Any questions? Well, if not, maybe I'll make comments, and that is that in some cases with exogenous viruses, the case of bovine leukemia virus, the situation is different in the sense that in a given infected animal you can find just a single integration site, though the situation may be different in different systems.

S.H. HUGHES: There are certainly complexities both in that system and the system with baboon endogenous virus integration sites which have not been resolved.

INITIATION OF MOLECULAR CLONING OF A TRANSFORMING ALLELE OF CHEMICALLY TRANSFORMED MOUSE CELLS[1]

Ben-Zion Shilo[2]
Chiaho Shih
Marnin Merrick
Ann Dannenberg
Mitchell Goldfarb
Robert Weinberg

Center for Cancer Research and Department of Biology
Massachusetts Institute of Technology
Cambridge, Massachusetts

I. INTRODUCTION

Among the lesions generated in a cell by chemical carcinogens, those which affect DNA structure have been thought to be of central importance in the subsequent development of a transformed, tumorigenic phenotype. This conclusion derives perhaps most directly from the work of Ames and others, in which carcinogenic potency of a large variety of organic chemicals has been shown to demonstrate a strong correlation with their respective mutagenic potencies (1-4). Having hypothesized a mutational origin of chemically-induced transformation, one is hard-pressed to further define the specific target genes involved and the phenotypic consequences which derive from their alteration.

The experiments reported here were designed to prove directly that the transformed phenotype of cells exposed to

[1]Supported by the Rita Allen Foundation, of which R.A.W. is a Fellow; by National Institutes of Health Grant Ca17537, and by Core Grant CA14051 to S. Luria.

[2]Is a Fellow of the Chaim Weizmann Postdoctoral Fellowships.

ISBN 0-12-633360-2

chemical carcinogens is encoded in a discrete DNA segment. Furthermcre, other work to be discussed here leads towards the molecular cloning of these transforming alleles from the descendants of cells transformed *in vitro* by carcinogens.

This work originated earlier in studies on the DNA genomes of murine retroviruses. These murine leukemia and sarcoma genomes are synthesized via reverse transcription in the cytoplasm of cells shortly after adsorption of a virus. Unintegrated linear and closed circular DNAs are extractable from infected cultures shortly after infection while viral genomes covalently integrated into the cellular genome are prepared from chronically infected cultures long after initial infection.

Those various forms of the viral DNA and their avian analogues have been characterized via a series of molecular biological techniques. Additionally, these DNAs were seen to be infectious as naked DNA molecules when applied to monolayer cultures using the technique of Graham and van der Eb (5). This technique allows a relatively efficient introduction of DNA into mammalian cells by applying the DNA to the cells in the form of a co-precipitate of DNA and calcium phosphate. Using this technique, we found that both unintegrated and integrated murine retrovirus DNAs showed a relatively high specific infectivity - as much as 10^6 biological units per microgram of viral DNA applied (6, 7).

These transfections included the transfer of DNA from cells containing a single-copy of a murine sarcoma provirus integrated per donor cell chromosome complement. In this case the sarcoma provirus was diluted with a million fold excess of cellular DNA sequences. Foci of transformed cells were readily observable 10-12 days after the application of donor DNA to the recipient mouse cell monolayer.

We reasoned that if single-copy viral transforming sequences could be readily transferred from cell to cell, then analogous experiments could be attempted with non-viral transforming alleles. Experiments were initiated attempting to passage the transformed phenotype from chemically transformed mouse fibroblasts to recipient NIH3T3 mouse fibroblast monolayer cultures.

II. RESULTS

The results of initial attempts at passaging the transformed phenotype have been reported in part previously (8). DNAs from 15 lines of mouse and rat cells were screened. These DNAs derived largely from *in vitro* transformed mouse fibroblast lines; a minority derived from chemically induced mouse and rat fibrosarcomas. Five of these 15 cell lines yielded DNA which induced significant numbers of foci upon application to recipient NIH3T3 monolayers (8).

The scoring of these foci was complicated by the tendency of the monolayer to yield small numbers of spontaneous overgrowths which are usually but not always distinguishable from genuine foci induced via transfection. To neutralize the effects of these spontaneous foci and the occasionally subjective criteria required to distinguish them from those induced by transfected DNA, we evaluated transfected cultures via double-blind protocols, in which the identity of a control or experimental culture dish was not known to the experimenter counting these foci.

Among these efficient donors of transforming DNA was a methylcholanthrene transformed C3H 10T1/2 cell line (MCA16) provided us by Drs. Rapp and Heidelberger. The DNA of this cell line induced foci with an efficiency of as much as 0.3 foci per microgram of transfected DNA. DNA extracted from these foci could serve in turn as donor DNA in a second cycle of transfection yielding foci once more with equivalent or even higher potency.

This DNA was treated with several restriction enzymes prior to transfection. Subsequent biological assays of the DNA revealed that SmaI, KpnI and EcoRI all inactivated the transforming potency of the transfected DNA while BamHI had no obvious effect. This suggested that the transforming allele was carried by one or several DNA segments which contained no BamHI sites in essential sequences. We consider it likely that the entire transforming allele is carried intact in one BamHI-derived fragment although proof of this will depend on successful cloning of the biologically active gene.

The results of these restriction enzyme cleavages made possible attempts at the molecular cloning of the entire segment containing the transforming allele. A variety of molecular cloning strategies were in principle possible in this endeavor. We chose an approach which had been developed for the cloning of yeast genes by Hicks, Fink and their colleagues (9). DNA extracted from the MCA16-5 donor cell culture was treated with BamHI endonuclease. In parallel, an equal mass of plasmid pBR322 DNA was treated with this enzyme. The two DNA preparations were mixed and the mixture of DNA fragments linked together by DNA ligase. The million different cellular BamHI fragments were thus joined randomly via their ends to linearized pBR322 genomes. Since the plasmid molecules were present in about 2-fold molar excess, each cellular DNA fragment was likely linked directly to at least one pBR322 segment.

This DNA was diluted with a 10 fold excess of carrier DNA from untransformed NIH3T3 cells, and transfected onto a recipient monolayer. Among the foci which derived from this transfection, three were grown up to mass cultures. Their DNA was prepared and analyzed by the Southern gel-filter transfer technique ("blotting").

The three DNAs whose analysis is shown in the three left most channels of Figure 1 were treated with endonuclease XbaI prior to their resolution by the Southern technique. Filters containing the adsorbed DNA fragments were incubated with radioactive pBR322 specific probe. The enzyme used to cleave the DNAs prior to electrophoresis (XbaI) does not cleave within the pBR322 sequence. Therefore, each of the bands visualized in these three channels of Figure 1 likely contains one or more pBR322 genomes linked to cellular DNA segments.

The existence of multiple pBR322 segments in these three transfected clones proves directly that the progenitors of each of these clones acquired DNA as a consequence of the transfection procedure. This biochemical demonstration of the transfection process follows analogous work done by others previously using the cloned thymidine kinase segment of herpes simplex virus as the marker of interest (10).

The large number of pBR322 segments acquired by each of these cells reflects a similar amount of donor DNA also present in each of these clones as a consequence of the transfection. Among these cell DNA segments is one containing the transformation allele whose presence has been selected during the derivation of these foci. In order to reduce the multiplicity of pBR322-linked segments present in a transfected cell, the DNAs of several of these foci were prepared and mechanically sheared prior to use in a second serial cycle of transfection.

Foci which derived from this second cycle of transfection have acquired the transformation allele once again, but in this case the allele has no longer been co-transferred together with a large number of pBR322 segments. Rather, only those pBR322 segments closely linked to the selected marker have survived this further cycle of transfection. The DNAs of the descendants of secondary transfectants are also analyzed in Figure 1. Two bands of hybridization are evident upon analysis of Kpn-cleaved DNA of a secondary transfectant (channel h). These two bands represent one virtually complete pBR322 genome, and one fragment of pBR322 whose incompleteness is evidenced by the small sized fragment seen after EcoRI cleavage of this DNA preparation (right most channel, channel i). These two plamids are in fact linked together via cell DNA sequences in one large DNA array as shown by the presence

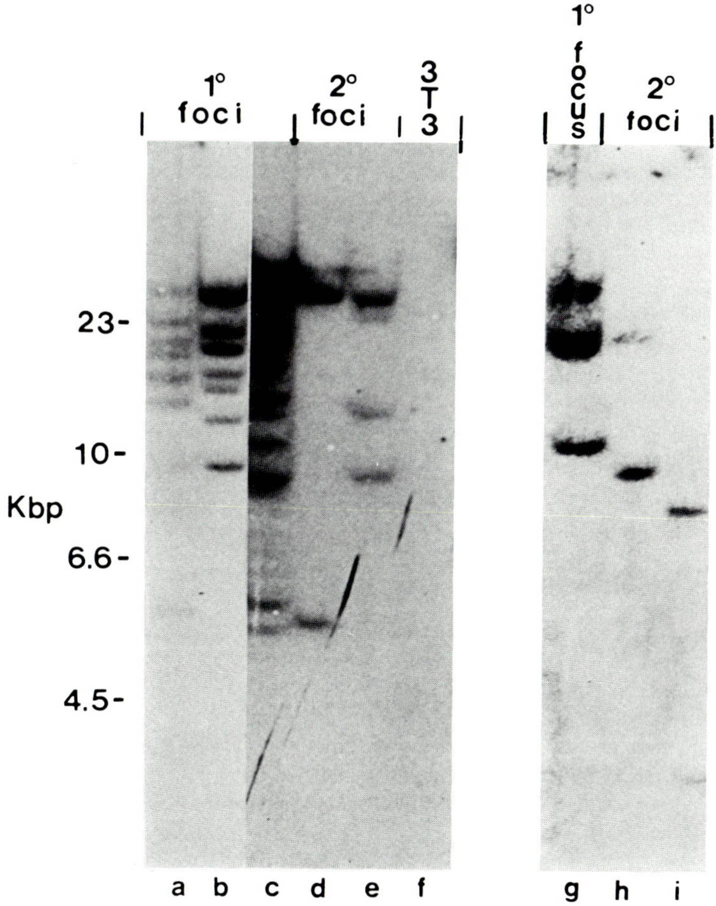

FIGURE 1. DNA derived originally from a methyl-cholanthrene transformed C3H 10T1/2 cell (MCA16, ref. 8) was cleaved with endonuclease BamHI, ligated to an equal mass of BamHI cleaved pBR322 and transfected into NIH3T3 cells. DNAs of three such foci were cleaved by Xba (channels a-c) or Kpn (channel f) prior to analysis by the Southern procedure using nick translated pB 322 as radioactive sequence probe. Secondary foci derived from DNA of channel c yielded DNAs analyzed after Xba treatment in channels d and e. Kpn cleaved DNA of primary focus (channel g) was used to generate a secondary focus whose Kpn-cleaved DNA is analyzed in channel h and whose EcoRI-cleaved DNA is analyzed in channel i. Analysis of NIH3T3 DNA is shown in channel f.

of only one large pBR322 containing DNA fragment seen after XbaI endonuclease cleavage of this DNA (data not shown).

The transformation allele likely resides adjacent to one of the pBR322 segments. Cleavage of this DNA by KpnI and treatment with DNA ligase allows the circularization of the KpnI derived fragments. These circularized molecules have been introduced into bacteria with the hope that one or both of the pBR322 segments would be functional in driving the replication of a chimeric plasmid in the bacteria. By selecting for the ampicillin marker which should remain intact in the pBR322 segments, we would hope to derive two chimeric plasmids whose size corresponds precisely to the size predicted by the two bands of hybridization predicted by Figure 1. One of these chimeric plasmids or a closely related derivative should contain most or all of the DNA sequences of the cellular transforming allele.

The success of this plasmid-driven amplification of transformation sequences is not assured. We intend to follow an alternative cloning strategy as well. Using the lambdaphage vector Charon 4A, we shall make a library of EcoRI-generated fragments derived from the DNA of the transfectant carrying the two pBR322 plasmids. By screening the components of this library with a radioactive pBR322 sequence probe, we intend to select the lambdaphage containing pBR322 together with linked cellular sequences.

Possession of such cloned transforming sequences will be of great utility and will allow the analysis of two questions of great interest in this biological system. First, what are the molecular differences between this transformation allele and its normal cellular allelic counterpart; second, whether the same or different cellular genes have been activated in the independent derivation of a series of other chemically transformed mouse cells.

REFERENCES

1. McCann, J., Choi, E., Yamasaki, E. and Ames, B.N., Proc. Natl. Acad. Sci. USA 72, 5135 (1975).
2. McCann, J., and Ames, B.N., Proc. Natl. Acad. Sci. USA 73, 950 (1976).
3. Bridges, B.A., Nature (London) 261, 195 (1976).
4. Bouck, N. and diMayorca, G., Nature (London) 264, 722 (1976).
5. Graham, F.L. and van der Eb, A.J., Virology 52, 456 (1973).

6. Smotkin, D., Gianni, A.M., Rozenblat, S. and Weinberg, R.A., Proc. Natl. Acad. Sci. USA 72, 4910 (1975).
7. Goldfarb, M.P. and Weinberg, R.A., J. Virology 32, 30 (1979).
8. Shih, C., Shilo, B., Goldfarb, M.P., Dannenberg, A. and Weinberg, R.A., Proc. Natl. Acad. Sci. USA 76, 5714 (1979).
9. Hicks, J.B., Hinnen, A. and Fink, G.R., Cold Spring Harbor Symp. Quant. Biol. 43, 1305 (1979).
10. Wigler, M., Sweet, R., Sim, G.K., Wold, B., Pellicer, A., Lacy, E., Maniatis, T., Silverstein, S. and Axel, R., Cell 16, 777 (1979).

DISCUSSION

R.S. ATHWAL: In a number of studies on chemically-transformed cells transformation phenotype is known to express as a recessive character. In view of this information how do you explain the expression of a recessive gene following DNA transfection when there is still a dominant gene present in the cell?

R.A. WEINBERG: The questioner alluded to a series of studies that suggest that transformed allele in transformed cells is genetically recessive. There are an equal number of studies which indicate it is dominant. I would say these experiments which we have done are clearly only compatible with its being a dominent allele in the genetic background of the NIH3T3 cells into which the DNA is introduced. To the extent this is generalizable (and I do not know that it is) then I would presume that the conclusions of recessivity are incorrect. But I emphasize that what we have done here is drawn some preliminary conclusions about a biological system which relates only to the genetic background of these mouse cells *in vitro* and may not in fact apply to all kinds of transformation in all cells in cultures and *in vivo*.

R.S. ATHWAL: My second question concerns the fact that in your experiments the transformation does not always occur with the same frequency. Could it be possible that the transformation is still recessive? What you might be getting is a replacement of a dominant gene in the cells, rather than integration of a second copy of the gene. Since integration into the homologous sites is not a very frequent phenomenon the frequency of obtaining transformed cells would be variable as well.

R.A. WEINBERG: The question was whether the transfection resulted in the replacement of a resident gene as a consequence of transfection. The mechanism of transfection in mammalian cells is vastly different from those that pertain in yeast and therefore there is no reason to believe that the transfecting DNA has any effect on the genetic apparatus which pre-exists in the recipient. I do not believe that a pre-existing is, for instance, evicted by the transfecting DNA and I do not believe that what we have observed gives hope for a case of recessivity of the allele.

D. MERLO: To what do you ascribe the appearance of two foci in the mock-infected cells?

R.A. WEINBERG: The two foci that come from the mock infected cells are spontaneous transformations of the recipient monolayer. As I told you, the monolayer tends to overgrow on occasion after transfection and we have occasionally seen one or two foci even upon transfer of *E. coli* DNA. Therefore the background level of one or two foci is something upon which one cannot build any experiments. Because of these background foci we insisted on doing these experiments all by double blind protocols to make sure that in the experimental samples there was a much higher level of focal transformation being transferred than in the control DNA samples. Cells will transform at a spontaneous rate *in vitro* if kept long enough in culture. They will look slightly different from those transformed via transfection.

D. GILLESPIE: Have you taken the sequence probe and examined the donor DNA for any rearrangements?

R.A. WEINBERG: No we have not done that and we are holding those kinds of experiments in abeyance until we have convincing proof that we have cloned what we would have liked to clone.

HUGHES: I would have thought naively that the act of transfecting even normal DNA might induce the kinds of reassortments or rearrangements necessary to at least rarely cause transformation. Have you pursued that and reached some conclusion?

R.A. WEINBERG: Well in fact the act of transfection may or may not represent a mild trauma for the cell. It does not seem to be an act which itself is capable of inducing transformation and one gets a small number of foci in control cultures which were just alluded to before, but it is not obvious that those foci arose as a consequence of spontaneous transformance of the culture which would have come up anyhow, or whether those small number of foci arose as a consequence of the introduction of foreign DNA into those cells. But clearly I am attempting to base our case, not on the difference between 5 foci and 2 foci, but on the difference between 1 or 2 or 0 foci coming from control cultures and 50 or 80 foci coming from experimental cultures. These small number of foci are potentially very interesting because they may represent in fact analogs of the chemical carcinogenesis events which are occurring spontaneously *in vitro*, but we are withholding our interest from them for the time being.

J.B. FLANEGAN: Do you know if the RNA from your different cell lines that is capable of causing transformation shows a different pattern of resistance to different restriction enzymes or do you see the same pattern of resistance?

R.A. WEINBERG: Dr. Flanegan asks whether DNA from the different cell lines is resistant to different restriction enzymes. We are just doing that now. Obviously one question would be whether the five different donor lines which we have played with and which are independently transformed, whether each of them has the same gene altered or whether each of the five has a different gene altered. That is, in principle, a question which is now unambiguously answerable by either the restriction enzyme analysis which was just proposed or more directly by cloning the allele of one cell line and asking whether when used as a hybridization probe, it will detect the transfer of the homologous allele from the other donors as well. We do not have an answer to that now but that is an exciting question to answer because it will address the question whether there are one or a multiplicity of targets in mouse fibroblasts which can serve as targets for the chemical carcinogen.

D. LEVY: How extensively have you characterized the phenotype of the donor and recipient cells in terms of their tumorgenicity and that sort of thing?

R.A. WEINBERG: The bulk of the tumor cells we have used focally overgrow, are refractile, grow in agar, and are tumorigenic. The vast majority of the recipient cells also focally overgrow, are tumorigenic and have similar phenotype. There is one exception to this, which is an interesting one. There is one donor which will form foci in monolayer but these cells will not grow in agar. When we transfect its DNA, interestingly enough, the recipient foci obviously form foci in monolayer but also will not grow in agar. So in that case we appear to have transferred one of the several transform phenotypes that is the trait of focal overgrowth in monolayer and not transferred the agar growth property which was not observed in the donor, nor is it observed in the recipient foci. That really is that only exception to the rule that the transformed cells will all grow in agar and all be tumorigenic in newborn mice.

CAPPED EUKARYOTIC MESSENGER RNAs ARE PRIMERS FOR INFLUENZA VIRAL RNA TRANSCRIPTION[1]

Robert M. Krug
Michele Bouloy[2]
Stephen J. Plotch

Memorial Sloan-Kettering Cancer Center
New York, New York

I. INTRODUCTION

One feature distinguishing influenza virus from other nononcogenic RNA viruses is that the functioning of the host nuclear RNA polymerase II is required for virus replication, specifically for viral RNA transcription (1-3). Our recent studies of viral RNA transcription have provided an explanation for this requirement for RNA polymerase II. We have shown that capped eukaryotic mRNAs strongly stimulate viral RNA transcription *in vitro* and donate their 5' terminal methylated cap and a short stretch of internal nucleotides to the 5' end of the viral RNA transcripts (4-6). A similar mechanism appears to operate *in vivo*, because we have found that the viral messenger RNAs (mRNAs) synthesized in the infected cell contain a short stretch of nucleotides at their 5' end, including the cap, that are not viral-coded (7). The need for newly synthesized host mRNA primers to stimulate viral RNA transcription would explain the requirement for the proper functioning of host RNA polymerase II in the infected cell. In addition, our results may provide an insight into cellular functions; in particular, they suggest that the fully methylated cap (cap 1) structure, m^7GpppX^m, which is found at

[1]Supported by NIH grants AI 11772 and CA 08748.
[2]Supported by NIH International Fellowship TW 02590

ISBN 0-12-633360-2

the 5' end of all mammalian cellular mRNAs (8), may have an important role in cellular processes other than protein synthesis. Here we will summarize some of the results of these recent studies.

II. RESULTS AND DISCUSSION

We had proposed a few years ago (9-11) that viral RNA transcription *in vivo* requires initiation by primer RNAs synthesized by RNA polymerase II and that the 5' terminal cap found on *in vivo* viral mRNA (12) is derived from these primer RNAs. These proposals were based on two sets of data: (i) the strong stimulation by a primer dinucleotide, ApG or GpG, of viral RNA transcription *in vitro* catalyzed by the virion-associated transcriptase (9,10,13); and (ii) the absence of detectable capping and methylating enzymes in virions (11).

Our later experiments essentially proved these proposals. We first identified primer RNAs in rabbit reticulocyte extracts, where they were shown to be globin mRNAs (4). We showed that β-globin mRNA, purified by polyacrylamide gel electrophoresis, stimulated viral RNA transcription about 80-fold and, on a molar basis, was about 2000 times more effective as a primer than ApG. Other capped eukaryotic mRNAs were also found to be extremely effective primers (4-6). The viral RNA transcripts primed by these eukaryotic mRNAs functioned as viral mRNAs in cell-free systems (4).

To determine whether the cap of the primer mRNA was physically transferred to the viral mRNA during *in vitro* transcription, we used as primer a mRNA containing radiolabel only in its cap (5). Globin mRNA containing ^{32}P only in its cap was prepared by enzymatically recapping β-eliminated globin mRNA in the presence of (α-^{32}P) GTP and S-adenosylmethionine (AdoMet). With this globin mRNA as primer in the presence of unlabeled nucleoside triphosphates, the resulting viral mRNA segments were shown to contain ^{32}P derived from the primer (Fig. 1A, lane 1). All of the ^{32}P in the viral mRNA segments was in cap structures, and none was in internal residues: DEAE-Sephadex chromatography of the RNase T2 digest of the viral mRNA segments showed that all of the ^{32}P eluted at a charge of -5 or -6 (Fig. 1B), where cap structures elute. A similar experiment was done with reovirus mRNAs containing ^{3}H label only in their caps (6). These mRNAs were synthesized *in vitro* with reovirus cores in the presence of (methyl-^{3}H) AdoMet. Again, the influenza viral mRNAs contained label (^{3}H) derived from the primer mRNAs, and this ^{3}H-label was shown to be in the cap structure m^{7}GpppG^{m}, identical to that

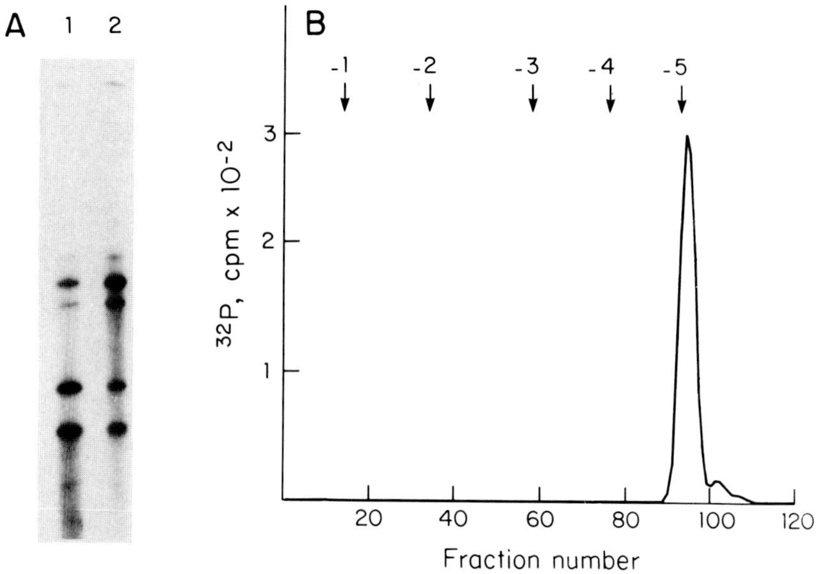

FIGURE 1. Transfer of the ^{32}P-labeled cap of globin mRNA to influenza viral mRNA during transcription *in vitro*. (A) Gel electrophoresis of deadenylated *in vitro* viral mRNA. The viral mRNA was synthesized in transcriptase reaction mixtures containing: lane 1, unlabeled nucleoside triphosphates and globin mRNA containing ^{32}P only in the cap; lane 2, (α-^{32}P) GTP as labeled precursor and unlabeled globin mRNA. (B) DEAE-Sephadex chromatography of the RNase T2 digest of the RNA eluted from the bands in lane 1. From reference 5.

found in the reovirus mRNA primers. Thus, the cap was indeed physically transferred to influenza viral mRNA from globin mRNA or reovirus mRNA primers.

To determine whether sequences in addition to the cap were transferred from a primer mRNA to influenza viral mRNA, we compared the size of the eukaryotic mRNA-primed viral mRNA segments to that of the ApG-primed viral mRNA segments. If the mRNA-primed segments contain additional sequences at their 5' ends, we would expect that these segments would be larger than those primed by ApG, as the latter segments initiate exactly at the 3' end of the virion RNA (vRNA) templates (14-17). First we compared the electrophoretic mobility of the globin mRNA-primed and ApG-primed viral mRNA segments after enzymatic deadenylation of the segments (5) (Fig. 2A). The globin mRNA-primed segments (lane 3) migrated slightly slower than the ApG-primed segments (lane 1). This difference in

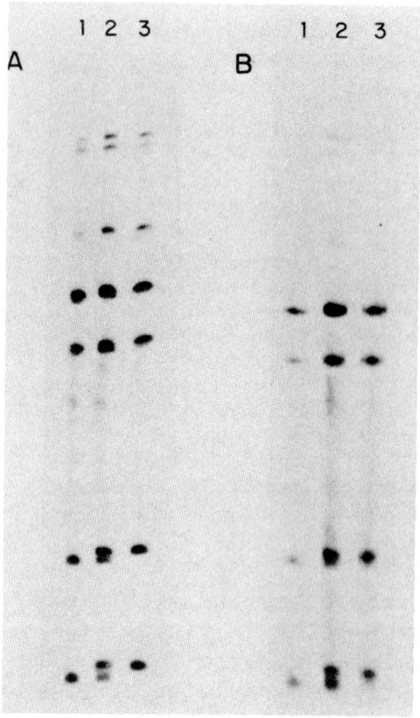

FIGURE 2. Gel electrophoresis of viral mRNA, after deadenylation, synthesized in reaction mixtures containing (α-^{32}P) GTP as labeled precursor and either ApG (lane 1) or globin mRNA (lane 3) as primers. Lane 2 contains a mixture of the viral mRNA of lanes 1 and 3. The deadenylylated viral mRNAs were electrophoresed directly (A) or after treatment with glyoxal (B). From reference 5.

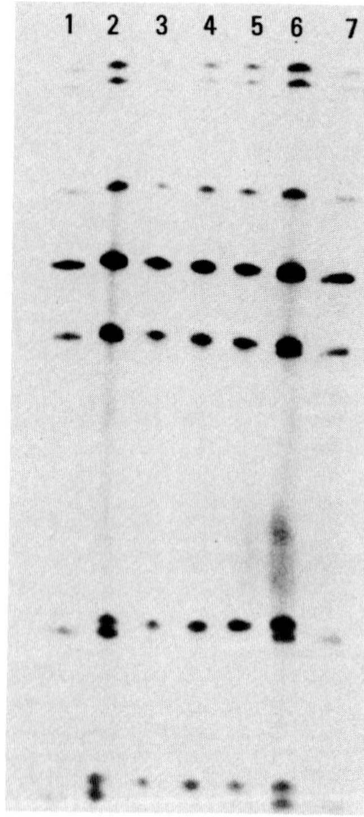

FIGURE 3. Comparison of the electrophoretic mobility of the influenza viral mRNAs primed by reovirus mRNA, globin mRNA, and ApG. The viral mRNAs, synthesized with (α-^{32}P) GTP as labeled precursor, were deadenylylated prior to electrophoresis. Lanes 1 and 7: ApG-primed viral mRNA. Lane 3: reovirus mRNA-primed influenza viral mRNA. Lane 5: globin mRNA-primed viral mRNA. Lane 2: mixture of ApG- and reovirus mRNA-primed influenza viral mRNAs. Lane 4: mixture of reovirus mRNA- and globin mRNA-primed influenza viral mRNAs. Lane 6: mixture of ApG- and globin mRNA-primed viral mRNAs. From reference 6.

mobility was confirmed by electrophoresing a mixture of these two RNAs (lane 2): doublets can be observed which are most evident at the position of the two smallest segments. These relative mobilities were unaffected by prior treatment with glyoxal to eliminate secondary structure in the RNAs (5,18) (Fig. 2B), or by prior decapping of the globin mRNA-primed segments (6). From the difference in mobility, we estimated that the globin mRNA-primed segments were 10-15 nucleotides larger than the ApG-primed segments. Other experiments established that these extra nucleotides were at the 5' end of the viral mRNA segments (5). A similar mobility difference was observed when a mixture of ApG-primed and reovirus mRNA-primed segments were analyzed by gel electrophoresis (6) (Fig. 3, lane 6). In fact, when the influenza viral mRNA segments primed by globin mRNA and by reovirus mRNA were mixed (lane 4), no doublets were evident. Thus, gel electrophoretic analyses indicated that approximately the same number of nucleotides, about 10-15, were transferred to influenza viral mRNA from either globin or reovirus mRNA primers.

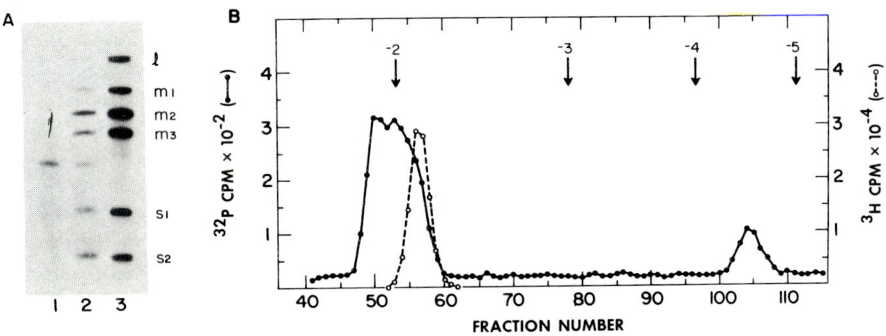

FIGURE 4. Transfer of the cap and internal nucleotides from uniformly ^{32}P-labeled reovirus mRNA to influenza viral mRNA during transcription in vitro. Experimental procedures were described in the text. (A) Gel electrophoresis of the influenza viral mRNA after deadenylation (lane 2). Lane 1: the ^{32}P-reovirus mRNA primer. Lane 3: deadenylylated influenza viral mRNA synthesized in a transcriptase reaction containing unlabeled globin mRNA and (α-^{32}P) GTP as labeled precursor. (B) DEAE-Sephadex chromatography of the RNase T2 digest of the s size influenza viral mRNAs of lane 2. From reference 6.

To obtain further evidence for the transfer of some internal nucleotides from the primer mRNA to influenza viral mRNA, we used uniformly ^{32}P-labeled reovirus mRNA as a primer in a transcriptase reaction mixture containing unlabeled CTP, ATP and UTP, and ^{3}H-labeled GTP (6). As shown in Fig. 4A, lane 2, the three m and two s influenza viral mRNA segments contained ^{32}P derived from the reovirus mRNA primer. When the s segments were eluted from the gel, digested with RNase T2, and the hydrolyzate analyzed by DEAE-Sephadex chromatography (Fig. 4B), ^{32}P-radiolabel was found in both internal residues at a charge of -2 and in the cap at -4.6. The number of chains of influenza viral mRNA synthesized can be estimated from the ^{3}H-radiolabel at -2. The results indicated that approximately one cap and 25 internal residues of reovirus mRNA were transferred to each chain of influenza viral mRNA. This estimation is in close agreement with that obtained from gel electrophoresis, especially considering the possible inaccuracies in the two types of analyses.

These studies did not allow us to identify which bases of the primer mRNA were transferred. To accomplish this, we (in collaboration with H. Robertson and E. Dickson of Rockefeller University) used ^{125}I-globin mRNA as primer and sequenced the ^{125}I-labeled region transferred to the viral mRNA. These results will be published elsewhere (H.D. Robertson, E. Dickson, S.J. Plotch and R.M. Krug, submitted for publication), and we will only present the conclusions here. We showed that the same set of ^{125}I-labeled oligonucleotides was present at the 5' end of all eight viral mRNA segments. The 5' terminus of β-globin mRNA has the sequence (19):

$$m^7Gpppm^6AmpC(m)pApCpUpUpGpCpUpUpUpUpGpApCp...$$

(with the numbers 1, 8, 13, 15 printed under residues 1, 8, 13 and 15)

and the predominant sequence at the 5' terminus of the viral mRNA molecules, representing 75% of the transferred oligonucleotides, was identical to the first 13 nucleotides (plus the cap) of this β-globin mRNA sequence. Because only the C-residues are labeled with ^{125}I, we cannot conclude with certainty that all 13 5' terminal nucleotides were transferred from β-globin mRNA. The results, however, certainly indicate that the β-globin mRNA donated at least the first eight, but no more than the first 14, 5' terminal bases to the viral mRNAs. It is almost certain that the ...UUUU... sequence (residues 9-12) originated from β-globin mRNA since there are no A residues in the 12-base 3' terminal sequence of the vRNA templates (14-17). One the other hand, the G-residue at position 13 of the viral mRNA could have arisen by transfer from the β-globin mRNA or as a result of transcription. Our data do not establish whether or not the A-residue at position 14 of the β-globin mRNA was transferred to the viral mRNA. Thus, the predominant sequence transferred from β-globin mRNA includes the first 12, 13 or 14 5' terminal nucleotides.

The minor ^{125}I-labeled RNase T1-resistant oligonucleotides found in the viral mRNAs indicated that shorter 5' terminal pieces of β-globin mRNA were at times transferred and also strongly suggested that the transferred sequences were linked to G as the first base transcribed. For example, 14% of the RNase T1 products were the sequences CUUUGp, CUUGp, CUGp and CGp, which can be presumed to derive their C- and U-residues from β-globin mRNA (residues 8-11) and which must obtain their G-residue by transcription. Thus, 14% of the time 5' terminal, cap-containing pieces of β-globin mRNA 8-11 bases in length were transferred. These results also indicate that β-globin mRNA was most probably cleaved to release major and minor 5' terminal fragments prior to their incorporation into the 5' termini of the influenza viral mRNAs.

The predominant sequence transferred from β-globin mRNA is not complementary at its 3' end to the 3' terminal common sequence of the vRNA templates, favoring a mechanism of priming that does not involve hydrogen-bonding between the primer mRNA and the template vRNA. Furthermore, other data eliminate the previous suggestions (5) that splicing within the β-globin mRNA could create a fragment about 15 bases in length which begins with the cap and ends with an AG derived from the interior of the β-globin mRNA molecule. We have found that capped, 5' terminal fragments of β-globin mRNA too short to contain an AG or any other sequence complementary to the 3' end of vRNA effectively stimulate viral RNA transcription (unpublished results). In addition, in collaboration with A. Shatkin's laboratory of Roche Institute, we have found that capped synthetic ribopolymers lacking AG are also very effective primers. For example, capped poly A and capped poly AU (1:1) were almost as effective primers as globin mRNA (A. Shatkin, A. LaFiandria, B. Broni and R.M. Krug, unpublished data).

In the absence of hydrogen-bonding between primer and template, the most likely mechanism for priming is that shown in Fig. 5. The primer mRNA is cleaved to produce 5' terminal fragments, and these fragments undergo a specific interaction with the transcriptase which causes stimulation of transcription coupled with the linking of the 5' terminal fragments to the first base transcribed. Thus, the fundamental interaction is between the primer mRNA (and/or its 5' terminal fragment) and one or more virion-associated proteins. One of the consequences of G, rather than A, being the first base transcribed is that the viral RNA transcripts would not necessarily contain an A complementary to the 3' terminal U of the template vRNA. In the example shown in Fig. 5, with the 5'

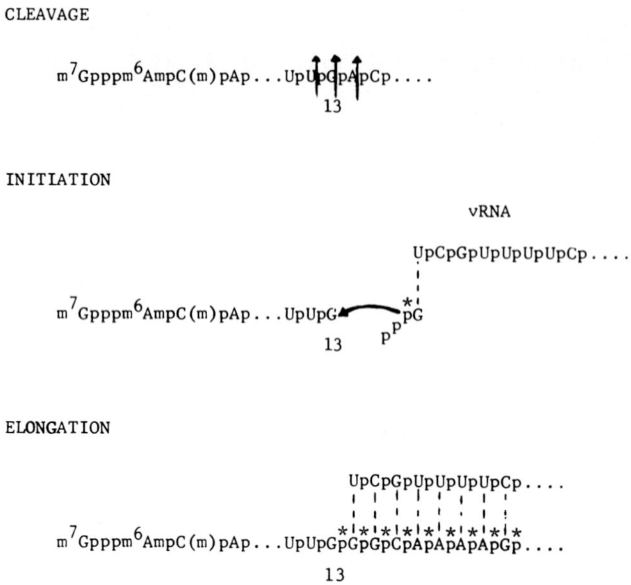

FIGURE 5. Postulated mechanism for the priming of influenza viral RNA transcription by β globin mRNA and other capped RNAs.

terminal fragments cleaved from β-globin mRNA at position G_{13}, the resulting viral mRNA molecules would contain a G at the position opposite the 3' terminal U of the vRNA template. The only way that an A could be found at this position would be if the 5' terminal fragment cleaved from the primer mRNA had a 3' terminal A. Preliminary experiments using globin mRNA as primer and (α-^{32}P)GTP as the only ribonucleoside triphosphate have yielded short 3' terminally labeled RNA fragments the size and other properties of which favor the type of priming mechanism shown in Fig. 5 (unpublished experiments).

The specific interaction between the primer mRNA and one or more virion-associated proteins most probably involves the

5' terminal methylated cap. Only capped mRNAs are active as primers (4-6). Removal of the m^7G of the cap of a mRNA by chemical (β-elimination) or enzymatic treatment eliminates all priming activity, and most of this activity can be restored by enzymatically recapping the β-eliminated mRNA (5). The cap must contain methyl groups, since reovirus mRNAs with 5' GpppG ends are not active as primers (6). Indeed, we have recently shown that each of the two methyl groups in the cap, the 7-methyl on the terminal G and the 2'-O-methyl on the penultimate base, strongly influences the priming activity of a mRNA (Bouloy _et al._, submitted for publication). Of particular interest is the effect of the 2'-O-methyl group. To demonstrate the importance of this group, we used several plant viral RNAs containing the monomethylated cap 0 structure, m^7GpppG. Brome mosaic virus (BMV) RNA 4 stimulated influenza viral RNA transcription only about 10-15% as effectively as globin mRNA. When the cap of BMV RNA 4 was enzymatically 2'-O-methylated, its priming activity was increased 14-fold. Qualitatively similar results were obtained with other plant viral RNAs: priming activity increased 3- to 20-fold following 2'-O methylation. This is the first instance in which the 2'-O-methyl group of the cap has been shown to have a strong and clear-cut effect on a specific function of a mRNA. Consequently, the fully methylated cap 1 structure, m^7GpppX^m, which is found in all mammalian cellular mRNAs and most animal viral mRNAs (8) is more stringently required for priming influenza viral RNA transcription than for translation in cell-free systems. In the influenza virion transcriptase reaction, recognition of the cap 1 structure could occur at both the cleavage and initiation phases of the reaction (see Fig. 5). Clearly, it will be of great interest to identify the virus-associated protein(s) which recognize the cap 1 structure, to establish whether this protein(s) is viral- or cell-coded, and to compare it to the cap-recognizing protein associated with eukaryotic ribosomes (20).

We also obtained strong evidence that priming of viral RNA transcription by capped host RNAs occurs in the infected cell (7). If such priming occurs _in vivo_, then the viral mRNA synthesized in the infected cell should contain 10-15 nucleotides at its 5' end, including the cap, which are not viral-coded. This is exactly what we found. First, when compared by gel electrophoresis (Fig. 6), the _in vivo_ viral mRNA segments were larger (migrated slower) than the ApG-primed _in vitro_ mRNA segments (7), identical to the mobility difference between the segments of eukaryotic mRNA-and ApG-primed _in vitro_ mRNA (see Figs. 2 and 3). In addition, when ^3H-methyl-labeled _in vivo_ viral mRNA was hybridized to vRNA, the 5' terminal cap structure of the mRNA was not protected

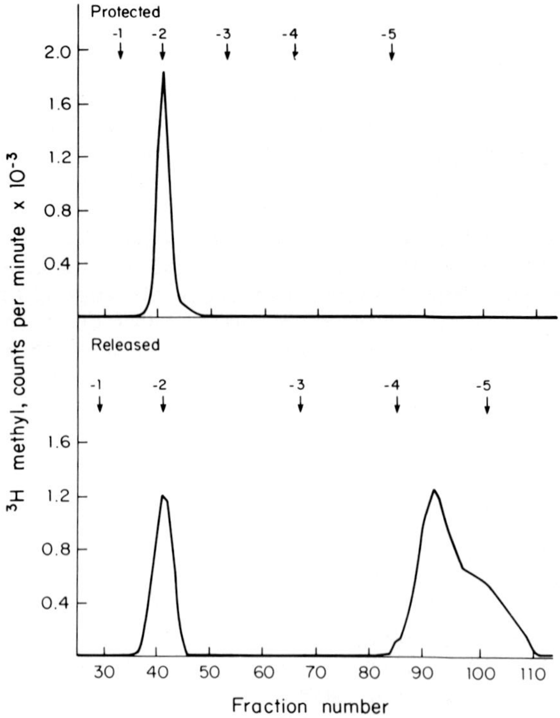

FIGURE 7. ^{3}H-methyl labeled species retained in (protected) or released from vRNA: in vivo viral mRNA hybrids by pancreatic RNase digestion. The in vivo viral mRNA contained the ^{3}H-methyl label. The hybrids were digested with pancreatic RNase, and the RNase-resistant double-strands and RNase-released material were collected. The double-strands (after heating and fast-cooling) and the released material were digested with RNase T2, and the hydrolyzates were analyzed by DEAE-Sephadex chromatography. From reference 7.

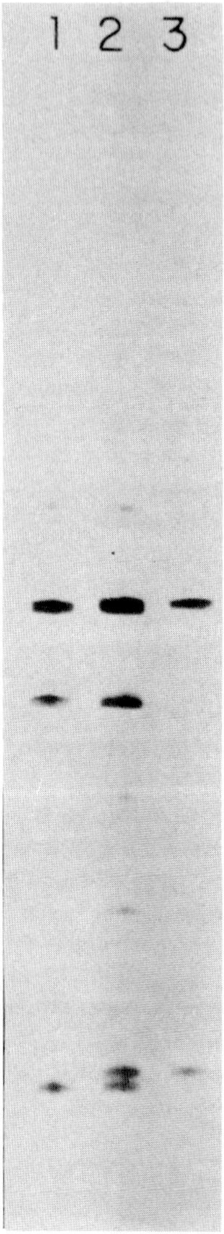

FIGURE 6. Comparison of the electrophoretic mobility of in vivo and ApG-primed in vitro viral mRNA. The two viral mRNAs were deadenylylated prior to electrophoresis. Lane 1: ApG-primed in vitro viral mRNA. Lane 3: in vivo viral mRNA. Lane 2: a mixture of the viral mRNAs of lanes 1 and 2. From reference 7.

against pancreatic or T1 RNase digestion (7) (Fig. 7). All of the cap (the -4.4 to 5.2 charge species) was released from the double-strands by the nuclease digestion. Only internal m^6A residues (charge -2) remained in the double-strands, although approximately one-third of these residues were released. As each molecule of in vivo viral mRNA contains an average of three m^6A residues (12), these results indicate that one of these m^6A's is in the 5' terminal sequence that is not viral coded. Thus, our results strongly suggest that host cell mRNAs and/or their precursors serve as primers for viral RNA transcription in the infected cell, and that they donate their cap and 10-15 internal nucleotides, one of which is m^6A, to the resulting viral mRNA molecules (7).

The synthesis of these host cell mRNA primers can be presumed to constitute the α-amanitin-sensitive step (RNA polymerase II function) required for viral RNA transcription. A critical, unanswered question is why new and continuous synthesis of these host mRNA primers is required. It may be that this requirement is due at least in part to the site of viral RNA transcription in the infected cell. The presence of m^6A residues in the viral mRNA (12) suggests that viral RNA transcription may have a nuclear phase, and other experiments measuring the steady-state level of viral mRNA (3,21) suggest that the nucleus is the site of at least primary transcription. If viral RNA transcription occurs in the nucleus, then the need for continued host cell mRNA synthesis may reflect the fact that the amount of available cellular mRNA and/or its precursors in the nucleus is limited and rapidly depleted. Clearly, it will be of great interest to establish definitively where both primary and amplified transcription occurs in the cell.

Another important question is whether these results with influenza viral RNA transcription shed any light on cellular functions. For example, is the recognition of methylated cap (cap 1) structures unique to influenza viral RNA transcription, or does it also occur during cellular transcription-related processes? One possibility is that the methylated cap structure is required for the proper processing of cellular heterogeneous nuclear RNA. An initial step in this processing, cleavage of heterogeneous nuclear RNA near its 5' end (22-24), is comparable to the cleavage of the primer RNA which occurs in the influenza virion transcriptase reaction (see Fig. 5). The cellular enzymes(s) may require the presence of the cap 1 structure, and it is in fact conceivable that the virion-associated activity is due to a cell-coded enzyme (possibly sequestered during a nuclear phase of viral RNA transcription). A later step in the processing of heterogeneous nuclear RNA, the ligation of the 5' terminal region onto distal regions of the RNA (22-24) could also involve cap recognition. This hypothesis, that the cap 1 structure has an important role in the processing of heterogeneous nuclear RNA, can be tested when the appropriate cellular enzymes(s) are identified and characterized. At the present time, the cap-recognizing protein(s) and cleavage enzyme(s) in influenza virions may serve as a useful model system.

ACKNOWLEDGMENTS

We thank Barbara Broni, Paul Simonelli and Mitchell Smilowitz for expert technical assistance.

REFERENCES

1. Lamb, R.A., and Choppin, P.W., J. Virol. 23, 816-819 (1977).
2. Spooner, L.L.R., and Barry, R.D., Nature 268, 650-652 (1977).
3. Mark, G.E., Taylor, J.M., Broni, B., and Krug, R.M., J. Virol. 29, 744-752 (1979).
4. Bouloy, M., Plotch, S.J., and Krug, R.M., Proc. Nat. Acad. Sci. USA 75, 4886-4890 (1978).
5. Plotch, S.J., Bouloy, M., and Krug, R.M., Proc. Nat. Acad. Sci. USA 76, 1618-1622 (1979).
6. Bouloy, M., Morgan, M.A., Shatkin, A.J., and Krug, R.M., J. Virol., in press (1979).
7. Krug, R.M., Broni, B.A., and Bouloy, M., Cell 18, 329-334 (1979).
8. Shatkin, A.J., Cell 9, 645-653 (1976).
9. Plotch, S.J., and Krug, R.M., J. Virol. 21, 24-34 (1977).
10. Plotch, S.J., and Krug, R.M., J. Virol. 25, 579-586 (1978).
11. Plotch, S.J., Tomasz, J., and Krug, R.M., J. Virol., 28, 75-83.
12. Krug, R.M., Morgan, M.M., and Shatkin, A.J., J. Virol. 20, 45-53 (1976).
13. McGeoch, D., and Kitron, N., J. Virol. 15, 686-695 (1975).
14. Skehel, J.J., and Hay, A.J., Nucleic Acids Res. 4, 1207-1219 (1978).

15. Both, G.W., and Air, G.M., Eur. J. Biochem. 96, 363-372 (1979).
16. Air, G.M., Virology 97, 468-472 (1979).
17. Robertson, J.S., Nucleic Acids Res. 6, 3745-3757 (1979).
18. McMaster, G.K., and Carmichael, G.G., Proc. Nat. Acad. Sci. USA 74, 4835-4838 (1977).
19. Lockard, R.E., and RajBhandary, U.L., Cell 9, 747-760 (1976).
20. Sonenberg, N., Rupprecht, K., Hecht, S., and Shatkin, A. J., Proc. Nat. Acad. Sci. USA 76, 4345-4349 (1979).
21. Barrett, T., Wolstenholme, A.J., and Mahy, B.W.J., Virology 98, 211-225 (1979).
22. Klessig, D.F., Cell 12, 9-21 (1977).
23. Berget, S.M., Moorce, C., and Sharp, P.A., Proc. Nat. Acad. Sci. USA 74, 3171-3175 (1977).
24. Darnell, J.E., Jr., Science 202, 1257-1260 (1978).

DISCUSSION

S. LANGBERG: What are the effects of the cap analog pm^7G on influenza viral RNA transcription?

R.M. KRUG: Experiments with cap analogs cannot be done in this system in exactly the same way that they have been done in protein synthesis. As you know, the cap analog pm^7G and also methylated cap structures of the form m^7GpppX^m inhibit the translation of capped mRNAs, but usually not that of uncapped mRNAs. For the analogous experiments with influenza viral RNA transcription, the only uncapped primers that we could use were the dinucleotides ApG or GpG, since uncapped mRNAs are inactive as primers. Optimal priming with ApG (or GpG) occurs at about 0.2 mM, a concentration 1000 times higher than that needed for optimal priming with a capped mRNA. At this high concentration, methylated cap structures and pm^7G themselves also exhibit low levels of priming, 5-10% of that of ApG. A similar low priming activity is observed with the compounds used as controls, unmethylatead cap structures (for example, GpppG) and pG. When added to globin mRNA-primed reactions, methylated cap structures and pm^7G at 0.1 to 0.4 mM did inhibit transcription 25-60%. However, the control compounds, GpppG and pG, caused a similar inhibition. These results were interpreted as simply reflecting the competition between a good primer (globin mRNA) and a poor primer (one of the small compounds) for the active site(s) of the transcriptase. A more specific effect was seen with ApG-primed transcription. Whereas the control compounds, GpppG and pG, inhibited the transcription primed by ApG, cap structures containing both methyl groups (for example, m^7GpppG^m) actually stimulated this transcription by 60-75%. Cap structures lacking the 2'-O-methyl group (m^7GpppG and m^7GpppA) also stimulated ApG-primed transcription, but only by 25%. The cap analog pm^7G stimulated ApG-primed transcription marginally, about 5-10%, while cap structures lacking the terminal 7-methyl group (for example, $GpppG^m$) behaved like the control compounds and inhibited ApG-primed transcription. Thus, ApG-primed transcription was stimulated specifically by cap structures containing 7-methyl G in 5'-triphosphate linkage with 2'-O-methylated A or G. These results suggest that this cap 1 structure is recognized by the transcriptase and causes an allosteric modification and activation of the enzyme. Actually, the fact that the cap is absolutely required for the priming activity of a mRNA and that both the 7-methyl and the 2'-O-methyl groups in the cap of a mRNA strongly influence its priming activity already indicated that one or more proteins in the virion transcriptase complex most probably recognizes the cap 1 structure.

S. LANGBERG: Is STNV RNA a good or bad primer?

R.M. KRUG: STNV RNA, which has a 5'-terminal diphosphate, is not active as a primer, but if it is enzymatically capped, it is then as good a primer as globin mRNA.

K.S. McCARTY: I was very interested in your initial observation that the beta globin mRNA was much better than the alpha globin mRNA as a primer. Are you suggesting that perhaps the virion-associated endonuclease did not cleave the alpha globin mRNA?

R.M. KRUG: We don't know why alpha globin mRNA is a relatively poor primer. In fact, it is one of the poorest mRNA primers we have encountered. We presume some of it is cleaved, however, because it does stimulate transcription about 20-fold. As you know, alpha globin mRNA is also not as good as beta globin mRNA in cell-free translation.

K.S. McCARTY: You didn't try to use the alpha globin mRNA after the transcriptase research for translation studies to see if it had been cleaved?

R.M. KRUG: No, we did not do that experiment. I do not think it would be a very clear experiment, because even with good primer mRNAs only a very small percentage of the primer mRNA is cleaved during the transcriptase reaction.

D. DRESSLER: A possibility that you did not mention is that the virus might be coupling the synthesis of its own messenger RNA molecules to the destruction of host cell messenger RNAs in order to gain control over cell metabolism. Is there any evidence that would support or rule out this hypothesis?

R.M. KRUG: Influenza does in fact shut off host protein synthesis, but so do several other RNA viruses which do not employ this method of making their messenger RNAs. Consequently, I do not think that this mechanism was devised specifically to inhibit host cell protein synthesis. This cannibalization of host cell messenger RNAs would, of course, lead to inhibition of protein synthesis if enough host cell messenger RNAs were involved.

R. KAVENOFF: I would like to know if you have looked at the stability of messenger RNAs with and without the cap. That is, does the cap make any difference in the degradation by nucleases associated with the virus?

R.M. KRUG: In terms of the stability of the bulk of the mRNA added to the reaction, the presence of the cap does not make any difference. You are probably referring to work in protein synthesis, where it has been reported that a 5' terminal methylated cap (m^7GpppX^m) or simply a 5' terminal blocked group (GpppX) on a mRNA stabilizes it against exonucleases present in several protein-synthesizing systems. One possible explanation for the inability of mRNAs lacking cap 1 structures to stimulate influenza viral RNA transcription is that these mRNAs are rapidly broken down by nucleases in the virus preparation. We therefore examined the stability of ^{3}H-methyl-labeled m^7GpppG^m-terminated (capped) and $pppG^m$-terminated (beta-eliminated) BMV RNA 4 during incubation with detergent-treated virus in the presence or absence of ribonuclease triphosphates. Most of both of these RNAs remained intact and sedimented at 12S, indicating that the absence of a capped 5' terminus did not render BMV RNA 4 more labile. Since the BMV RNA 4 molecules in these experiments were labeled at their 5'-end, we would have detected exonuclease activity. In terms of the specific endonucleolytic cleavage to generate the 5' terminal fragments which participate in the priming reaction, we have not yet determined the relative efficiency of cleavage of capped and uncapped mRNAs.

R. KAVENOFF: What is this endonucleolytic degradation like? Is it quite extensive?

R.M. KRUG: No, it is not. In a typical reaction (50 µl), we add about 12 pmoles of globin mRNA and make about 0.2 pmoles of influenza viral messenger RNA. Consequently, the amount of product molecules is only 2 percent of the primer molecules added, and only a small percentage of the primer molecules are specifically cleaved. We have obtained preliminary evidence for the production of these specific 5' terminal fragments, but they are difficult to detect against a background of nondegraded material and of fragments generated by nonspecific nucleases associated with the virus.

M.F. SINGER: Before I ask my question, let me be sure whether you are indicating that the addition of the first G to the 5' terminal capped fragment of the primer was directed by the template virion RNA.

R.M. KRUG: While we cannot be sure, our present working hypothesis is that the addition of the first G is template directed - either by the 3' terminal U or the penultimate G of the virion RNA. The surprise of course is that the first base added is not an A.

M.F. SINGER: Have you speculated at all on the possibility that the cap may in fact be required to get the addition of this unexpected base, G rather than A?

R.M. KRUG: That is a possibility, but it cannot be tested because uncapped RNAs do not participate in the reaction. We feel that one possibility is that the reason A is not added first is that the 3' terminal U is not copied at all. What I did not mention in my talk is that we have determined whether other ribonucleoside triphosphates other than G yield 3' terminally labeled fragments of globin or other messenger RNAs in the absence of the other three ribonucleoside triphosphates. A and U do not work, but C does, suggesting that the G at the third position at the 3' end of the virion RNA (3' UCG...) also can direct incorporation. With another virus, Semliki Forest virus, the 3' terminal base of the template is apparently not copied during the synthesis of the viral messenger RNA.

P.A. SHARP: Isn't it atypical to have an m^6A within the first 15 nucleotides at the 5' end of capped cellular messenger RNAs? Does this suggest that the virus is utilizing a subset of cellular messenger RNAs containing an m^6A near its 5' end for utilization as primers?

R.M. KRUG: It is unusual to find an m^6A so close to the 5' end of host messenger RNAs. There is an alternative possibility however: the methylation of A could occur during or after the 5' terminal sequences of the host messenger RNA are transferred to the viral messenger RNA. In this regard, it is important to emphasize that the presence of m^6A in the mRNA is not required for priming and transfer of primer sequences, because globin messenger RNA and reovirus messenger RNAs are effective primers in vitro and do not contain m^6A.

P.A. SHARP: Then, does the virus contain an enzyme which methylates adenosine?

R.M. KRUG: We do not know yet. About two years ago, we carried out a series of experiments looking for capping and methylating enzymes in the virion. None were detected, but these experiments were not specifically designed to detect a small amount of methylation of internal A residues.

P.A. SHARP: In influenza virus-infected cells, do you see small cap-containing fragments of heterogeneous nuclear RNA generated by the virus?

R.M. KRUG: We really have not yet done definitive experiments looking for such fragments generated in vivo.

A RECURRING SV40 VARIANT CONTAINING MONKEY DNA SEQUENCES

Maxine F. Singer, Edward L. Kuff, Theresa N.H. Lee, Thomas F. McCutchan, Joseph Papamatheakis and Ronald E. Thayer

Laboratory of Biochemistry
National Cancer Institute
Bethesda, Maryland

Ernest Winocour

Department of Virology
Weizmann Institute of Science
Rehovot, Israel

I. INTRODUCTION

It has been evident for some time that African green monkey kidney cells infected with simian virus 40 (SV40) are actively engaged in recombination. After a few serial passages of initially plaque-purified SV40 through such cells an abundance of altered viral genomes can readily be detected within infected cells and in viral particles. The resulting viral stocks are generally a mixture of wild type virus and different variants some of which may be defective. Among the variant genomes are deletions, duplications, inversions, recombinations with monkey sequences, and tandem reiterations of SV40 or SV40-monkey DNA segments (see Fareed and Davoli, 1977 and Kelly and Nathans, 1977 for reviews).

Several examples of the last mentioned group of variants have been characterized in varying detail (Oren *et al.*, 1978; Gutai and Nathans, 1978a and b; Wakamiya *et al.*, 1979; McCutchan *et al.*, 1979). Typically, they consist of between 3 and 10 tandem repetitions of basic repeat units comprising viral and monkey DNA segments arranged in a closed circular duplex DNA which approximates the size of the wild type genome. The basic repeat units in different variants always

ISBN 0-12-633360-2

include the SV40 replication origin but differ in the amount of the wild type SV40 genome that is retained and in the size and nature of the incorporated monkey DNA segments. Such circularly redundant DNA molecules are not unique to SV40-infected monkey cells. They have also been observed in polyoma-infected mouse cells (Griffin and Fried, 1977) and are structurally analogous to the altered mitochondrial genomes of the "petite" mutants of yeast (Faugeron-Fonty et al., 1979). Furthermore, such molecules are generated upon transfection of monkey cells with segments of the SV40 and polyoma genomes (Fareed and Davoli, 1977).

This paper will focus on one particular defective variant that was first detected some time ago after four serial passages of plaque purified SV40 strain 777 on the BSC-1 line of African green monkey kidney (AGMK) cells (Rozenblatt et al., 1973; Oren et al., 1978; Wakamiya et al., 1979).

The 1210 base pair (bp) long basic repeat unit of this variant (termed CVP8/1/P2) is shown schematically in Figure 1A: the entire nucleotide sequence is known. The repeat unit contains 427 contiguous bp of SV40 DNA; of these 392 are exactly as they are in the wild type 776 genome (Fiers et al., 1978; Reddy et al., 1978), one (residue 1177) differs from the published sequence for 776 in being an A·T bp rather than a G·C, and 34 (residues 954 through 987) are joined (at residue 987) in an inverted orientation to wild type sequences that are 153 bp away in the wild type genome. The variant also contains 783 bp of monkey DNA: of these 156 are homologous to the highly repeated α-component DNA of African green monkey (Rosenberg et al., 1978; Maio, 1971) while the remainder represent DNA sequences that are present in very low copy number in the monkey genome.

One of the outstanding questions about variants such as CVP8/1/P2 is whether they represent unique products of nonspecific, chance recombinations or specific and recurring structures reflecting important events during the infectious cycle. The question has remained largely unanswered because of the absence of any very sensitive means to detect a particular variant within a population consisting of a number of different variants as well as wild type genomes.

One possible approach is to use those DNA segments in a known variant that are homologous to very low copy number monkey genomic sequences as probes for the detection of related variants. Earlier, some such experiments were carried out using as probes RNA prepared, for example, by random cleavage of variant cRNA and subsequent removal of SV40 and α-component sequences by exhaustive hybridization to wild type SV40 and monkey DNA. These experiments suggested that variants containing the same infrequent monkey DNA sequences as CVP8/1/P2 were present in several independently passaged stocks of SV40 (Oren *et al.*, 1976; Frenkel *et al.*, 1974; Rosenberg *et al.*, 1977). More recently, detailed knowledge of the structure of CVP8/1/P2 has permitted the preparation of highly purified probes by excision of specific segments with restriction endonucleases and also by purification of those segments by molecular cloning in *E. coli*. In particular, the segment of monkey DNA sequence in CVP8/1/P2 that extends from the endo R·*Bam*H1 site at residue 364 to the endo R·*Hind*III site at residue 679 (Fig. 1A) has been purified, labeled with ^{32}P *in vitro*, and used as a probe (Winocour *et al.*, 1979; Papamatheakis *et al.*, 1980a and b). Using this probe, termed fragment C, the emergence of variants essentially identical to CVP8/1/P2 has been observed repeatedly. The provenance and characterization of these new isolates will be reviewed in this paper. Further, evidence supporting the view that the new isolates represent independent events will be summarized and the implications of these findings discussed.

II. RESULTS AND DISCUSSION

A. Identification of CVP8/1/P2-Type Variant in Plaques Initiated by Wild Type Virus

1. *Isolation by in situ Plaque Hybridization (Winocour et al., 1979)*. The spread of wild type virus through a cell monolayer will kill all the cells before a rare defective variant can replicate enough to be detectable in the total progeny. Consequently an infectious center modification (Winocour *et al.*, 1979) of an *in situ* plaque hybridization procedure (Villarreal and Berg, 1977) was developed and used to detect variants containing fragment C. In brief, cells infected at high multiplicity with a stock of plaque-purified wild type SV40 strain 777 grown from a limiting dilution were mixed with a 20-fold excess of uninfected cells and plated at high density. The monolayers that formed were covered with agar and after some days at 37° the agar was removed and the monolayers transferred to nitrocellulose filters. The filters were hybridized, after appropriate treatment, to the fragment C probe. Eight positive plaques were detected after screening a total of 1.5×10^6 infected cells. All the positive plaques arose when the cells initially infected with SV40 were subconfluent (growing): no positive plaques were seen when 2×10^6 confluent cells were infected and screened. One of the positive plaques, termed 290, was recovered from the corresponding agar overlay and a stock was prepared. Assay of the stock by the *in situ* plaque hybridization test now showed the presence

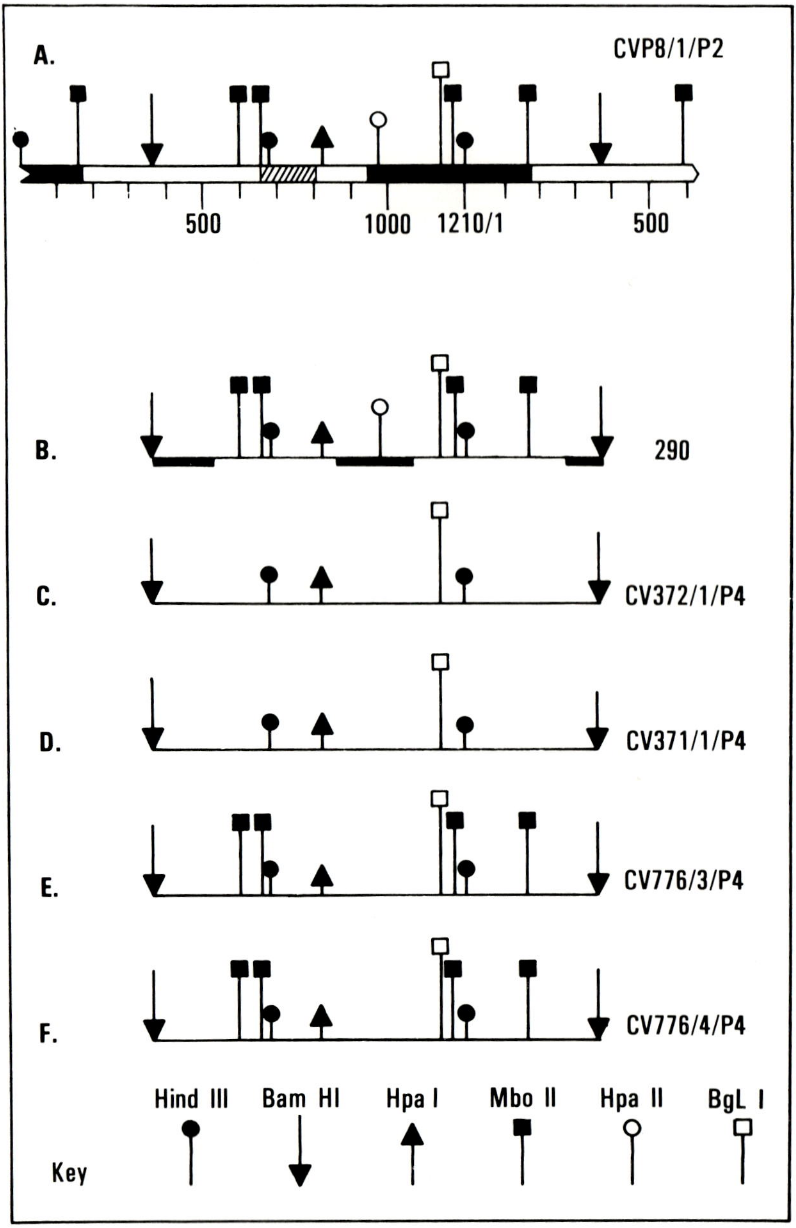
A.
CVP8/1/P2
500
1000
1210/1
500
B.
290
C.
CV372/1/P4
D.
CV371/1/P4
E.
CV776/3/P4
F.
CV776/4/P4
Hind III
Bam HI
Hpa I
Mbo II
Hpa II
BgL I
Key

FIGURE 1

LEGEND for FIGURE 1. Characterization of the structure of several independently isolated SV40 variants. A. Structure of the major repeating unit of variant CVP8/1/P2 DNA. The complete nucleotide sequence is known (Wakamiya *et al.*, 1979). An entire repeat unit (residues 1 to 1210) is shown plus 600 base pairs of the next repeat. Only those restriction endonuclease sites pertinent to the present work are indicated. The dark bars represent SV40 sequences; the open bars are monkey sequences that occur infrequently in the monkey genome; the slashed bars are sequences homologous to highly repeated α-component DNA of the monkey genome. B. Structure of *Bam*H1 fragment purified from variant 290 (Papamatheakis *et al.*, 1979a). DNA I purified from cells infected with stock 290 (see text) was cleaved with endo R·*Bam*H1 and the resulting DNA fragments were ligated to endo R·*Bam*H1 treated pBR322 (Bolivar *et al.*, 1977) and used to transfect *E. coli* K12, strain RR1. A colony that hybridized with ^{32}P-labeled fragment C, α-component DNA, and SV40 DNA (Thayer, 1979) was purified and used to prepare the recombinant plasmid. The inserted segment was characterized by restriction endonuclease digestion as indicated. The portions indicated by the thickened lines were subjected to nucleotide sequence analysis (Maxam and Gilbert, 1977). C. Structure of *Bam*H1 fragment purified from CV372/1-passage 4 (Papamatheakis *et al.*, 1979b). DNA I was purified from cells infected with the lysate obtained from the fourth serial passage of series CV372/1 (see Table II) and purified by molecular cloning and analyzed as described in B. Procedures for D, E, and F were as described for B and C. D. Structure of *Bam*H1 fragment purified from passage series CV371/1. E. Structure of *Bam*H1 fragment purified from passage series CV776/3. F. Structure of *Bam*H1 fragment purified from passage series CV776/4.

of a replicating variant containing the fragment C sequence in 15 out of 10^5 infected cells thus suggesting that the stock was enriched for the initially detected variant. The variant was recloned by mixing 10^5 cells infected with the stock with 2 x 10^6 uninfected cells and dispersing the cells into 100 miniwells. When the resulting lysates were hybridized with fragment C, 16 showed strongly positive reactions, 6 reacted weakly and 78 gave no reaction. Stocks were prepared from three of the strongly positive miniwells and the variants were characterized in detail.

2. Characterization of the 290 Variant (Papamatheakis et al., 1980a). All three isolates were carried through most of the procedures described below and gave essentially identical results, suggesting that the miniwell procedure and subsequent preparation of stocks had yielded material representing the initial 290 isolate.

DNA I was isolated from cells infected with the miniwell stocks and analyzed by gel electrophoresis, restriction endonuclease digestion, Southern blotting, and hybridization with ^{32}P-labeled fragment C as well as with ^{32}P-labeled α-component probe and ^{32}P-labeled SV40 DNA. The results indicated that the 290 DNA I consisted of a mixture of wild type SV40 and variants. A single type of variant which hybridized with all three probes appeared to be abundant in the mixture. Furthermore, all the properties of that variant were consistent with its being very similar to CVP8/1/P2. In particular, a fragment of approximately 1200 bp in length that hybridized with all three probes was produced upon treatment with endo R·BamHl.

In order to obtain pure samples of the 290 variant for more detailed analysis and to avoid molecular rearrangements that might occur upon further passaging in monkey cells, the approximately 1200 bp long BamHl fragment was cloned in an *E. coli* host-vector system (pBR322, *E. coli* K12 strain RR1, P-3 physical containment). Analysis of the cloned fragment confirmed the presence of all three types of DNA sequences and showed that the restriction endonuclease map of 290 was essentially identical to that of the major repeating unit of CVP8/1/P2 (see Fig. 1B). The primary nucleotide sequence of portions of the cloned 290 Bam H1 fragment were determined (Maxam and Gilbert, 1977) (Fig. 1B). The data confirmed the identity of 290 and CVP8/1/P2 from about residue 280 through 500 (with the exception that the G·C bp at residue 426 is missing in 290) and 860 through 1060. Thus, 290 contains the same 153 bp deletion of wild type SV40 sequences and the same inversion of 34 bp as does CVP8/1/P2. Furthermore, the joint between SV40 sequences and monkey sequences at residues 953/954 is identical in the two independent isolates.

B. Emergence of CVP8/1/P2-Type Variants upon Passaging Wild Type SV40 at 33°

1. Preferential Replication of CVP8/1/P2 at 33° (Singer and Thayer, 1980). Ever since its initial detection it has been clear that the CVP8/1/P2 type variant is replicated efficiently, relative to its wild type helper, in BSC-1 cells (e.g., see Oren *et al.*, 1978 and Wakamiya *et al.*, 1979). All these observations were made in cells maintained at 37°. At least in part the preferential replication of CVP8/1/P2 at 37° must reflect the presence of 4 copies of the SV40 origin of replication sequence (Lee and Nathans, 1979). Recent experiments summarized in Table I have indicated that the variant is replicated even more efficiently, relative to its wild

TABLE I. Preferential replication of variant CVP8/1/P2 at 33°

Time after infection (hr)	Percent of CVP8/1/P2 in DNA I	
	33°	37°
24		14
36		14
48	53	19
59		29
72	61	50
96	65	
120	63	
120*	0	

LEGEND for TABLE I. In this experiment confluent monolayers of BSC-1 cells were infected with a mixture of variant CVP8/1/P2 and wild type SV40 strain 777 (moi of 10 of wild type). The plates were incubated at either 33° or 37° and at the indicated times were pulse-labeled with 10 μCi per ml of ^{3}H-thymidine. After 2 hr DNA I was purified from lysates (Hirt, 1967) of the infected cells. The percent of DNA I resistant to cleavage by endo R·EcoRI was measured by sedimentation in alkaline sucrose gradients. CVP8/1/P2 contains no endo R·EcoRI site (Wakamiya *et al.*, 1979) while wild type SV40 contains a single such site. In the line marked * infection was with wild type virus alone. Data are taken from Singer and Thayer, 1980.

type helper, at 33° compared to 37°. The preferential rate of replication of the variant compared to wild type is most marked at early times in the infectious cycle. Different times were chosen for pulse labeling at the two temperatures in order to compensate for the delayed time course of viral infection at 33°. Maximal rates of DNA synthesis were attained by 48 hr postinfection at 37° and by 72 hr post-infection at 33°.

2. Multiple Serial Passages of Plaque-Purified Wild Type Virus at 33°. The many multiple serial passages of SV40 that have been described in the literature were carried out at 37°. In view of the preferential replication of CVP8/1/P2 at 33° it might be expected that related variants would be amplified in stocks passaged at 33°, should such variants occur. Table II summarizes the data obtained from nine independent passage series at 33°, using three different thrice-plaque purified viral stocks to initiate the passaging and two different cell types. ^{3}H-labeled DNA I was purified from cells infected with the passaged stocks and screened for variants containing α-component sequences by hybridization to filters containing SV40 or monkey DNA. The results are expressed as "hybridization index"; the percent of ^{3}H-labeled DNA I hybridizing to monkey DNA divided by the percent hybridizing to SV40 DNA. DNA I isolated from cells infected with the yield of the first high multiplicity passage of each series appeared to be wild type DNA alone, free of detectable variant DNAs either by restriction endonuclease digestion or by the hybridization screen: most of the hybridization indices were 0.01 or less. On the other hand, as shown in Table II, DNA I isolated from cells infected with the yield of the fourth high multiplicity passage of each series hybridized significantly with the monkey DNA except in one case. Additional hybridization analysis of the DNA I progeny of the 4th passage stocks of passage series CV776/3, CV776/4, CV371/1, CV371/3, CV372/1, and CV372/3 indicated the presence of fragment C sequences covalently linked to α-component and SV40 DNA sequences (Papamatheakis et al., 1980b; Singer and Thayer, 1980). Restriction endonuclease analysis of these DNA I samples revealed the presence of wild type DNA as well as a number of different variants in each. Digestion with endo R·BamH1 demonstrated that in each case one abundant variant species yielded a fragment about 1200 bp in length. The implication of these results, namely that high multiplicity passage at 33° often yielded variants similar to CVP8/1/P2, among others, has been confirmed by analysis of the BamH1 fragments after purification by molecular cloning in E. coli as described above for variant 290 (Papamatheakis et al., 1980b). The results are summarized in Fig. 1.

TABLE II. Presence of variants containing monkey α-component sequences in SV40 stocks serially passaged four times at 33°

Independent Passage Series	Cell Type	Hybridization Index
CV776/2	BSC-1	0.04
CV776/3	BSC-1	0.17
CV776/4	AGMK	0.08
CV371/1	BSC-1	0.29
CV371/2	BSC-1	0.05
CV371/3	AGMK	0.07
CV372/1	BSC-1	0.21
CV372/2	BSC-1	0.02
CV372/3	AGMK	0.10

LEGEND for TABLE II. The initiating viral stock is designated by the name of the thrice-purified plaque from which it was made (e.g., CV776). Several passage series were independently initiated with each stock and each such series is designated by a number after the slash: thus CV776/3 is the third independent series initiated with stock CV776. The type of cell used for each series is given. BSC-1 is an established line of African green monkey kidney cells. AGMK designates primary African green monkey kidney cells. ^{3}H-labeled DNA I was isolated from cells infected with the yield of the fourth serial passage in each series. The hybridization index is the percent of the ^{3}H-DNA I that hybridized to a filter containing total monkey DNA divided by the percent that hybridized to a filter containing SV40 DNA. The data are taken from Singer and Thayer, 1980.

C. Provenance of the CVP8/1/P2 Type Variants

1. The CVP8/1/P2 Type Variant was not Detectable as a Contaminant in Plaque Purified SV40 Stock. The data summarized here indicate that the CVP8/1/P2 type defective has emerged on at least seven (290, CV371/1/P4, CV371/3/P4, CV372/1/P4, CV372/3/P4, CV776/3/P4 and CV776/4/P4) independent occasions in addition to the original isolation. The CVP8/1/P2 type defective was first detected after 4 serial passages of a plaque-purified strain 777 virus (Lavi *et al.*, 1973; Rozenblatt *et al.*, 1973). The *in situ* plaque-hybridization experi-

ment that gave rise to variant 290 was also initiated with strain 777 virus: the plaque-hybridization experiment itself demonstrated that the initiating 777 stock was not grossly contaminated with the variant. Thus, no positive plaques were observed when 2 x 10^6 confluent cells were infected and screened by the infectious center-*in situ* plaques hybridization procedure (Winocour *et al.*, 1979) although eight positive plaques were counted when 1.5 x 10^6 subconfluent cells were infected and screened. These data are consistent with the possibility that the variant emerged in growing infected cells rather than being present in the initiating stock. The plaque-purified stocks termed CV371 and CV372 were also of the strain 777 type (by restriction endonuclease analysis) while stock CV776 was a plaque purified strain 776 virus (confirmed by restriction endonuclease analysis). These stocks were, as indicated above, free of CVP8/1/P2-type variants as measured by restriction endonuclease digestion patterns and lack of hybridization to monkey DNA of progeny DNA I. Furthermore, about 10^6 subconfluent BSC-1 cells infected with each of these stocks were screened by the infectious center-*in situ* plaque hybridization procedure using fragment C as a probe and no positive plaques could be identified (Singer and Thayer, 1980). Thus, the frequent occurrence of this variant does not reflect a simple inadvertent contamination of the viral stocks by the variant. Similarly, the cells used for the passages are unlikely to be a source of the variant by simple contamination. The *in situ* screening procedures were carried out on the BSC-1 cell lines used to isolate 290 as well as some of the variants accumulated at 33°. Three of the 33° passage series in BSC-1 cells (CV776/3, CV371/2 and CV372/2) were carried out in a laboratory in another building at the NIH using a cell line carried independently in that laboratory for some years. Similarly, the three 33° passage series carried out on primary AGMK cells were also carried out in that laboratory. Media and supplies belonging to that laboratory were used for those passages.

2. The Possible Origin of CVP8/1/P2 Type Variants. Previous discussions of the origin of SV40 variants containing covalently linked viral and host DNA have assumed that the variants arise by recombination between the two genomes and reiteration of a recombined fragment to yield a replicable genome of a size that can readily be packaged (reviewed by Kelly and Nathans, 1977; Fareed and Davoli, 1977). If independent recombinational events, amplified by preferential replication, are the source of repeated isolations of variants of the CVP8/1/P2 type at 33°, then those events must involve a high degree of specificity.

Still, and in spite of the evidence to the contrary reviewed above, the possibility that extremely rare copies of this variant persist in viral stocks derived from thrice purified plaque isolates (even when made at limiting dilutions) cannot be rigorously excluded. It is also possible that such a molecule exists in low copy number in rare monkey cells and can be rescued upon infection with SV40. One aspect of the structures outlined in Fig. 1 supports this last interpretation. As noted in the Introduction, residue 1177 in CVP8/1/P2 differs from the published sequence for wild type strain 776 DNA in being an A·T rather than a G·C base pair. This change introduces an additional endo R·MboII site (Roberts, 1978) into the molecule. Recent experiments (Papamatheakis *et al.*, 1980b) have demonstrated that this endo R·MboII site is also found in strain 777 DNA (and is, as expected, missing in 776). Thus it was not surprising to find that endo R·*Mbo*II site in CVP8/1/P2, and 290, both of which emerged after infections with strain 777. The presence of the endo R·*Mbo*II site is readily discernible. Cleavage of the 1210 bp *Bam*H1 monomer fragment with endo R·*Mbo*II yields, among others, a 540 bp long fragment (residue 648-1188) if the recognition site around residue 1177 is present: if the site were absent the 540 bp fragment would be replaced by one 731 bp long (residues 648-1210/1-169). However, the same endo R·*Mbo*II site occurs in the cloned *Bam*H1 fragments from CV776/3/P4 and CV776/4/P4 (Fig. 1) both of which were isolated from serial passages initiated with strain 776. These data suggest that the variants did not arise by simple recombination between the infecting 776 and the host DNA. Thus, it seems possible that the variant genome preexists, albeit very infrequently, either in the cells or in viral stocks. Replication of the variant might be stimulated by SV40 infection and the packaging of the variant genome into SV40 capsids would ensure its spread and amplification. Experiments designed to test the various hypotheses concerning the provenance of the variant are currently being carried out.

ACKNOWLEDGMENTS

The authors are grateful to Kevin Dougherty for his help with some of the experiments and to Ms. May Liu for her expert preparation of the manuscript.

REFERENCES

Bolivar, F., Rodriguez, R.L., Greene, P.J., Betlach, M.C., Heyneker, H.L., Boyer, H.W., Crosa, J.H., and Falkow, S. (1977). Gene 2, 95-113.

Fareed, G.C., and Davoli, D. (1977). Ann. Rev. Biochem. 46, 471-522.

Faugeron-Fonty, G., Culard, F., Baldacci, G., Goursot, R., Prunell, A., and Bernardi, G. (1979). J. Mol. Biol. 134, 493-537.

Fiers, W., Contreras, R., Haegeman, G., Rogiers, R., Van de Voorde, A., Van Heuverswyn, H., Van Herreweghe, J., Volckaert, G., and Ysebaert, M. (1978). Nature 273, 113-120.

Frenkel, N. Lavi, S., and Winocour, E. (1974). Virology 60, 9-20.

Griffin, B.E., and Fried, M. (1977). J. Mol. Biol. 117, 473-495.

Gutai, M.W., and Nathans, D. (1978). J. Mol. Biol. 126, 259-274.

Gutai, M.W., and Nathans, D. (1978). J. Mol. Biol. 126, 275-288.

Hirt, B. (1967). J. Mol. Biol. 26, 365-369.

Kelly, T.J. Jr., and Nathans, D. (1977). Adv. Virus Res. 21, 85-173.

Lavi, S., Rozenblatt, S., Singer, M., and Winocour, E. (1973). J. Virol. 12, 492-500.

Lee, T.N.H., and Nathans, D. (1979). Virology 92, 291-298.

Maio, J.J. (1971). J. Mol. Biol. 56, 579-595.

Maxam, A., and Gilbert, W. (1977). Proc. Natl. Acad. Sci. U.S.A. 74, 560-564.

McCutchan, T., Singer, M., and Rosenberg, M. (1979). J. Biol. Chem. 254, 3592-3597.

Oren, M., Kuff, E., and Winocour, E. (1976). Virology 73, 419-430.

Oren, M., Lavi, S., and Winocour, E. (1978). Virology 85, 404-421.

Papamatheakis, J., Kuff, E., Winocour, E., and Singer, M. (1980a). Manuscript in preparation.

Papamatheakis, J., Lee, T.N.H., Thayer, R., and Singer, M. (1980b). Manuscript in preparation.

Reddy, V.B., Thimmappaya, B., Dhar, R., Subramanian, K.N., Zain, B.S., Pan, J., Ghosh, P.K., Celma, M.L. and Weissman, S.M. (1978). Science 200, 494-502.

Roberts, R.J. (1978). Gene 4, 183-193.

Rosenberg, M., Segal, S., Kuff, E., and Singer, M. (1977). Cell 11, 845-857.

Rosenberg, H., Singer, M., Rosenberg, M. (1978). Science 200, 394-402.
Rozenblatt, S., Lavi, S., Singer, M., and Winocour, E. (1973). J. Virol. 12, 501-510.
Singer, M.F., and Thayer, R.E. (1980). Submitted, 1980.
Thayer, R.E. (1979). Anal. Biochem. 98, 60-63.
Villarreal, L.P., and Berg, P. (1977). Science 196, 183-185.
Wakamiya, T., McCutchan, T., Rosenberg, M. and Singer, M. (1979). J. Biol. Chem. 254, 3584-3591.
Winocour, E., Singer, M., Kuff, E. (1979). Cold Spring Harbor Symposium on Quantitative Biology, in press.

DISCUSSION

P.A. SHARP: If SV40 pre-exists in the cell in this configuration, wouldn't it have been picked up on Southern blot of CV1 cell DNA?

M.F. SINGER: I would be interested in knowing of such experiments. There are several reasons for thinking that unless one did them under very special conditions one might not have seen them. For example, the pieces of SV40 might be rather small. One problem that was pointed out to me this week is that the SV40 sequences around the origin might snap back.

P.A. SHARP: I would like to address a question to Mike Botchan. Have you obtained any evidence in monkey cells for segments as long as 100 nucleotides that would hybridize to SV40 sequences?

M. BOTCHAN: No, there are none, but we have seen in libraries of human DNA, or monkey DNA, sequences which hybridize faintly to the SV40 origin of replication. We believe that the extent of hybridization that we have seen in the DNA isolated from these plaques, represents homologies much shorter than 100 nucleotides - probably 25 to 50. We only see this kind of hybridization in a Davis type screen of plaques where we get an enormous amplification of a signal in a very small area. We can't see it in a Southern blot of genome DNA.

D. GILLESPIE: If you use SV40 with mutations in those regions that have been preserved, are they invariably lost in the variant which emerges?

M.F. SINGER: In the two cases where we started with strain 776, we got back variants with the MboII sites which we had thought was diagnostic of strain 777.

J. ROGERS: Does the host sequence in these variants contain the ubiquitous sequence homologous to the SV40 origin and the RNA polymerase III transcription units?

M.F. SINGER: There are no sequences in the host DNA part of the variant that are homologous to SV40 sequences.

M. BOTCHAN: Dr. Rogers is referring to a small 14 base pair repetitive sequence, which is near RNA polymerase III binding sites and seems to be homologous to the SV40 origin. There are many such sequences in the monkey genome but I have looked at the sequences from this repeating unit and they don't seem to share homology. However the insert in an SV40 variant described by Dhruva and Subramanian (poster session) does share significant homology with that sequence.

P.A. SHARP: If you had to reconstruct this genome, how many unique combinations of SV40 and cell DNA are required?

M.F. SINGER: You mean how many recombination events? I do not know whether the monkey segment is contiguous in the monkey genome.

TWO YEAST tRNA GENES CONTAINING INTERVENING SEQUENCES

Hyen S. Kang, Richard C. Ogden and John Abelson
Department of Chemistry
University of California, San Diego
La Jolla, California 92093

ABSTRACT

The nucleotide sequences of one $tRNA^{Trp}$ gene and one $rTNA_3$ gene from *Sacharomyces cerevisiae* have been determined. In each case, the coding portion of the gene is interrupted in the anticodon region by an intervening sequence of about 30 base pairs. Precursors to $tRNA^{Trp}$ and $tRNA_3^{Leu}$ accumulate at the non-permissive temperature in the yeast *rna1* mutant. RNA sequence analysis of *in vivo* ^{32}P-labelled precursors indicate that they have mature termini and contain all the expected oligonucleotides from the intervening sequence suggesting that these precursors are colinear with the genes. Plausible secondary structure models can be made which fit a proposed generalized structure for tRNA precursors with intervening sequences.

INTRODUCTION

The biosynthesis of some tRNAs in *Saccharomyces cerevisiae* involves RNA splicing. It is well established that certain yeast tRNA genes are not colinear with the corresponding mature tRNAs. For example, each of three $tRNA^{Tyr}$ genes contains a fourteen base pair intervening sequence (1) and four $tRNA^{Phe}$ genes contain intervening sequences of either 18 or 19 bp (2). It has been demonstrated that generation of the mature tRNA sequence in these cases occurs by splicing of a precursor RNA that *is* colinear with the gene (3,4). An investigation of the activities involved in yeast tRNA splicing *in vitro* and

ISBN 0-12-633360-2

characterization of RNA intermediates has been reported (5,6).

We have sequenced single genes for yeast $tRNA^{Trp}$ and $tRNA_3^{Leu}$ and have shown that they too contain intervening sequences in the anticodon region. Precursors to $tRNA^{Trp}$ and $tRNA_3^{Leu}$ accumulate in the yeast <u>rna1</u> mutant ts136 (3,7) and standard oligonucleotides mapping has indicated that these precursors contain mature 5' and 3' termini, some base modification, and the intervening sequence. Analyses of the sequences, base modification and splicing pathway of pre-$tRNA^{Trp}$ and pre-$tRNA_3^{Leu}$ along with the other identified precursor tRNAs accumulating in ts136 have been recently summarized (8,9).

MATERIALS AND METHODS

In vivo ^{32}P-labelled precursors to $tRNA^{Trp}$ and $tRNA^{Leu}$ were isolated as described previously (3,10) and analyzed by standard procedures (11). Construction and isolation of plasmic DNAs containing the $tRNA^{Trp}$ and $tRNA_3^{Leu}$ genes have been described previously (12,13). Restriction endonucleases, DNA polymerase (Klenow), RNases, ^{32}P-orthophorphoric acid and α-^{32}P-deoxyribonucleotides were obtained commercially. Cellulose acetate strips (Schleicher and Schuell) and DEAE paper (Nucleopore Corporation and Whatman) were used for RNA fingerprinting. Modified nucleotides were identified by the method of Nishimura (14).

Isolation of Restriction Fragments and DNA Sequencing. DNA restriction fragments carrying the respective tRNA genes were separated by agarose gel electrophoresis and identified by the method of Southern (15) using tRNA probes end-labelled with γ-^{32}P-ATP and T4 polynucleotide kinase. A partial restriction map was thus determined in the region of the tRNA gene. Restriction enzyme digestions were performed according to standard procedures. For sequencing purposes, Hinf I restriction fragments were terminally labelled using DNA polymerase (Klenow) with α-^{32}P dCTP, dATP and dTTP for one hour at 37° and subsequently digested with Hind III or Hae III. Terminally labelled fragments were separated by polyacrylamide gel electrophoresis and eluted. The strands were separated and sequenced according to the method of Maxam and Gilbert (16). Taq I restriction fragments were terminally labelled with DNA polymerase and α-^{32}P-dGTP and dCTP and the appropriate fragments sequenced as described above.

RESULTS

<u>Restriction Mapping and DNA Sequence Analysis.</u> The mature tRNA^{Trp} sequence (17) indicates the presence of a single Hinf I site and two Taq I sites. A partial restriction map constructed using these and other restriction enzymes is summarized in Figure 1. Figure 1A shows the construction of pJB20m from pBR313 and Hind III cleaved total yeast DNA as described previously (12,13). A partial restriction map of the yeast insert and the portion of the tRNA^{Trp} gene is shown in Figure 1B. As expected, ^{32}P-labelled yeast tRNA^{Trp} hybridized to two Hinf I restriction fragments from pJB20m in a Southern plot. One band, approximately 2 kb in length, could be reduced to 1.4 kb after cutting with Hind III. For sequencing purposes, Hinf I fragments were end-labelled, cut with Hind III, and the 1.4 kb fragment isolated by gel electrophoresis and sequenced (16). The other Hinf I fragment (400 bp) can be cut with Hae III to a 150 bp fragment which contains the 5' end of the gene. Following end labelling as above, the 400 bp fragment was isolated by gel electrophoresis and digested with Hae III. The two labelled bands were separated by electrophoresis on a 7.5% polyacrylamide gel and the smaller band (150 bp) was sequenced. The intervening sequence was confirmed by isolating the 1 kb EcoRl/Hinc II restriction fragment containing the tRNA^{Trp} gene, cutting with Taq I, end labelling with α-^{32}P-dGTP and α-^{32}P-dCTP and separating the bands by polyacrylamide gel electrophoresis. The intervening sequence is totally contained on the fragment 45 bp long, both strands of which were sequenced. Figure 1C shows the intragenic restriction sites and the sequencing strategy employed to determine the nucleotide sequence of the flanking regions, coding region and intervening sequence.

The preliminary results of a similar analysis of pJB2k, the plasmid containing the tRNA^{Leu} gene, are shown in Figure 2. As can be seen in Figure 2A, the plasmid contains two fused yeast Hind III fragments. The two Hinf I sites expected from the mature tRNA$_3^{Leu}$ sequence (18) are shown in Figure 2B. The Hinc II site

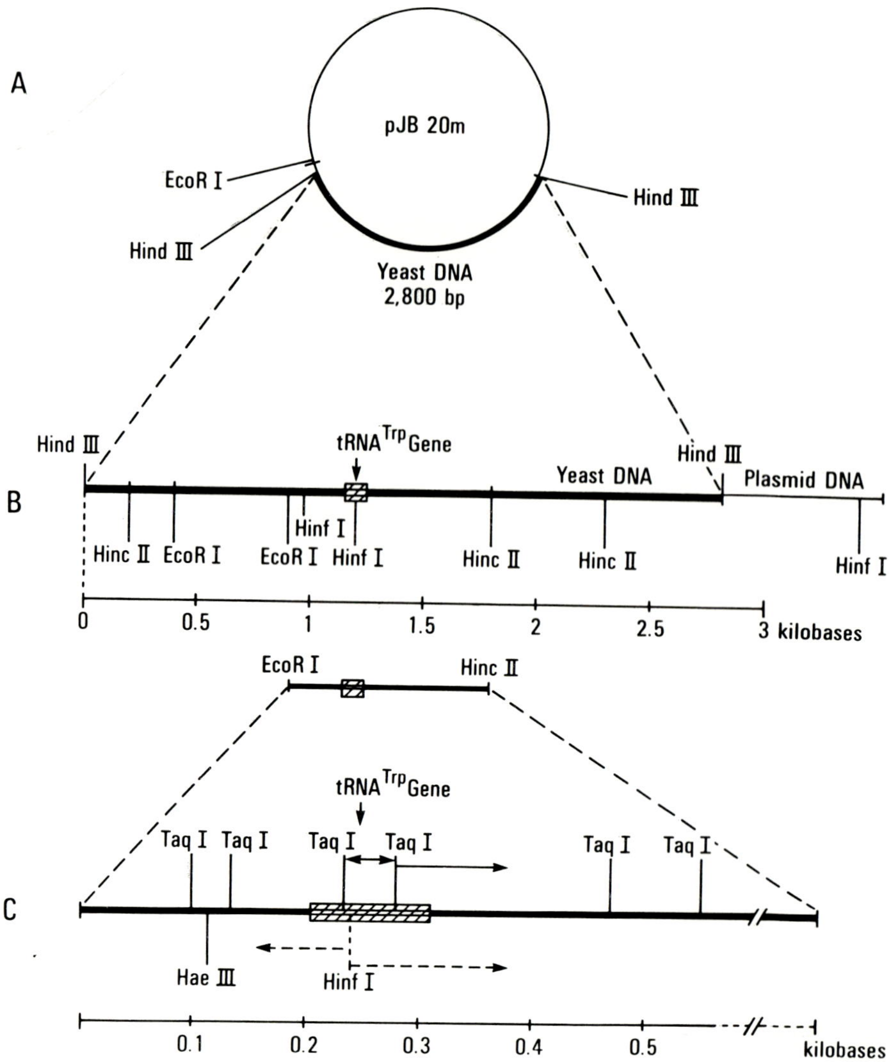

Fig. 1. (A) Schematic diagram of the plasmid containing the yeast $tRNA^{Trp}$ gene. (B) Partial restriction map of the yeast DNA insert. (C) Partial restriction map of the region containing the tRNA gene showing the intragenic restriction sites and other sites used for sequencing. The arrows indicate the direction and approximate extent of the sequence determination.

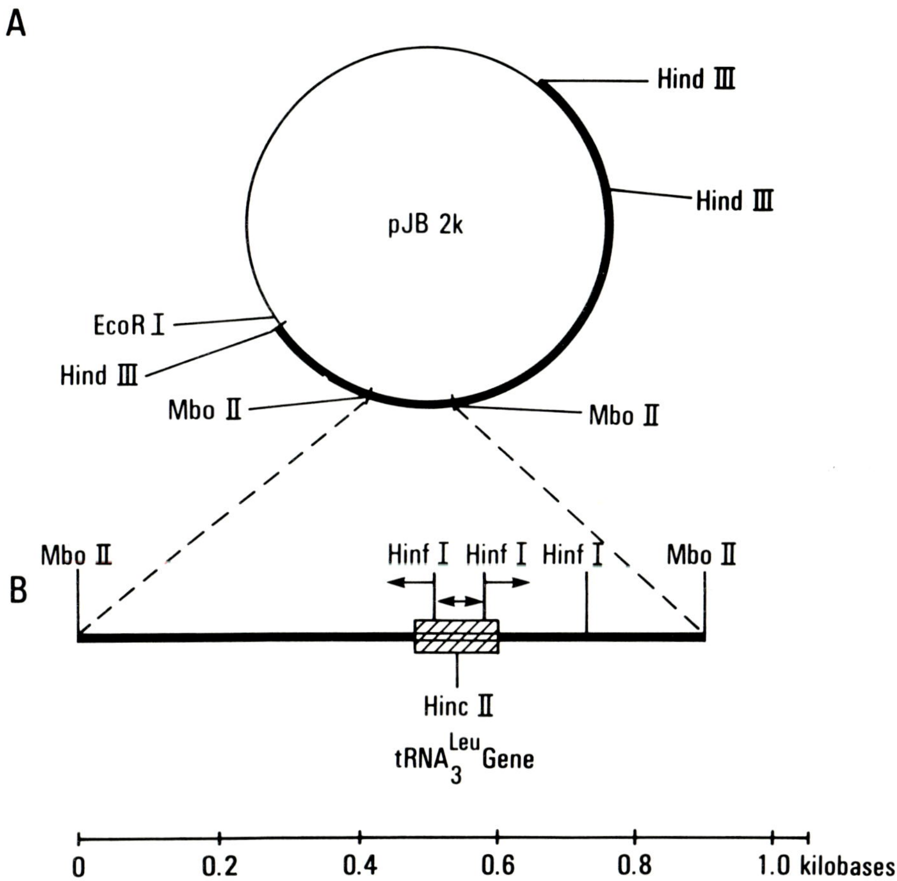

Fig. 2. (A). Schematic diagram of the plasmid containing the yeast tRNA ^{Leu} gene showing the two fused Hind III fragments. (B) Partial restriction map of the Mbo fragment showing the restriction sites used in the sequence determination. The arrows show the direction of the various sequence determinations. The Hinc II site (also Hpa I site) is in the intervening sequence.

within the $tRNA_3^{Leu}$ gene derives from the middle of the intervening sequence. pJB2k DNA digested with Mbo II gives a 900 bp fragment which contains the $tRNA_3^{Leu}$ gene as judged by a Southern blotting experiment. After isolation by agarose gel electrophoresis, this fragment was cut with Hinf I, labelled using DNA polymerase (Klenow) and the four labelled bands separated by polyacrylamide gel electrophoresis. The intervening sequence is contained on a fragment 70 bp long. The appropriate fragments were strand separated and sequenced (16) as shown in Figure 2B.

Figure 3 shows typical sequencing gels of the intervening sequence and junction regions for the two genes. The complete sequences of the coding region, intervening sequence and flanking regions of the two genes are shown in Figure 4 together with the mature tRNA sequences. The DNA sequences confirm the mature tRNA sequences (17,18) and in addition show an intervening sequence in each gene in the region of the anticodon (overlined in the tRNA sequence). The precise position of the intervening sequence cannot be deduced from the DNA sequence but has been determined for both genes by analysis of the intermediates in the RNA splicing pathway (6). As observed in other yeast tRNA genes, the 3' terminal CCA_{OH} sequence of the mature tRNA is not encoded within the genes. Adjacent to the 3' end of the coding region, both genes show tracts of 6 ($tRNA^{Trp}$) and 9 ($tRNA^{Leu}$) A residues on the transcribed strand in a generally AT rich region. There is no significant homology between the two genes in either the intervening sequence (AT rich in both cases) or in the 5' non-coding region. The overall architecture of the two genes will be discussed below.

<u>RNA Sequence Analysis</u>. Precursors to $tRNA^{Trp}$ and $tRNA_3^{Leu}$ were purified by two-dimensional polyacrylamide gel electrophoresis (Figure 5). Following identification and isolation, as described previously (3,10), the precursors were analyzed by standard fingerprinting procedures (11). In each case, the oligonucleotide expected from the mature sequence anticodon loop is absent and several additional unique RNase T1 and RNase A products are observed. Analysis of these oligonucleotides showed that they derive from the intervening sequence and the junction regions between the coding region and the intervening sequence. All the data including molar yield and sequence and

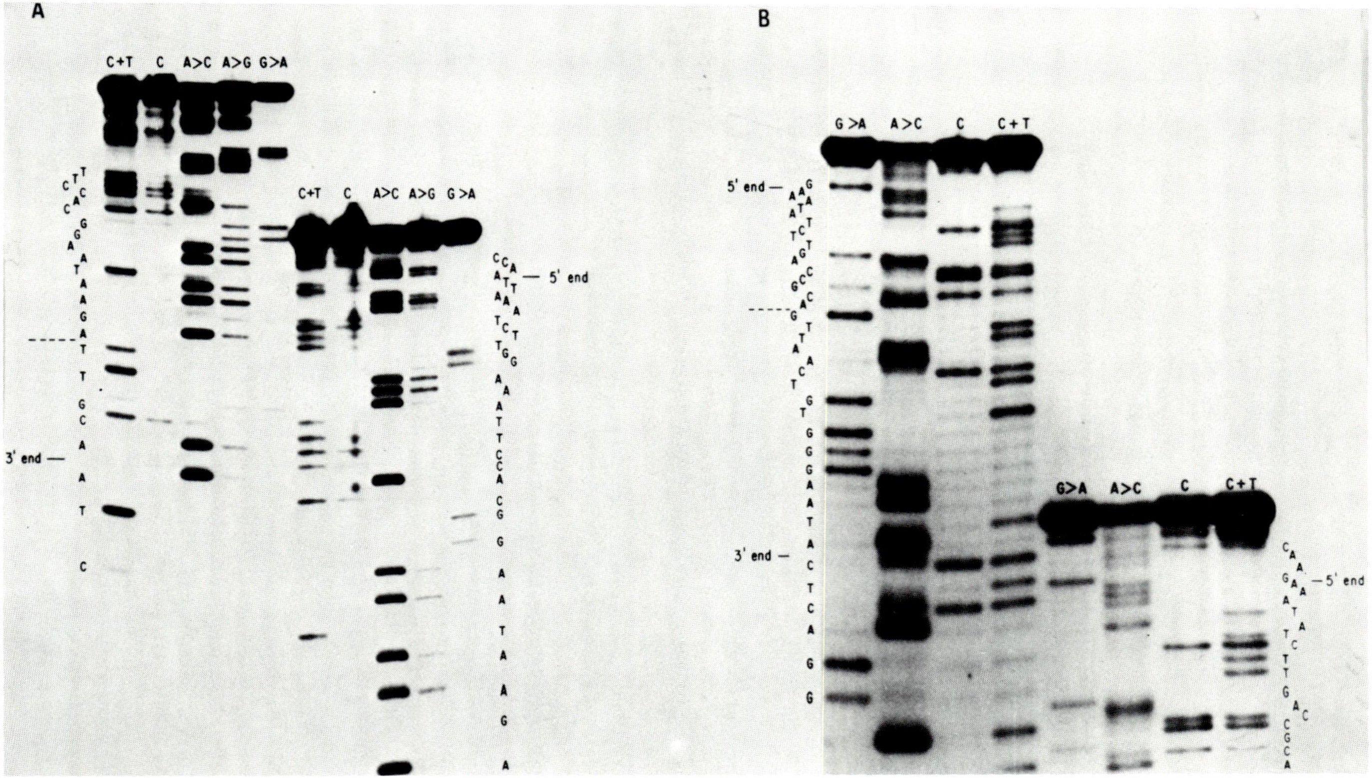

Fig. 3. Sequencing gels (16) of the intervening sequences of the two tRNA genes. (A) The $tRNA^{Trp}$ gene. (B) The $tRNA_3^{Leu}$ gene. The ends of the intervening sequences (corresponding to the precursor tRNA sequence), as determined previously (6) are indicated. The dashed line indicates the position of overlap of the two loadings.

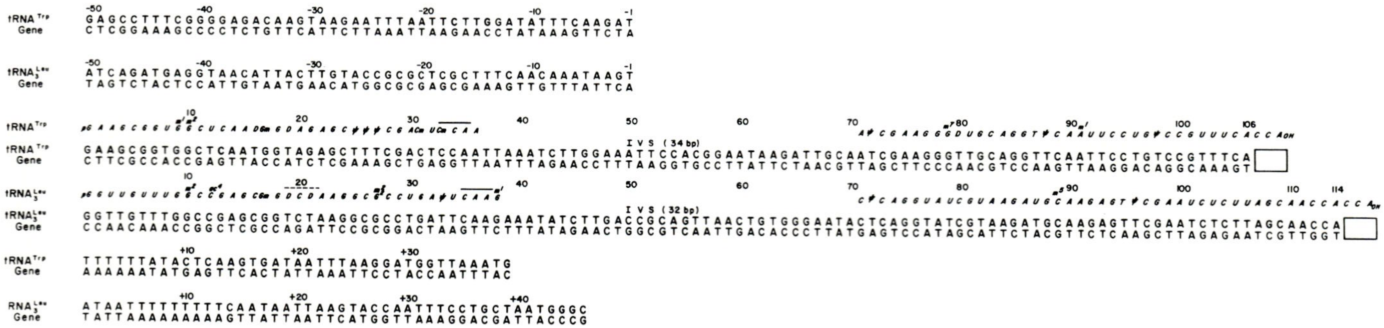

Fig. 4. DNA sequences of the yeast $tRNA^{Trp}$ and $tRNA_3^{Leu}$ genes including their 5' and 3' flanking regions. The primary sequence of each tRNA is shown above the respective gene. The anticodon is overlined in each case. The box at the 3' end of each gene indicates the postion of the uncoded CCA_{OH} terminus.

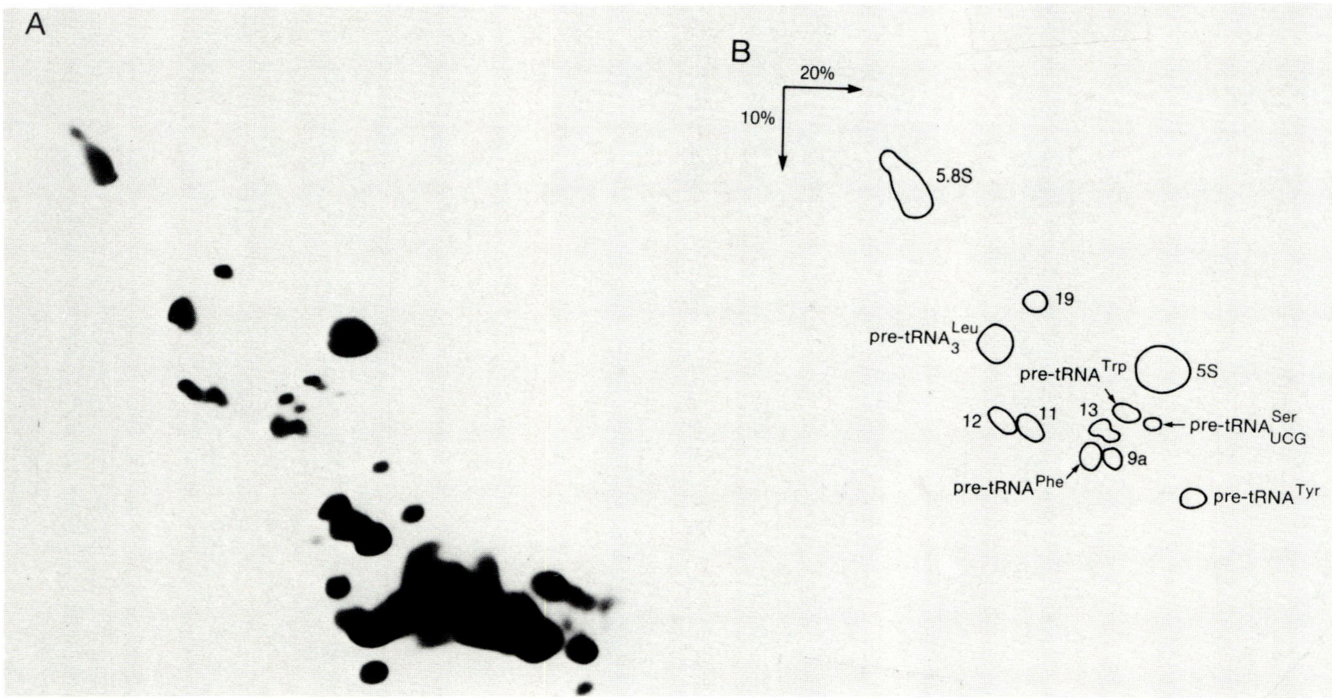

Fig. 5. Two dimensional poluacrylamide gel electrophoresis of *in vivo* ^{32}P-RNA isolated from the yeast *rna1* mutant. (A) Autoradiogram of the two-dimensional separation. (B) Diagrammatic representation and identification of the ^{32}P-pre- tRNAs.

composition analysis for the RNase T1 and RNase A oligonucleotides are consistant with the precursors having mature 5' and 3' termini and being colinear with the sequenced genes. Figure 6 summarizes the data for pre-$tRNA^{Trp}$ and Figure 7 summarizes the data for pre-$tRNA_3^{Leu}$.

DISCUSSION

We have sequenced two cloned yeast tRNA genes for $tRNA^{Trp}$ and $tRNA_3^{Leu}$ including their 5' and 3' flanking regions and demonstrated that each gene contains an intervening sequence. Colinear transcripts of these genes with mature termini and some other post-transcriptional modifications accumulate at the non-permissive temperature in the yeast rnal mutant, ts136. The *in vivo* precursor tRNAs have previously been shown to be substrates for the yeast tRNA splicing system (3). It is evident that the single gene sequenced for each tRNA splicing system (3). It is evident that the single gene sequenced for each tRNA appears to be homologous with the respective precursors from ts136. The fingerprints of the purified precursors to, however, show other oligonucleotides present in submolar yield (for example P10, P11 in Figure 7D) which may represent minor heterogeneities within the intervening sequence as demonstrated in the cases of pre-$tRNA^{Tyr}$ and pre-$tRNA^{Phe}$ (3).

The intervening sequences found in these two genes are larger than previously found in tRNA genes (1,2,19) although we do have preliminary evidence of a still larger intervening sequence in a precursor tRNA of unknown aminoacyl specificity. The implications of this size heterogeneity as regards precursor structure and a common yeast tRNA splicing system have been discussed (8,9). Plausible secondary structure models which incorporate the structural similarities already reported (3,7-9,19) are presented in Figure 8. In this particular model, we hypothesize that in the case of pre-$tRNA^{Trp}$ and pre-$tRNA_3^{Leu}$ additional nucleotides can be accomodated in a hairpin thus maintaining the generalized structure in regions of the splice points and anticodon. It should be emphasized that other base pairing schemes, in which splice points occur in single-stranded regions do exist although preliminary structure probes on pre-$tRNA^{Tyr}$ (4) and pre-$tRNA^{Ser}$ (19) favor the type of model shown. Additional

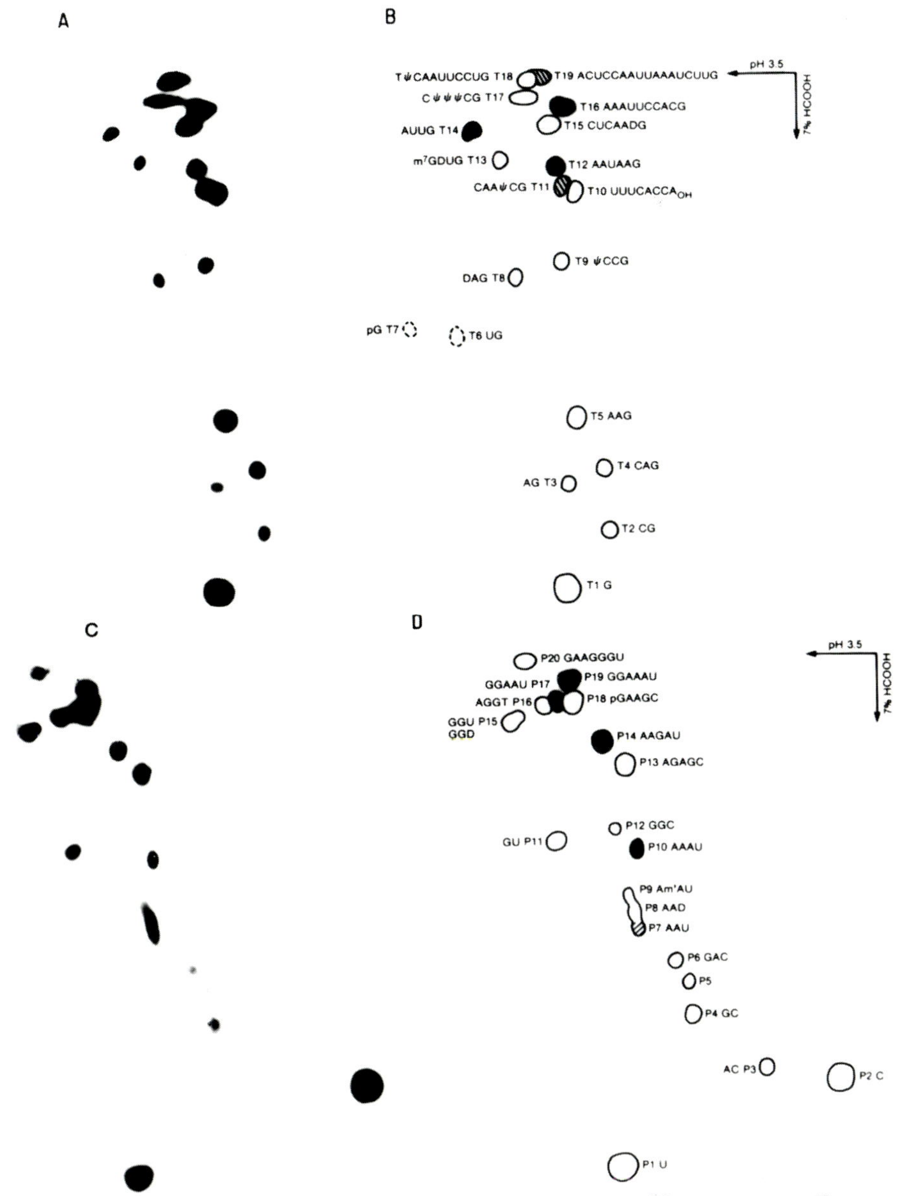

Fig. 6. RNA fingerprint analysis of *in vivo* ^{32}P-pre-tRNA^{Trp}. (A) Autoradiogram of the RNase T1 digestion products. (B) Schematic representation of the subsequent oligonucleotide analysis. (C) Autoradiogram of the RNase A digestion products. (D) Analysis of the RNase A oligonucleotides. Open symbols indicate oligonucleotides common to both pre-tRNA and tRNA. Closed symbols denote those oligonucleotides uniquely derived from the intervening sequence. Hatched symbols designate overlap oligonucleotides which span the junction between the tRNA and intervening sequence. In some cases the sequence of an loigonucleotide was confirmed by comparison with known tRNA or DNA sequence.

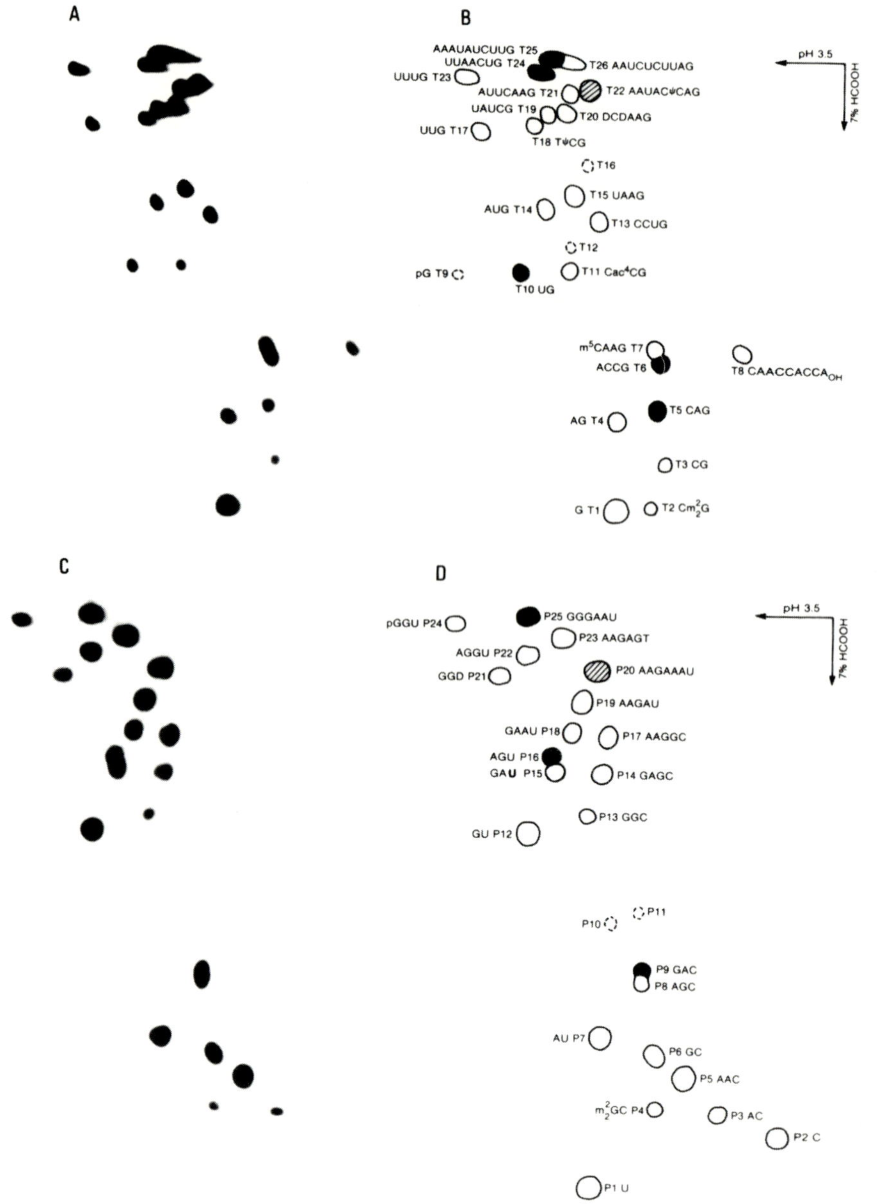

Fig. 7. RNA fingerprint analysis of in vivo ^{32}P-pre-tRNA$^{Leu}_3$. (A) Autoradiogram of the RNase T1 digestion products. (B) Analysis of the RNase T1 oligonucleotides. (C) Autoradiogram of the RNase A digestion products. Nucleotides P1 and P2 were run off to improve resolution of the larger oligonucleotides. (D) Analysis of the RNase A oligonucleotides. Symbols are as in Figure 6.

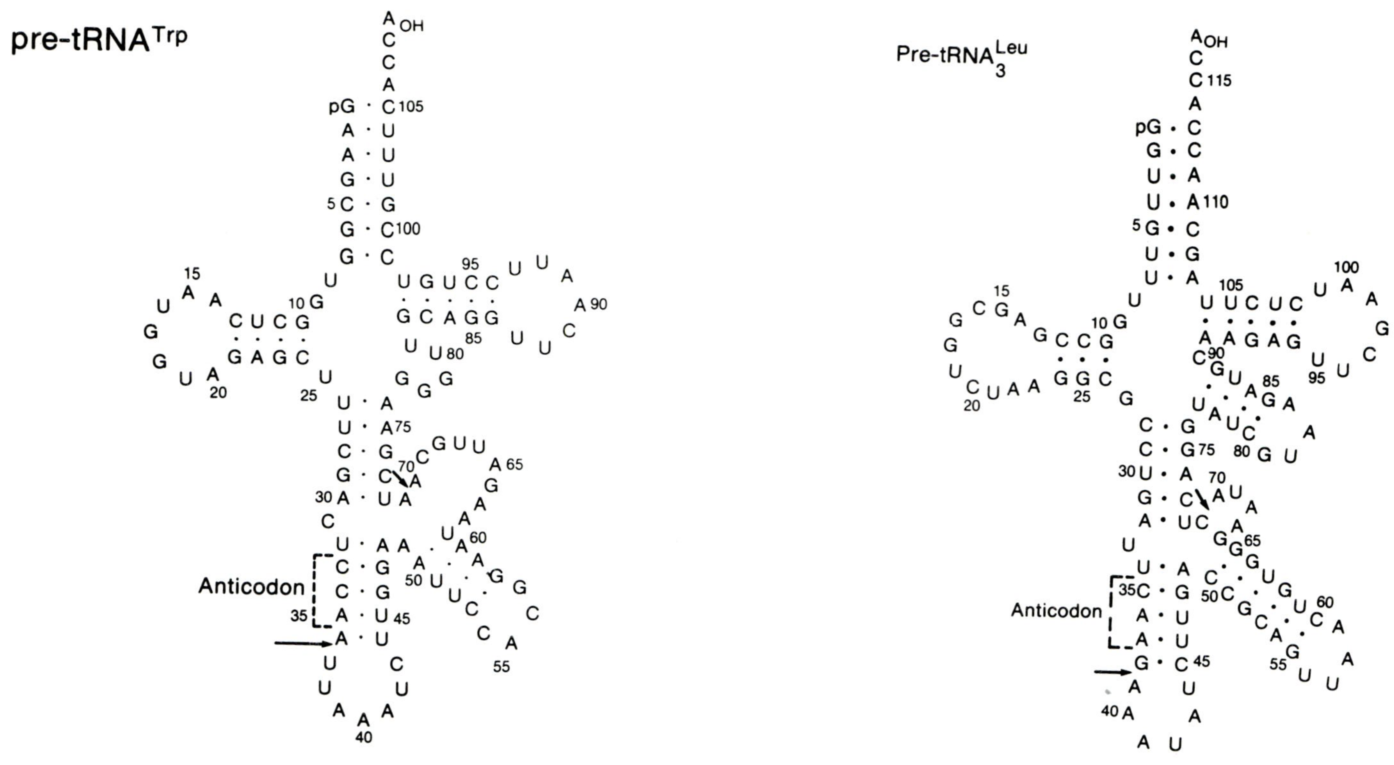

Fig. 8. Secondary structure models for pre-tRNA^{Trp} and pre-tRNA$_3^{Leu}$. The models were constructed to conform to the possible generalized structure for tRNA precursors with intervening sequences. The arrows indicate the ends of the intervening sequences.

sequence data and the variety of chemical and enzymatic techniques now available for exploring RNA structure will help to elucidate the precursor structures.

Homology between the two genes exists beyond the coding region at the 3' end but not at the 5' end. We have investigated *in vitro* transcription of the two sequenced genes in the *Xenopus* germinal vesicle system (10, J. Johnson et al., manuscript in preparation). In each case, one of the transcription products is a 5' and 3' extended precursor. This products is trimmed in both *Xenopus* and yeast extracts in the absence of ATP to give the precursor with mature termini containing the intervening sequence. This is spliced in an ATP-dependent reaction in either system to yield the mature tRNA sequence. In both cases, fingerprint analysis of the largest observed transcript suggestss a short 5' extension (a 5'-triphosphate has been detected at position -2 for the $tRNA^{Trp}$ gene - O. Schmidt, personal communication). At the 3' end, the RNAs extend into the common stretch of U residues. The relevance of these observations to possible sequences involved in polymerase binding, initiation and termination in the heterologous system are discussed elsewhere (J. Johnson et al., manuscript in preparation).

ACKNOWLEDGEMENTS

This work was supported by NIH Grant No. CA10984.

REFERENCES

1. Goodman, H.M., Olson, M.V. and Hall, B.D. (1977) Proc. Natl. Acad. Sci., U.S.A. 74, 5453-5457.
2. Valenzuela, P., Venegas, A., Weinberg, F., Bishop, R. and Rutter, W.J. (1978) Proc. Natl. Acad. Sci., U.S.A. 75, 190-194.
3. Knapp, G., Beckmann, J.S., Johnson, P.F., Fuhrman, S.A. and Abelson, J.N. (1978) Cell 14, 221-236.
4. O'Farrell, P.Z., Cordell, B., Valenzuela, P., Rutter, W.J. and Goodman H.M. (1978) Nature 274, 438-445.
5. Peebles, C.L., Ogden, R.C., Knapp, G., and Abelson, J. (1979) Cell 18, 27-35.
6. Knapp, G., Ogden, R.C., Peebles, C.L., and Abelson, J. (1979) Cell 18, 37-45.

7. Hopper, A.K., Banks, F., and Evangelidis, V. (1978) Cell 14, 211-219.
8. Ogden, R.C., Knapp, G., Peebles, C.L., Kang, H.S., Beckmann, J.S., Johnson, P.F., Fuhrman, S.A., and Abelson, J., and Schimmel, P. Eds. Cold Spring Harbor Laboratory. In Press.
9. Kang, H.S., Ogden, R.C., Knapp, G., Peebles, C.L. and Abelson, J. (1979) in Eukaryotic Gene Regulation. ICN-UCLA Symposia on Molecular and Cellular Biology. Academic Press. In Press.
10. Ogden, R.C., Beckmann, J.S., Kang, H.S., Abelson, J., Soll., D. and Schmidt, O. (1979) Cell 17, 399-406.
11. Brownlee, G.G. (1972) in Laboratory Techniques in Biochemistry and Molecular Biology. T.S. Work and E. Work, Eds. (New York: American Elsevier Publication Co., Inc.) pp. 67-99.
12. Beckmann, J.S., Johnson, P.F. and Abelson, J. (1977) Science 196, 205-208.
13. Beckmann, J.S., Johnson, P.F., Knapp, G., Sakano, H., Fuhrman, S.A., Ogden, R.C., and Abelson, J. (1979) in Nonsense Mutations and tRNA Suppressors. Celis, J.E., and Smith, J.D., Eds., pp. 207-234, Academic Press.
14. Nishimura, S. (1972) in Progress in Nucleic Acid Research and Molecular Biology, Davidson, I.N., and Cohn, W.E., Eds., 12, 49-85.
15. Southern, E.M. (1975) J. Mol. Biol. 98, 503-517.
16. Maxam, A.M., and Gilbert, W. (1977) Proc. Natl. Acad. Sci. U.S.A. , 74, 560-564.
17. Keith, G., Roy, A., Ebel, J.P., and Dirheimer, G. (1971) FEBS Lett. 17, 306-308.
18. Kowalski, S., Yamane, T. and Fresco, J.R. (1971) Science 172, 385-387.
19. Etcheverry, T., Colby, D., and Guthrie, C. (1979) Cell 18, 11-26.

DISCUSSION

R.W. CHAMBERS: I have two questions John. Is there any acceptor activity of the pre-tRNA's - the ones with the mature ends containing the insert?

J. ABELSON: We have not done that experiment but O'Farrell et al. did do the experiment for the pre-tRNA ^{Tyr} and detected no acceptor activity for the precursor.

R.W. CHAMBERS: Has anyone done any mis-acylation studies?

J. ABELSON: We have not. I do not know if anyone in the Goodman, Rutter group has done anything on them.

R.W. CHAMBERS: All of the tRNA's that you have listed there have purines just after the anti-codon, but they are all unmodified. Is there any data to show whether or not there is cutting and splicing with tRNA's containing modified purines.

J. ABELSON: We do not know that because no precursor occurs with a modified nucleotide in that position.

R.W. CHAMBERS: I guess what I am really asking is this: All of those pre-tRNA's you showed end up without a modification in the purine. There is another group that ends up with a modification. Do the latter pre-tRNA's belong to the group that do not contain the intervening sequences?

J. ABELSON: Of course, the $tRNA_2^{Ser}$ + $tRNA^{Ser}_{UCA}$ are the most striking cases of that. This is a set of tRNA's which do have a hyper-modified nucleoside, isopentenyl adenosine, and do not have intervening sequences. I think that probably the presence of an intervening sequence has nothing to do with whether or not that base is modified, although one might have supposed that it was a feature of the modification reaction. We do know that tRNA's whose modification has been removed can be modified in the mature transfer RNA, so it is unlikely that this is a mechanism for example in the synthesis of isopentenyl adenosine or the Y base in $tRNA^{Phe}$.

I.B. DAWID: Does the ligase make hybrid molecules between halves of different tRNA's?

J. ABELSON: We have tried to do this in several cases and it does not work. We conclude that probably a mature tRNA-like structure is required by the ligase and that is supported by the finding that the ligase is quite strongly inhibited by mature tRNA.

A. BOLLON: Does the presence of the respective tRNA synthetase interfere with any of the splicing events?

J. ABELSON: I doubt it since the splicing reaction goes perfectly well in the crude extract where those enzymes are present, but the experiment has not been done with an excess of purified tRNA synthetase.

I. HERSKOWITZ: You alluded to the fact that your studies ight aid in understanding messenger RNA processing. Is there evidence other than for mitochondrial mRNA's that yeast RNA's have intervening sequences?

J. ABELSON: You will hear of some evidence for splicing in the mitochondrial system this afternoon. In addition we have obtained evidence for the presence of an intervening sequence in the actin gene of yeast. Ray Ng was able to clone the yeast actin gene because it cross-hybridizes with a Dictyostelium actin gene which had been cloned by Rick Firtel in San Diego. There is a single actin gene in yeast. Ng's sequence studies revealed that this gene contains at its 5 prime end an intervening sequence of 310 base pairs. This intervening sequence is between the codons in actin and his sequencing of the actin mRNA confirms that the second and third messenger RNA does not contain the intervening sequence, so yeast must have a splicing mechanism as in other eukaryotic cells for the processing of nuclear messenger RNA. This had not been discovered before because only a few of the genes studied were ones that were not selected by being able to function in E. coli. I think if a gene can function in E. coli it is likely that it does not contain an intervening sequence. So that is where we are on that.

R.W. CHAMBERS: Just a comment concerning the question I asked earlier. There have now been 8 tRNA's looked at by modification with HSO_3^-. 5 of those 8 show very clearly that the anticodon is involved in recognition for aminoacylation. I think it is interesting that there is no activity with precursor molecules in view of the hypothetical tertiary structure you showed where the anticodon presumably would be inaccessible, at least distorted, for the recognition process involved in aminoacylation.

SYNTHESIS AND PROCESSING OF THE MOUSE BETA GLOBIN mRNA PRECURSOR

Jeffrey Ross[1]
Alan J. Kinniburgh[1]
Janet Mertz[1]

Department of Oncology
University of Wisconsin
Madison, Wisconsin

I. INTRODUCTION

The experiments described here concern the transcription and processing of RNA from structural genes in mammalian cells. The synthesis of globin mRNA in nucleated mouse erythroid cells has been exploited as a model system to investigate specific mRNAs, rather than total cell mRNA, which consists of thousands of structurally distinct molecules. Ambiguous results and interpretations that might arise as a result of the size heterogeneity and sequence complexity of total mRNA (1) can be minimized in experiments focusing on specific mRNAs.

Experiments described several years ago indicated that the mouse alpha and beta globin genes are each transcribed into separate precursor RNAs (2-5) that are approximately 1.5- and 3-fold larger than cytoplasmic (mature) alpha and beta globin mRNAs, respectively. These experiments were performed by analyzing pulse-labeled RNA extracted from cultured nucleated erythroid cells, under conditions in which artifacts related to mRNA aggregation were minimized. Subsequent experiments revealed that the processing of the large precursors occurred in several steps, involving distinct RNA molecules intermediate in size between the precursor and the mature mRNA (6).

---

[1]Supported by NIH grants CA-07175, CA-23076, and CA-22484.

ISBN 0-12-633360-2

These experiments raised two major questions: 1) what are the structures of the precursors?; 2) by what pathway are the precursors processed to mature mRNAs? Here we describe experiments designed to characterize the major structural features of the beta globin mRNA precursor and the mechanism by which it is processed.

MATERIALS AND METHODS

Procedures for the preparation and culture of mouse fetal liver cells, labeling and isolation of RNA, nucleic acid hybridization, and polyacrylamide gel analyses were performed as previously described (2,6).

RESULTS

Initial experiments from this laboratory indicated that the beta globin mRNA precursor contains approximately 1800 nucleotides, subdivided into 3 components: structural or cytoplasmic sequences corresponding to those found in the mature mRNA (690 nucleotides), poly A (150 nucleotides), and RNA sequences present in the nuclear precursor but not in the cytoplasmic RNA (960 nucleotides). The RNA sequences unique to the precursor are cleaved from it during the post-transcriptional processing reactions. Several experiments were performed to determine the structural organization of these three components. One approach was to form R loops between the precursor RNA and cloned DNA derived from cytoplasmic beta globin mRNA (7). Such R loops should consist of mRNA-specific nucleotides and poly A, both of which anneal to the cloned DNA segment, plus precursor-specific RNA, which does not anneal to the DNA. In control experiments cytoplasmic globin mRNA formed a smooth R loop circle with cloned cDNA (Figure 1, top), because the DNA and RNA sequences were totally complementary. In contrast, two regions of non-hybridized RNA were observed in R loops with precursor RNA (Figure 1, bottom). One region formed a "head", a stretch of RNA emerging from one end of the R loop and identified in independent experiments as the 5' terminal segment of the precursor. The second non-annealed region formed an internal loop near the middle of the DNA-RNA portion of the R loop. This structure is diagnostic of an intervening sequence (IVS) within the RNA. Biochemical experiments confirmed independently that the beta globin mRNA precursor in fact contained two intervening sequences, a larger

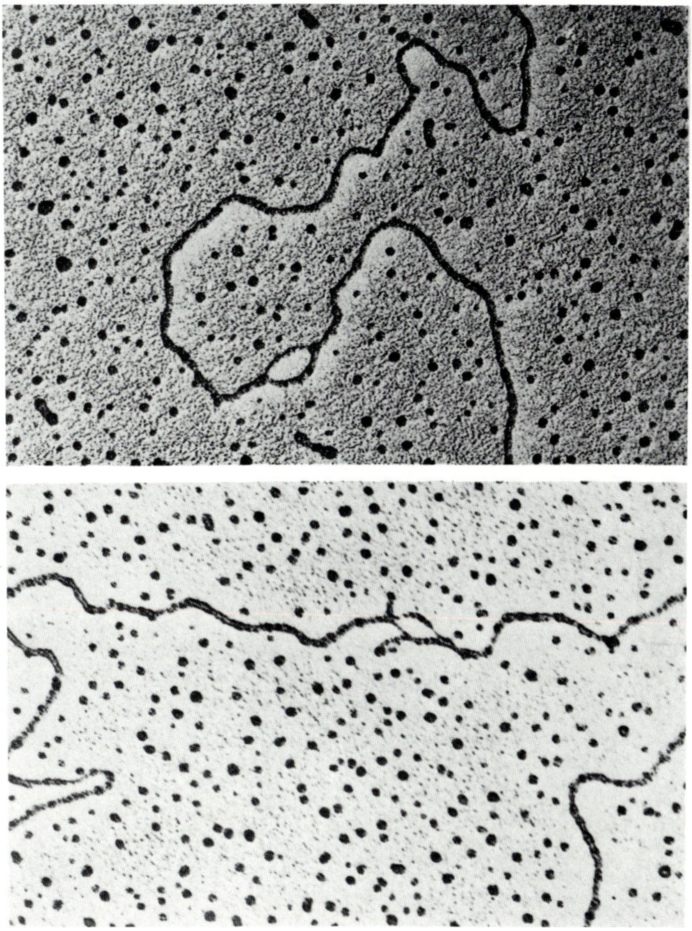

mRNA

PRECURSOR

Fig. 1. R loops between cytoplasmic beta globin mRNA sequences cloned into bacteriophage lambda and cytoplasmic globin mRNA (top) or the beta globin mRNA precursor (bottom).

one corresponding to that observed in the R loop of Figure 1 plus a second, smaller IVS (too small to be observed in the electron microscope) between the large R loop and the 5' terminus (see reference 7 for details). We conclude that the precursor contains two IVS as does the beta globin gene (8,9), plus extra nucleotides at the 5' terminus (see Figure 5 for a summary).

To determine the molecular weights of the processing intermediates, pulse-labeled mouse fetal liver total cell RNA was electrophoresed in a formamide-containing polyacrylamide gel, and RNA from each slice was eluted and hybridized to

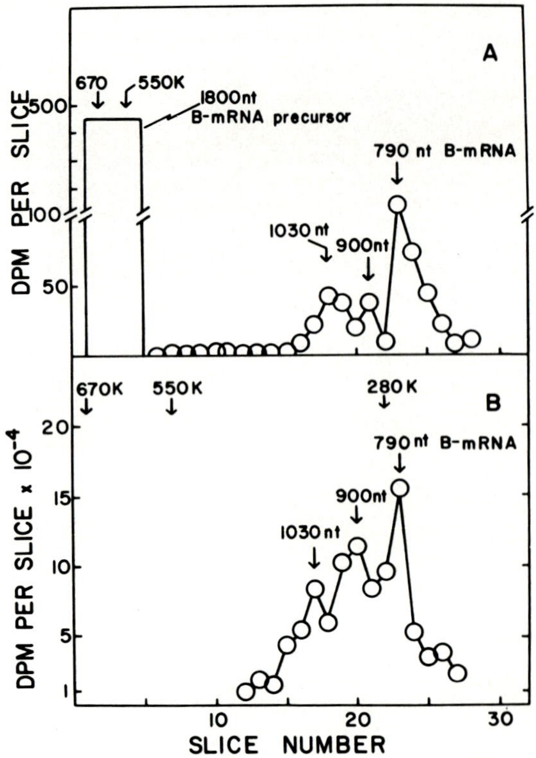

Fig. 2. Analysis of pulse-labeled mouse fetal liver cell beta globin mRNA processing intermediates. RNA was electrophoresed in a formamide-containing polyacrylamide gel, and slices were analyzed by hybridization to cloned beta-specific DNA or filters (10). Panels A and B are separate experiments. The precursor region of the gels was pooled into a single fraction. © The MIT Press

cloned beta globin DNA immobilized on filters. The precursor (1800 nucleotides) plus two classes of smaller RNA, containing approximately 1000 and 900 nucleotides, and mature beta globin mRNA (790 nucleotides) were observed (Figure 2) (10). Pulse-chase experiments were attempted, to determine conclusively if the 1000 and 900 nucleotide RNAs were obligate processing intermediates, but the precursor turns over so rapidly that it is not feasible to label it without labeling the smaller RNAs as well. Therefore, kinetic experiments demonstrating a flow of label from precursor to intermediates to mRNA were not feasible. However, since there is no apparent wastage of globin-specific RNA, as determined by kinetic experiments on total pulse-labeled globin RNA sequences (2), it seems quite likely that these RNAs are processing intermediates.

This conclusion was strengthened by the following biochemical analysis, which revealed that these RNAs were structurally related to mature beta globin mRNA. The intermediate size RNAs were purified, and the structure of each RNA was analyzed by the RNase protection method described by Kinniburgh, Mertz, and Ross (7). The data show that the 1,000 nucleotide RNA contains the same three structural globin mRNA segments that are found in the precursor, plus two IVS (Figure 3). Therefore, although 800 nucleotides have been excised from the precursor to generate this smaller RNA, there are still two IVS in this RNA, located at the same positions as the original IVS in the precursor. We conclude that a portion, but not all of the larger IVS is removed during the first processing step. The IVS nucleotides that are not excised by this cleavage reaction are ligated back together, creating a small IVS where the larger IVS once existed. We designate this process as a partial splicing reaction. Partial splicing of the smaller IVS may occur as well.

The splice pattern of the 900 nucleotide intermediate was analyzed in the same manner, and five RNA peaks were observed (Figure 4A). RNAs I through III correspond to the mRNA segments of the precursor and the 1000 nucleotide intermediate. The RNA in peak V contains 790 nucleotides and represents fully spliced beta globin mRNA molecules which contaminate the 900 nucleotide RNA (see Figure 1B). The RNA in peak 4 contains approximately 340 nucleotides, which is the expected size of an RNA generated by the fusion of fragments II and III. Appropriate controls demonstrated that these fragments were not generated by random cleavages (Figure 4C). These data indicate that the 900 nucleotide fraction consists of two RNAs, one with two IVS and three mRNA segments, the other with one IVS, one mRNA segment (I), and a fusion product of two mRNA segments, II + III.

A model describing the major transcriptional and processing events required to generate mature beta globin mRNA from the precursor is shown in Figure 5. Although there are some published data suggesting that there exist beta globin precursors larger than 1800 nucleotides (4), we and others have been unable to find such large molecules (3,5,6), and one report indicates that the high molecular weight "precursors"

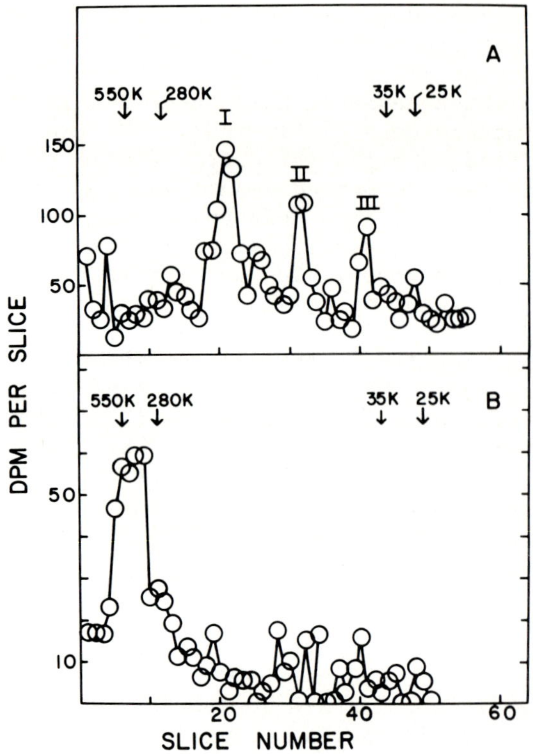

Fig. 3. Analysis of the 1000 nucleotide intermediate RNA by the RNase protection method (7). Panel A: Radioactive, purified 1000 nucleotide RNA was annealed to unlabeled cDNA, treated with RNase A, and RNase-resistant fraction was electrophoresed. Panel B: Similar protocol, but RNase omitted.

previously identified are the result of an artifact of the cDNA cellulose hybridization method (11).

We have also been unable to find conclusive evidence for globin gene transcription in non-erythroid cells. Several laboratories have reported that globin mRNA nucleotide sequences exist in non-erythroid cells, suggesting that the globin genes are transcribed in these cells, but implying that the transcripts are not stable (12,13). On the other hand, data from experiments with cultured cells (14) and with

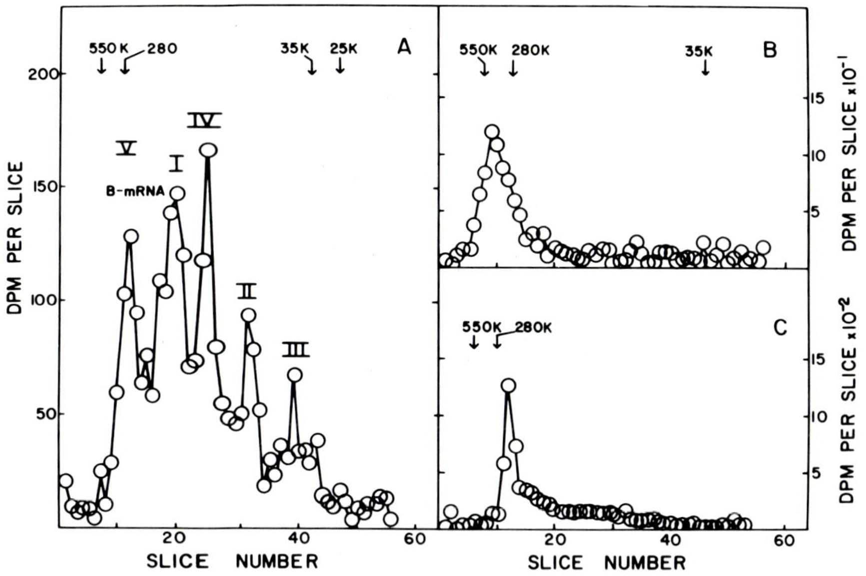

Fig. 4. Analysis of the 900 nucleotide RNAs by RNase protection (7). Panel A: Purified 900 nucleotide RNA annealed to cDNA, treated with RNase A, and the RNase-resistant material electrophoresed. Panel B: As per panel A, but RNase omitted, Panel C: Cytoplasmic (mature) globin mRNA treated as per panel A.

in vitro nuclear transcription systems (15) support the notion of control of globin gene expression at the transcriptional level. In initial experiments from our laboratory, we detected little, if any, globin mRNA nucleotide sequences in nuclear RNA or total cell RNA from BALB/3T3 fibroblasts (JR, unpublished data). To investigate this question with cells closer to the hematopoietic lineage (but not with animal tissue, in which even minor contamination with reticulocytes would yield false positive artifacts), nuclear RNA was

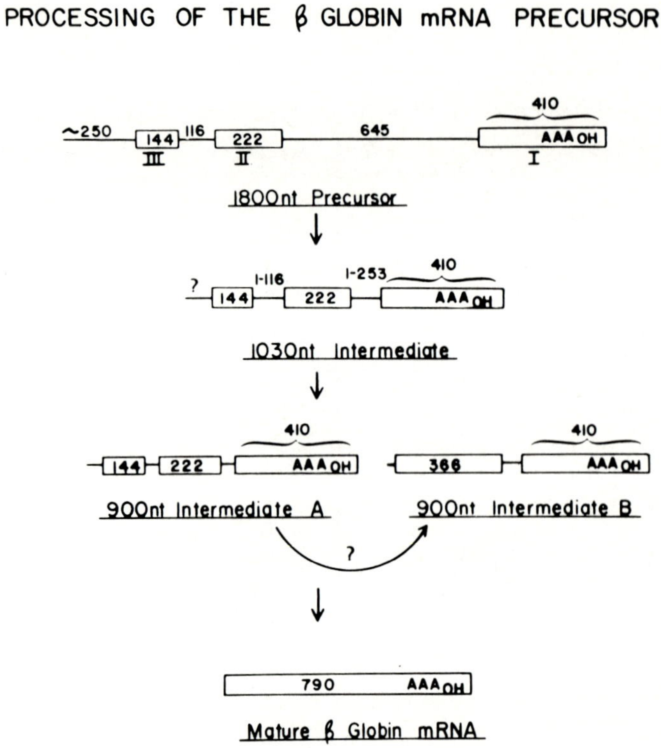

Fig. 5. Model for the processing of the mouse beta globin mRNA precursor.

isolated from a line of Abelson virus-transformed lymphoid cells (18-4, obtained from Dr. R. Risser). Large quantities of nuclear RNA were hybridized with highly radioactive globin cDNA (prepared with alpha plus beta globin mRNA), and the percent hybridization was compared with that of authentic globin mRNA. The results indicate that there is less than one full length globin mRNA molecule per nucleus (Table 1). The maximum extent of hybridization with RNA dissolved to the limit of its solubility was only 6% (data not shown). Such low hybridization could be the result of rare sequences in other RNA molecules that are shared with small portions of alpha and beta globin mRNAs. Therefore, we are unable to find conclusive evidence for globin gene transcription in non-erythroid cells.

TABLE I. ANALYSIS OF LYMPHOID CELL RNA BY ANNEALING TO GLOBIN ^{3}H-cDNA

RNA	Hybridization (%)	Number of full-length globin mRNA molecules hybridized
Globin mRNA (1.2 pg)	6	1.2×10^7
18-4 (332 μg; from 3.8 X 10^8 nuclei	6	1.2×10^7

DISCUSSION

The major finding from these experiments is that the mouse beta globin mRNA precursor is processed to functional mRNA by stepwise removal of intervening sequence nucleotides (see Figure 5). The first processing step involves cleavage of the 1800 nucleotide precursor to a 1000 nucleotide intermediate. Most of the 800 nucleotides are excised by partial splicing of the larger IVS. A portion of the smaller IVS and some or all of the 5'-terminal RNA might be removed in the same reaction. A second processing step shortens this intermediate by 100 nucleotides. One of the 900 nucleotide intermediates contains 2 IVS, and it probably arises from the 1000 nucleotide intermediate by additional partial splicing reactions. The second 900 nucleotide RNA is a partially ligated molecule, consisting of fused 5' terminal globin segments, approximately 100 nucleotides of the larger IVS, and the 3'terminal mRNA fragment with its poly(A). It is not clear from our data if the 900 nucleotide molecules A and B (Figure 5) are kinetically related, or if they are each independently derived from the 1000 nucleotide intermediate.

These data demonstrate that some IVS are themselves composed of one or more IVS. As a result, processing of this precursor involves stepwise cleavage-ligation reactions, all but the last of which generate new IVS. These results raise the following questions: 1. Is more than one enzyme required for each processing reaction? 2. What are the nucleotide sequences at which the initial cleavage-ligation reaction takes place? 3. Could a mutation in an IVS affect the rate or extent of processing of mRNA precursors? This question is particularly intriguing, in view of the possibility that beta+ thalassemia is due to abnormal processing of the beta globin, but not the alpha globin mRNA precursor.

ACKNOWLEDGMENTS

We are grateful to Lorri Steffen and Shi-Da Yu for excellent technical assistance and to Karen Denk for typing the manuscript.

REFERENCES

1. Lewin, B., Cell 4, 11 (1975).
2. Ross, J., J. Mol. Biol. 106, 403 (1976).
3. Curtis, P. J., and Weissman, C., J. Mol. Biol. 106, 1061 (1976).
4. Bastos, R. N., and Aviv, H., Cell 11, 641 (1977).
5. Kwan, S.-P., Wood, J. G., and Lingrel, J. B., Proc. Nat. Acad. Sci., USA 74, 178 (1977).
6. Ross, J., and Knecht, D. A., J. Mol. Biol. 119, 1 (1978).
7. Kinniburgh, A. J., Mertz, J. E., and Ross, J., Cell 14, 681 (1978).
8. Jeffreys, A. J., and Flavell, R. A., Cell 12, 1097 (1977).
9. Tilghman, S. M., Tiemeier, D. C., Seidman, J. G., Peterlin, B. M., Sullivan, M., Maizel, J. V., and Leder, P. Proc. Nat. Acad. Sci., USA 75, 725 (1978).
10. Kinniburgh, A. J., and Ross, J. Cell 17, 915 (1979).
11. Haynes, J. R., Kalb, F. V., Rosteck, P., and Lingrel, J. B. FEBS Letters 91, 173 (1978).
12. Humphries, S., Windass, J., and Williamson, R. Cell 7, 267 (1976).
13. Perlman, S. M., Ford, P. J., and Rosbash, M. M. Proc. Nat. Acad. Sci., USA 74, 3835 (1977).
14. Lowenhaupt, K., Trent, C., and Lingrel, J. B. Dev. Biol. 63, 441 (1978).
15. Orkin, S. H., and Swerdlow, P. S. Proc. Nat. Acad. Sci., USA 74, 2457 (1977).

THE ROLE OF RNA SPLICING IN REGULATION OF VIRAL GENE EXPRESSION

G. Khoury, J. Alwine, R. Dhar, N. Goldman, P. Gruss, G. Jay, S. Segal and I. Seif

Laboratory of Molecular Virology
National Cancer Institute, National Institutes of Health
Bethesda, Maryland 20205

We have been interested for some time in eukaryotic gene regulation, particularly at the level of RNA synthesis and processing. The model system we have chosen to study is the lytic cycle of SV40. The advantages of using this virus have become clear over the past several years. These include a small double stranded DNA genome which is easily manipulated and mutagenized. Furthermore, the biology of the virus, both in the lytic interaction with permissive African green monkey kidney cells and in the abortive cycle in non-permissive (transforming) rodent cells is known in greater detail than that of any other animal virus (see 1,2). Finally, we now know the entire nucleotide sequence of the viral DNA, and extensive sequences of the spliced viral mRNA's (3,6).

The use of an animal virus to investigate the regulatory signals involved in gene expression, however, embodies the underlying assumption that the nucleotide sequences which control viral and eukaryotic genes are identical. At the level of mRNA synthesis and processing, a number of observations lead us to believe that transcription of SV40 genes provides an accurate model for transcription of genes in eukaryotic organisms:

1. SV40 is a small DNA virus which carries into the infected cell no known enzymes. Thus it is entirely reliant on the cellular "machinery", at least for the expression of its early genes.
2. SV40 transcription is initiated by alpha amanitin-sensitive cellular polymerase II.
3. Post-transcriptional processing of SV40 mRNA involves the same steps as does the processing of cellular mRNAs--capping, polyadenylation, splicing, and internal methylation.

During the past few years we have been involved in studies of post-transcriptional processing at the level of mRNA splicing. A quantitative and qualitative analysis of the viral RNA synthesized in cells infected by SV40 and its mutants has permitted us to draw a number of conclusions regarding the potential mechanism and function of splicing

ISBN 0-12-633360-2

events. In addition to the analysis of the primary RNA sequence of preferred donor and acceptor splice sequences (7,8), we have obtained evidence that suggests that RNA conformation plays an essential role in splicing reactions (9, our unpublished results). We present the results of studies suggesting that splicing may be a required function for the stabilization of a number of viral, and presumably, eukaryotic mRNAs (10-12). We also describe an approach in progress, which employs recombinant molecules to assess genetic regulatory signals.

RESULTS AND DISCUSSION

Nucleotide Sequence at Splice Sites

Initial studies by Breathnach et al. (13) suggested that nucleotide sequences around splice sites were not random, but followed a general form. This "rule" was expanded upon as described below.

An analysis of the primary sequence of regions involving splice sites was undertaken to provide an insight into the mechansim of RNA splicing. We determined the DNA sequences that corresponded to the early and late splice sites of human papova-virus BKV transcripts. A comparison of these sequences with those of other viral and eukaryotic mRNAs (see 7) has indicated a number of similarities, which are discussed below. We refer to the RNA nucleotides (or their equivalents in the DNA of the genome) involved in the ligation process, as donor (D) and acceptor (A) nucleotides. D and A bear a 5'→3' location to one another in the spliced transcript. We refer to the regions that surround D and A in the genome or the primary transcript, as donor site (DS) and acceptor site (AS).

Sequences Characteristic of Donor and Acceptor Sites

A list of the nucleotide sequences associated with various eukaryotic splice sites is presented in Fig. 1. As the intervening sequences (IVS) vary considerably in size, they have been only partially represented in this Figure, except for the smallest IVS of SV40 and BKV, which are only 31 nucleotides long and are presented in the entirety (marked by a central horizontal bar). We have analyzed the 20-30 nucleotides at the donor and acceptor sites because deletion mutants of SV40 suggest that the nucleotides close to the excision points are sufficient for the splicing reactions (C.-J. Lai et al., in prep.). In the cases reported in Fig. 1, the same sequences are present at both donor and acceptor sites; these repeated sequences would allow splicing

	D donor	GT	AG	A acceptor		
Ovalbumin	AAGCTCAAAAG	GTAGGCAACTCTC	TTCTGCTGTTTGCTCTAG	ACAACT		
	AAATAAATAAG	GTGAGCCTACAGT	TATTATTTTCAATTACAG	GTTGTT		
	TTGAAGCTCAG	GTACAGAAATAAT	CTTTCTCTTTGTATTCAG	TGTGGC		
	CAATCCTGCCA	GTAAGTTGCTCTA	AATACTCTTGCTTTACAG	GAATAC		
	TCAGACAAATG	GTAAGGTAGAACA	TCCTATTCATTCTTAAAG	GAATTA		
	GAGTGACTGAG	GTATATGGGCATA	TTTTTTTGGTTGCTCCAG	CAAGAA		
	GCCTTGAGCAG	GTATGGCCCTAGA	ATTCTCATTTCCTTGCAG	CTTGAG		
Hemoglobin	GCCCTGGGCAG	GTTGGTATCCAGG	CTATGTTTCCCTTTTTAG	GCTGCT	β	Mouse
	AGAACTTCAGG	GTGAGTCTGATGG	TCTTCCATATTCCCACAG	CTCCTG	β	
	TCAACTTCAAG	GTATGCGCTGGGA			α	
	GCCCTGGGCAG	GTTGGTATCCTTT	CTGTTTTCATTTTCTCAG	GCTGCT	β	Rabbit
	AGAACTTCAGG	GTGAGTTTGGGGA	TTCTTCTTTTTCCTACAG	CTCCTG	β	
Immunoglobin	TCTCTGCTCAG	GTCAGCAGCCTTT	TTTCTTACCTGTTTGCAG	GAGCCA	λ2	Mouse
	TCTCAGCTCAG	GTCAGCAGCCTTT	TTTCTCATCTGTTTGCAG	GGGCCA	λ1	
	GACTGTCCTAG	GTGAGTCACTCCT	TCTTACTTCATCCTGCAG	GCCAGC	λ1	
			ATATACTTTTTCTTGTAG	CCAAAA	γ1	
	CAAGAAAATTG	AGAGGACATATAG	ATAGCTTTCTCTCCACAG	TGCCCA	γ1	
	CATATGTACAG	GTAAGTCAGTC	TTTCTTCTTCATCCTTAG	TCCCAG	γ1	
	CAAAACCAAAG	GTGAGAGCTGCAG	CTCACTCACAG	GCAGAC	γ1	
Adenovirus	CAGCTGTTGGG	GTGAGT		CTCGCG	Fiber	
			GTGTAG	GTACTC		
	ACAGTCGCAAG	GTAGGC	CAACAG	TTTCCA		
	CAAAACAGGAG	GTGAGC	TTGCAG	ATGAAA		
			CGTCCATGTCGCCGCCAG	AGGAGC	Hexon	
SV40	ATGCAACTGAG	GTATTTGCTTCTT	ATTGTTTGTGTATTTTAG	ATTCCA	D4-A4	
			TGTGTTTGTGTATTTTAG	ATTC	dl 1440	
			TCTGCTTGTGTATTTTAG	ATTC	dl 1410	
			AACTGCTGTGTATTTTAG	ATTC	dl 884	
	AAAGCTATAAG	GTAAATATAAAAT	ATTGTTTGTGTATTTTAG	ATTCCA	D5-A4	
	CGCCTCAGAAG	GTACCTAACCAAG	TTTTGTCTTTTATTTCAG	GTCCAT	D1-A2	
	CGCCTCAGAAG	GTACCTAACCAAG	TTTTTGTGTTTGTTTTAG	AGCTTT	D1-A1	
	AAAGTTAACTG	GTAAGTTTAGTCT —	— TTTTGTCTTTTATTTCAG	GTCCAT	D3-A2	
	AAAGTTAACTG	GTAAGTTTAGTCT	TGTTGCCTTTACTTCTAG	GCCTGT	D3-A3	
	CCTCCGTTAAG	GTTCGTAGGTCAT	TTTTGTCTTTTATTTCAG	GTCCAT	D2-A2	
BKV	GTAGCTCAGAG	GTTTGTGCTGATT	TAATTATTTTTTTTATAG	GTGCCA	"D4-A4"	
	AAAGCTTTAAG	GTAACTAACTTAT	TAATTATTTTTTTTATAG	GTGCCA	"D5-A4"	
	AAGACTCCTAG	GTAAGTAATCCCT —	— TTTTTTTTGTATTTCCAG	GTTCAT	"D3-A2"	
	AAGACTCCTAG	GTAAGTAATCCCT	TGTTGCCTTTACTTCTAG	GCCTGT	"D3-A3"	
	CTTCAGTGAAA	GTTGGTAAAACCT	TTTTTTTTGTATTTCCAG	GTTCAT	"D3-A2"	
			TTTTTATTTTTATTTTAG	AGCTTT	"A1"	
Polyoma virus	CCGGCTTCCAG	GTAAGAAGGCTAC			Early	
	GTATAATCCAA	GTAAGTATCAAGA				
			CGGCCTATATTCTTACAG	GGCTCT		
	AG	GTAAGT	——YYxYAG			
		GTA G	(−)AG			
		GTA T				
		GT AG				
		GT GT				

Figure 1. A comparison of eukaryotic and viral mRNA splice sites (see 7). The sequences surrounding the donor (D) and the acceptor (A) nucleotides are presented in their DNA form. The "GT-AG" rule of Breathnach et al. (13) is illustrated at the top and our suggestions for preferred donor and acceptor sites at the bottom (Y = pyrimidine). Because of their length or incomplete sequencing, most IVS are presented in part (illustrated by a central gap in the sequence); two IVS are presented in their entirety and the gap is filled by a horizontal bar (SV40 and BKV D3-A2 splices). The putative BKV and polyoma virus splice sites have been sequenced only at the DNA level. Sources for the splice site sequences are listed in Ref. 7.

to occur in several different frames while still generating the same spliced product. For example, the spliced mouse beta-globin RNA (14) 5'-GCAGGC-3' can be generated from the primary transcript 5'-C↓A↓G↓G↓U . . . U↓A↓G↓G↓C-3' by any one of the four pairs of endonucleolytic cleavage reactions designated by the pairs of arrows. Nevertheless, Breathnach et al. (1978), having compared a number of splice sites, propose that the IVS are bounded by GT at the 5' end and by AG at the 3' end. In other words, the donor nucleotide is followed by the dinucleotide sequence GT and the acceptor nucleotide is preceded by AG. The only exception to this rule at present is an IVS for a cloned mouse immunoglobulin heavy chain, which contains an AG at the 5' end instead of GT (15). Proposed transcripts of BK and polyoma (Py) virus appear to conform to this rule (Fig. 1).

The nucleotide junctions formed after splicing include G-G, G-A, G-U, G-C, A-G. In each case the donor nucleotide is a purine. It is furthermore a G, with the exceptions of an ovalbumin donor, a suggested donor of BKV and of polyoma virus (7).

A more detailed comparison (Fig. 1) indicates that the donor site at the 5' GT end of the IVS takes at least one of four forms: ↓GTAxG, ↓GTAxxT, ↓GTxAG, or ↓GTxxGT, with only one exception (↓GTACAG in ovalbumin). We refer to these sequences as the four groups of donor sites. These four groups derive from two basic forms: ↓GTxxG and ↓GTA.

Considering that the dinucleotide that precedes the 5' end of the IVS is generally an AG (see Fig. 1), and with the four groups in mind, an optimal donor site might be AG↓GTAAGT. If the dinucleotide prior to the excision point (arrow) is not an AG, then there is a greater tendency to include a maximum of remaining sequence ↓GTAAGT; if the 5' IVS deviates significantly from ↓GTAAGT, the donor AG is more likely to be present as in the ovalbumin splice sequence described above (AG↓GTACAG).

At the acceptor site, the 3' end of the IVS is pyrimidine rich. With few exceptions it ends in PyPyXPyAG↓. Within the 16 nucleotides preceding the 3' dinucleotide AG↓, dipurines other than AA are rarely present. Furthermore, no AG appears within the 13 nucleotides preceding the 3'-terminal AG↓. The absence of dipurines (in particular AG) preceding the AG↓ at the 3' end of the IVS may serve to minimize the likelihood of erroneous splices.

From the description of donor and acceptor sequences outlined above, some similarities of one to another are apparent. This is the case within the frequently observed junction repeat AG↓GT. Nevertheless, none of the acceptor sites, listed in Fig. 1, can be categorized in any of the four groups of donor sites, and none of the donor sites listed has an AG preceded by a pyrimidine stretch, which is characteristic of an acceptor site. This suggests that the repeats may be present because of similarity in recognition of donors and acceptors by the splicing enzymes, but that there are additional nucleotide signals that establish a clear distinction between donor and acceptor. These signals are located primarily within the IVS. Consequently, they are removed during the splicing process. These preferred sequences for splicing are based on the limited number of splice sites presently available. It is clear that within the near future a large number of additional sequences will become available and these will be used both to test and extend the "rules" presented above. In preliminary studies with hybrid viral RNAs, for example, recent observations suggest that acceptor sequences varying slightly from the "rules" can be used in the absence of the optimal (normal) acceptor (Khoury et al, unpublished).

The value of these splicing rules is at least two-fold. The canonical sequences presented by ourselves (7) and others (13,8,16) makes certain predictions about those putative enzymes needed for recognition of a splicing complex. It has recently been suggested that small RNA's may provide for a splicing structure by acting as a strut which can pair with sequences around a splice junction (16-20). In fact, there appears to be a rather good match between some of these small RNAs and the junctional sequences described above (16-18).

Secondly, using these rules, one can predict splice sites based on the DNA sequence, and the approximate size of spliced RNA segments. The latter can be obtained from electronmicroscopic heteroduplex analysis or S1 nuclease mapping (21). An example of such a prediction is presented below. Given the DNA sequence (22-24) of the early region of polyoma virus (Py) and the approximate sizes of spliced RNA segments of the three early Py mRNAs (R. Kamen, personal communication), we and R. Kamen have determined the presumed splice sites (Fig. 2).

The location of two donor sequences (D1 and D2) and two acceptor sequences (A1 and A2) which can be used in all three possible combinations defines three early Py mRNAs:

1. Large T mRNA uses D1, and splices to A1. This transition goes from reading frame _1_ to reading frame _2_, a large open frame until just before the polyadenylation site. This mRNA encodes a 100,000 dalton polypeptide, the Py large T or 100K protein.

2. Middle τ mRNA involves a splice from D2 to A2. An mRNA begins in frame _1_, enters frame _3_ after the

```
HPA II
 5/4   420      YD1                   440                    460
XCC.GGC.TTC.CAG.GTA.AGA.AGG.CTA.CAT.GCG.GAT.GGG.TGG.AAT.CTA.AGT.ACC.AAA.GAC.ACC.
    gly-phe-gln-val-arg-arg-leu-his-ala-asp-gly-trp-asn-leu-ser-thr-lys-asp-thr-

       480                          500                          520
TTT.GGT.GAT.AGA.TAC.TAC.CAG.CGG.TTC.TGC.AGA.ATG.CCT.CTT.ACC.TGC.CTA.GTA.AAT.GTT.
phe-gly-asp-arg-tyr-tyr-gln-arg-phe-cys-arg-met-pro-leu-thr-cys-leu-val-asn-val-

       540                          560                          580
AAA.TAC.AGC.TCA.TGT.AGT.TGT.ATA.TTA.TGC.CTG.CTT.AGA.AAG.CAA.CAT.AGA.GAG.CTC.AAA.
lys-tyr-ser-ser-cys-ser-cys-ile-leu-cys-leu-leu-arg-lys-gln-his-arg-glu-leu-lys-

       600                          620                          640
GAC.AAA.TGT.GAT.GCC.AGG.TGC.CTA.GTA.CTT.GGA.GAA.TGT.TTT.TGT.TTT.GAA.TGT.TAC.ATG.
asp-lys-cys-asp-ala-arg-cys-leu-val-leu-gly-glu-cys-phe-cys-phe-glu-cys-tyr-met-

       660                          680                          700
CAA.TGG.TTT.GGA.ACA.CCA.ACC.CGA.GAT.GTG.CTG.AAC.CTG.TAT.GCA.GAC.TTC.ATT.GCA.AGC.
glu-trp-phe-gly-thr-pro-thr-arg-asp-val-leu asn-leu-tyr-ala-asp-phe-ile-ala-ser-

       720                          740                          760 D2
ATG.CCT.ATA.GAC.TGG.CTG.GAC.CTG.GAT.GTG.CAC.AGC.GTG.TAT.AAT.CCA.AGT.AAG.TAT.CAA.
met-pro-ile-asp-trp-leu-asp-leu-asp-val-his-ser-val-tyr-asn-pro-ser-lys-tyr-gln-

       780                HAE III   800         A1               820 A2
GAG.GGC.GGG.TGG.GTA.TTT.ACG.GCC.TAT.ATT.CTT.ACA.GGG.CTC.TCC.CCC.TAG.AAC.GGC.GGA.
glu-gly-gly-trp-val-phe-thr-ala-tyr-ile-leu-thr-gly-leu-ser-pro-stop

       840               HAE III
GCG.AGG.AAC.TGA.GGA.GAG.CGG|CCX.
```

FIGURE 2

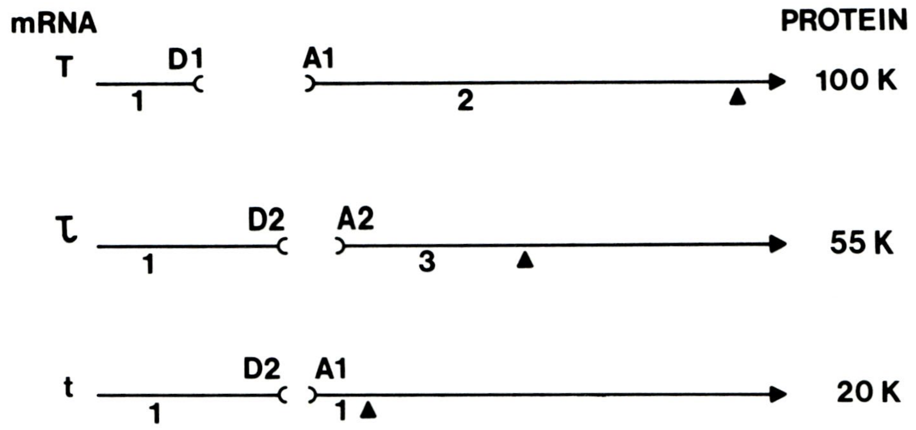

Figure 2. Putative splice sites for early polyoma virus (Py) mRNAs. At the top of the figure is the mucleotide sequence from the proximal portion of the early Py gene region (24). Based on the "splicing rules" previously presented (7) and the size of early Py exons from S_1 nuclease mapping (R. Kamen, personal communication), we suggest two optimal donor sequences and two optimal acceptor sequences. Three possible combinations of these donors (D_1 and D_2) and acceptors (A_1 and A_2) would define three potential early Py m RNAs. Based on the sequence, termination codon sites, a common initiator AUG, and the early protein sizes, we suggest that D_1-A_1, D_2-A2, and D_2-A_1 define the large-T mRNA, middle τ mRNA and small-t mRNA, respectively. (See text for further description.)

splice, and terminates in the middle of this distal exon (body) encoding a 55K protein, Py middle T antigen.

3. The mRNA for small t antigen has a small splice (D2-A1) with re-entry of frame 1. Just distal to A1, reading frame 1 has a termination codon resulting in the production of a 20K protein, Py small t antigen.

Although these splice sites which do fit the above rules have yet to be confirmed, the likelihood is high that they are correct. Not only are the sizes of spliced segments in accord with R. Kamen's S_1 nuclease data, but the size of the predicted Py proteins agree with published results (25-27).

The Use of Viral Mutants to Investigate RNA Processing

A number of studies have been performed by ourselves (9,10,12) and other laboratories (11,28-30) to assess the regions of the SV40 genome required for the generation of stable mRNAs. An approach which we have used to attack this problem involves the use of deletion mutants of SV40. The ability or inability of these mutants to generate stable viral-coded mRNA or proteins provides an indication of the role of the deleted sequences.

Late SV40 Deletion Mutants

Generally, the late SV40 deletion mutants employed have been non-viable, and are grown with helper viruses (e.g. an early ts mutant such as ts A58) under complementing conditions (high temperature). To summarize the results of studies which have been published, it seems fair to say the deletion mutants in coding sequences, well removed from splice sites, induce the synthesis of stable truncated mRNA molecules and in most cases, their corresponding truncated polypeptides in infected monkey kidney cells (10,29,31-34). Deletions which remove splice junctions as well as intervening and flanking sequences, are "polar" mutants as described by Cole et al. (29). They generate no stable late mRNA (10), thus giving rise to an epistatic effect on a distal gene. This observation provided evidence that splicing may play a role in the generation of stable mRNA. Another line of evidence to suggest that splicing is critical to the generation of stable mRNA derives from the studies of Hamer and Leder (see 11 for summary). These investigators made recombinant molecules between SV40 and a mouse beta globin gene. When a particular recombinant molecule was missing the late SV40 splice sites, no stable mRNA was generated unless the globin insert contained its own splice site.

An extension of these studies involved the construction of a late SV40 deletion mutant from a reverse transcript of 16S late SV40 mRNA, which precisely removes the intervening sequence of this transcript at the DNA level (12). The deletion site in the resulting "intron minus" mutant was confirmed at the DNA level by nucleotide sequence analysis. The prediction for this mutant's biological activity, assuming that the late transcriptional promoter can function, is that capsid protein VP1 should be made. In complementation studies, however, this mutant was unable to complement a tsB mutant defective in VP1. Analogous studies, performed by B. Howard and P. Berg (personal communication), yielded similar results. Furthermore, no VP1 protein could be detected in cells inoculated with the DNA of this intron minus mutant. Finally, we could detect no stable mRNA after an infection with this mutant, suggesting that at least for the SV40 16S late mRNA, and perhaps as a more general rule, splicing may be in some way essential for generating stable transcripts (12).

Splicing could be coupled to a transport system which shuttles transcripts from the nucleus to the cytoplasm. This model is attractive in part because virtually all cytoplasmic transcripts are spliced while a number of nuclear mRNAs have been found which appear to represent primary products of transcription, or partially spliced RNA molecules. Furthermore, splicing has not yet been found for prokaryotic mRNAs which have no need for traversing a nuclear membrane. Any splicing theory should eventually encompass recent experimental observations that certain mRNAs (viz. histone transcripts and the mRNA for adenovirus protein IX; U. Petturson et. al., personal communication, are not spliced. Furthermore, it must deal with the additional data, perhaps of evolutionary significance, that many yeast mitochondrial mRNAs are spliced (35-37).

Early SV40 Mutants

Among the most interesting early SV40 mutants are the so-called early viable deletions (38-42). These reside in the coding sequence for small t-antigen, but at the same time, in the intervening sequence of large-T antigen. Since small-t antigen is not required for growth of SV40 in tissue culture, and since these mutants can normally splice large-T mRNA, the mutants are viable. A number of interesting observations derive from the study of the mRNAs of these mutants (12,43,44). First, the ability to create the normal large-T mRNA splice, in spite of a series of variably-sized deletions within its intervening sequence, suggests that those nucleotides absent from the various deletion mutants are not essential to the large-T mRNA splicing enzyme complex.

A second observation is based on the effect which these deletions exert on the efficiency of the small-t mRNA splicing reaction. The effect depends on the site of the deletion relative to the small t-mRNA splice junction. Mutants which eliminate the small t-mRNA donor splice junction, as expected, prevent the production of small-t mRNA. Two deletion mutants, dl-2000 and dl-2007, which reside in a central portion of the intervening sequence, have little or no effect on small-t mRNA production (12,43). A large number of deletion mutants, some as far as 200 nucleotides from the small-t donor site (i.e. in 5'-flanking sequences) still result in the production of reduced quanties of small-t mRNA. Although it has not been rigorously demonstrated that the early viable SV40 deletions directly affect either the splicing process or mRNA stabilization, at least one model suggests that these deletions affect the ability of a precursor RNA to efficiently form an active complex with the putative splicing enzyme. This model would predict that secondary or tertiary RNA structure, in addition to primary nucleotide sequence, play a role in determining the efficiency (i.e. frequency) with which a particular splicing reaction will proceed.

Although this hypothesis is not easily tested at present, there are some experimental data which bear on question of alterations in template structure affecting splicing efficiency.

In addition to the studies of late and early SV40 deletion mutants described above, two sets of experiments with early SV40 mRNAs may be relevant. The assumption has been made that a single early SV40 primary transcript is processed (spliced) in one of two ways, either to small-t mRNA or large-T mRNA. It was observed by Berk and Sharp (45) that the large-T mRNA was about three times as abundant as small-t mRNA.

Recently, it has been shown that the ratio of large-T mRNA to small-t mRNA is dependent on the incubation temperature (J. Alwine et. al., in preparation). At low temperature (32°), the large T:small t mRNA ratio is lower than 3:1, while at a high temperature (41°), the ratio is greater than 3:1. Furthermore, the host cell plays a role in determing the relative amounts of large-T and small-t mRNA produced. While in AGMK cells at 37° this ratio may be 3:1, under similar conditions in mouse cells there is considerably more small-t mRNA (S. Segal and E. May, unpublished data). Presumably, host cell factors which interact with the primary transcript can determine the efficiency with which various splices can be made.

Recombinant DNA Molecules to Assay Splicing and Gene Expression

To further investigate the biochemical signals important in gene regulation, we have made recombinant molecules between SV40 and eukaryotic genes. These experiments have several advantages:

1. Mutations can be placed at strategic locations in the recombinant eukaryotic DNA segment, well removed from the signals required for propagation of SV40.
2. The heterologous DNA segment, and the RNA and protein which it encodes, are easily assayed in the presence of the SV40 helper virus macromolecules. This is a distinct advantage over looking for the products of SV40 deletion mutants.
3. The eukaryotic genes inserted into SV40 can potentially be studied for regulatory features which may be unique to the production of their RNA or proteins (e.g. hormone responsiveness, post-translational modification, etc.).

A particular example of such a set of experiments presently ongoing in our laboratory is described below.

In collaboration with Drs. Lomedico, Efstratiadis and Gilbert at Harvard, we have made recombinant molecules between SV40 and the rat insulin genes. There are two rat insulin genes, rI_1 and rI_2, both of which have an intervening sequence in the 5'- noncoding sequence (46-48). They differ primarily in that rI_1 is missing an intervening sequence in the coding region which is present in rI_2.

Both of these genes, which produce slightly different proteins, have been shown to be active _in vivo_. Furthermore, there is now rather compelling evidence that rI_1 evolved from rI_2 through the precise loss of an intron (Lomedico, Efstratiadis and Gilbert, personal communication).

In one set of experiments (Gruss et. al., in preparation), we have set rI_1 into the late region of SV40. The segment of SV40 deleted to accomodate the insertion of rI_1 included the known SV40 late splice sites, extending approximately from 0.72 to 0.14 map units, in a clockwise direction. Thus, splicing of RNA was presumed to be dependent on the intact rI_1 splice junction. Preliminary analysis of the RNA (not shown) suggests that in fact, the rI_1 splice junction was active and presumably led to the generation of stable RNA molecules (cf ref's 10-12,34). It also appeared from this RNA analysis that the rat insulin polyadenylation signal, which was situated 800 to 900 nucleotides proximal to the late SV40 polyadenylation signal, was used preferentially. Although there was an indication that the rat insulin promoter might be active, confirmation of this conclusion will rest on the analysis of 5'-ends of insulin-containing RNA molecules.

By immunoprecipitation, we were able to show that infection of AGMK cells by the recombinant SV40-rI_1 led to the production of rat pro-insulin (Gruss et. al., in preparation). The identification of this polypeptide was based on co-migration of the ^{35}S-labelled immunoprecipated protein with a bovine pro-insulin marker of similar size. The insulin immunoreactivity was also confirmed by radioimmune assay. These findings suggest that the rat prepro-insulin hydrophobic leader peptide can be cleaved off in AGMK cells, but that these cells do not have the enzymatic machinery to remove the internal C-peptide (see 49).

In future studies, we will be interested both in the events which determine this proteolytic cleavage, as well as the nucleotide signals which regulate the synthesis and processing of the insulin-containing mRNAs.

REFERENCES

1. Kelly, T.J. and Nathans, D. (1977). Academic, New York, Vol. 21, pp. 86-173.
2. Acheson, N.H. (1976). Cell 8, 1-12.
3. Reddy, V.B., Thimmappaya, B., Dhar, R., Subramanian, K.N., Zain, B.S., Pan, J., Ghosh, P.K., Celma, M.L. and Weissman, S.M. (1978). Science 200:494-502.
4. Fiers, W., Contreras, R., Haegman, G., Rogiers, R., van de Voorde, A., van Heuverswyn, H., van Herreweghe, J., Volckaert, G. and Ysebaert, M. (1978). Nature (London) 273:113-120.
5. Ghosh, P.K., Reddy, V.B., Swinscoe, J., Lebowitz, P. and Weissman, S.M. (1978). J. Mol. Biol. 126:813-846.
6. Reddy, V.B., Ghosh, P.K., Lebowtiz, P., Piatak, M. and Weissman, S.M. (1979). J. Virol. 30, 279-296.
7. Seif, I., Khoury, G. and Dhar, R. (1979). Nucleic Acids Res. 6, 3387-3398.
8. Dawid, I.B. and Wahli, W. (1979). Devel. Biol. 69, 305-328.
9. Khoury, G., Gruss, P., Dhar, R. and Lai, C.-J. (1979). Cell 18, 85-92.
10. Lai, C.-J. and Khoury, G. (1979). Proc. Natl. Acad. Sci. 76:71-75.
11. Hamer, D. and Leder, P. (1979). Cell 18, 1299-1303.

12. Gruss, P., Lai, C.-J., Dhar, R. and Khoury, G. (1979). Proc. Natl. Acad. Sci. 76, 4317-4321.
13. Breathnach, R., Benoist, C., O'Hare, K., Gannon, F. and Chambon, P. (1978). Proc. Natl. Acad. Sci. 75:4853-4857.
14. Konkel, D.A., Tilghman, S.M. and Leder, P. (1978). Cell 15:1125-1132.
15. Sankano, J., Rogers, J.H., Huppi, K., Brack, C., Traunecker, A., Maki, R., Wall, R. and Tonegawa, S. (1979). Nature (London) 277:627-633.
16. Rogers, J. and Wall, R. (1980). PNAS in press.
17. Lerner, M.R., Boyle, J.A., Mount, S.M., Wolin, S.L. and Steitz, J.A. (1980). Nature 283, 220-224.
18. Murray and Holliday (1979). FEBS Lett. 106, 5-7.
19. Zain, S., Gingeras, T., Bullock, P., Wong, G. and Gelinas, R.E. (1979). J. Mol. Biol. 135, 413-433.
20. Monstein, R., and Philipson, L. in press.
21. Berk, A.J. and Sharp, P.A. (1977). Cell 12:721-732.
22. Friedmann, T., Esty, A., LaPorte, P. and Deininger, P. (1979). Cell 17, 715-724.
23. Soeda, E., Arrand, J.R., Smolar, N. and Griffin, B.E. (1979). Cell 17, 357-370.
24. Hattori, J., Carmichael, G.G. and Benjamin, T.L. (1979). Cell 16:505-513.
25. Ito, Y., Spurr, N. and Dulbecco, R. (1977). Proc. Natl. Acad. Sci. U.S.A. 74, 1259-1263.
26. Schaffhausen, B.S., Silver, J.E. and Benjamin, T.L. (1978). Proc. Natl. Acad. Sci. U.S.A. 75, 79-83.
27. Hutchinson, J.A., Hunter, T., and Eckhart, W. (1978). Cell 15, 65-77.
28. Hamer, D. and Leder, P. (1979). Nature 28, 35-40.
29. Cole, C.N., Landers, T., Goff, S.P., Manteuil-Bruttag, S. and Berg, P. (1977). J. Virol. 24:277-294.
30. Mulligan, R.C., Howard, B.H. and Berg, P. (1979). Nature 277, 108-112.
31. Villareal, L.P., White, R.T. and Berg, P. (1979). J. Virol. 29, 209-219.
32. Subramanian, K. (1979). Proc. Natl. Acad. Sci. USA 76, 2556-2560.
33. Rundell, K., Collins, J.K., Tegtmeyer, P., Ozer, H.L., Lai, C.-J. and Nathans, D. (1977). J. Virol. 21:636-646.
34. Goff, S., and Berg, P. (1980). J. Mol. Biol., in press.
35. Slonimski, P.P., Pajot, P., Jacq, C., Foucher, M., Perrodin, G., Kochko, A. and Lamouroux, A. (1978b). New York: Academic Press, pp. 339-369.
36. Borst, P. and Grivell, L.A. (1978). Cell 15, 705-723.
37. Hanson, D.K., Miller, D.H., Mahler, H.R., Alexander, N.J. and Perlman, P.S. (1979). J. Biol. Chem. 254, 2480-2490.

38. Shenk, T., Carbon, J. and Berg, P. (1976). J. Virol. 18:664-671.
39. Sleigh, M.J., Topp, W.C., Hanich, R. and Sambrook, J.P. (1978).
40. Feunteun, J., Kress, M., Gardesl, M. and Monier, R. (1978). Proc. Natl. Acad. Sci. 75:4455-4459.
41. Thimmappaya, B. and Shenk, T. (1979). J. Virol. 30, 668-673.
42. Volkaert, G., Feunteun, J., Crawford, L.V., Berg, P. and Fiers, W. (1979). J. Virol. 30:674-682.
43. May, E., Kress, M. and May, P. (1978). Nucleic Acids Res. 5:3083-3099.
44. Khoury, G., Alwine, J.A., Dhar, R., Gruss, P., Lai, C.-J., Segal, S., and Seif, I. (1980). Cold Spring Harbor Symp. Zuant. Biol. 44, in press.
45. Berk, A.J. and Sharp, P.A. (1978). Proc. Natl. Acad. Sci. USA 75, 1274-1278.
46. Villa-Komaroff, L., Efstratiadis, A., Broome, S., Lomedico, P., Tizard, R., Nabor, S.P., Chick, W.L. and Gilbert, W. (1978). Proc. Natl. Acad. Sci. USA 75, 3727-3731.
47. Cordell, B., Bell, G., Tischer, E., De Noto, F.M., Ullrich, A., Pictet, R., Rutter, W.J., and Goodman, H.M. (1979). Cell 18, 533-543.
48. Lomedico, P., Rosenthal, N., Efstratiadis, A., Gilbert, W., Kolodner, R. and Tizard, R. (1979). Cell 18, 545-558.
49. Tager, H.S. and Steiner, D.F. (1974). Ann. Rev. Biochem. 43, 509-538.

DISCUSSION

H. AVIV: What is the evidence for preferring the recognition sites of both the promoter and Poly (A) addition of insulin over that of SV40?

G. KHOURY: The evidence is that the S1 gel band which I pointed out was distinct and dark. It corresponded to the size which one would expect if the insulin polyadenylation site was used. A much fainter band which migrates higher on the gel corresponds in size to the RNA expected if the SV40 polyadenylation site is used.

P. ZAMECNIK: I am intrigued by your description of the adapter RNA which holds the edges of the spliced area together. In a facetious way I would like to suggest a trivial name for it. You might call it "Band-Aid" RNA because it holds together the edges of the splice.

G. KHOURY: I am intrigued by the name which you provide for it, but since it was not our concept, and in fact it is just now being pursued in a number of laboratories, I hesitate to give any name to it except to cite, as I have, the people who I think are actively working on it and are interested in it.

P.A. SHARP: If you summarize your data, Paul Berg's data and Hamer's data, would you say that the lack of intervening sequence affects RNA synthesis at the stage of transport or of stability? Can you distinguish between the two?

G. KHOURY: Based on the data we have I think that one cannot distinguish between the two and in fact there may not be a distinction. It is possible that transport may be needed for stabilization; in other words the nucleus for some reason for certain RNA's may provide an unhealthy environment. But I do not think that there are any data to distinguish between these two possibilities.

J.D. SHARP: Have you looked at the glucose or glucagon dependence of insulin production in these cells?

G. KHOURY: No.

TRANSCRIPTION OF ANIMAL VIRUS GENES IN VITRO[1]

James L. Manley[2]
Hiroshi Handa[3]
Shu-Ying Huang
Malcolm L. Gefter
Phillip A. Sharp

Department of Biology and Center for Cancer Research
Massachusetts Institute of Technology
Cambridge, Mass.

I. INTRODUCTION

It is well established that three separate classes of RNA polymerases exist in mammalian cells (1). Of these, one class (RNA polymerase II) has been implicated as the class of enzyme responsible for mRNA synthesis. This assignment is based primarily upon studies with the drug α-amanitin. Thus the same very low concentration of the toxin which selectively inhibits purified RNA polymerase II when assayed *in vitro* also inhibits the synthesis of hn RNA, the presumptive mRNA precursor in isolated nuclei (2). The enzyme has a very high molecular weight (over 500,000) and may contain as many as nine or more different subunits (3).

---

[1]Research supported by grants from the American Cancer Society (NP-6) and National Institutes of Health (AI13357).

[2]Supported by NIH grant CA09255 to the Dept. of Biology, Massachusetts Institute of Technology.

[3]Supported by the Leukemia Society of America, Inc.

ISBN 0-12-633360-2

In addition to RNA polymerase II, a complex array of enzymes is required to complete the synthesis of a mature mRNA molecule. An enzyme, or enzymes, is needed to add the '5' "cap" structure (4) found on all mammalian mRNAs examined to date. Additional enzymes are required to methylate the 5' cap structure, as well as to carry out internal methylations, such as occur at the N6 position of some adenosine residues (5). Another set of enzymes are necessary to create the 3' ends of mature mRNAs. These must include at least (a) specific endonuclease(s) and a poly (A) polymerase (6,7). Finally, an unknown number of enzymes must exist to carry out the "splicing" reactions necessary to remove the intervening sequences found in most mRNA precursors (8,9).

In most cases, the enzymes which catalyze these reactions have not been identified. This is in large part due to the lack of suitable substrates with which to assay these enzymes. An ideal substrate would be an RNA species synthesized _in vitro_ which corresponds to the primary transcript of a particular well-characterized gene. However, synthesis of such a molecule requires the availability of an _in vitro_ transcription system which is capable of accurately initiating transcription. Such systems derived from bacterial cells have been well-characterized. Only recently, though, have systems derived from mammalian cells begun to be available. We described a system consisting of nuclei isolated from adenovirus serotype 2 (Ad2)-infected HeLa cells which is capable of accurately initiating transcription of the Ad2 late transcription unit (10). This system had two disadvantages for enzymological studies. First, in addition to newly synthesized RNA, these nuclei contain large amounts of endogenous RNA, including partially processed precursor molecules as well as mature mRNAs. Second, the presence of the nuclear membrane presents a barrier to further purification of any activity which might be detected. Recently, Weil et al. (11) described a completely soluble system which consists of highly purified RNA polymerase II, a high speed supernatant derived from the cytoplasm of uninfected HeLa cells, and exogenously added DNA. This system is also capable of initiating synthesis of the Ad2 major late transcription unit. This development represented a break-through in the study of transcription in eukaryotic cells. However we felt that such a system had two important drawbacks for the study of all the enzymes involved in mRNA synthesis. First, large amounts of purified RNA polymerase II are required. Second, the relative purity of this system suggests that many of the other enzymes needed for mRNA synthesis might not be present. In order to circumvent these problems, we have attempted to develop a simple whole-cell lysate transcription system which would not be dependent upon the addition of purified RNA polymerase II for activity. Here we describe the characterization of such a system, and show that it supports accurate initiation of transcription on exogenously added Ad2 DNA as well as DNA from the monkey virus, SV40.

II. METHODS

Most of the methods used in this work have been described in detail previously. In particular, the preparation of the HeLa cell lysate used for *in vitro* transcription, the conditions for RNA synthesis used in the *in vitro* reaction mixtures and the purification and analysis of the RNA produced, are described by Manley et al. (12) or in references cited therein. Details of techniques not previously described are included in the text.

III. RESULTS

Preparation of an Adenovirus DNA Template

The mRNAs encoded by the human adenoviruses have been extensively studied over the last several years (13,14). Not only do they undergo all the processing reactions that cellular mRNAs do but in many cases these reactions were first detected in the adenovirus system. Specifically, as referred to above, the major late transcription unit has been quite well characterized, and has served as a model for several studies dealing with transcription initiation. Figure 1 diagrams the first 5,000 nucleotides of the long (∿25 kb) primary transcript. The capped 5' end is encoded at position 16.5 (15). The three segments of the 228 nucleotide tripartite leader are complementary to DNA sequences located at coordinates 16.5, 19.5, and 26.5 (8,9).

In order to test whether the RNA polymerase II contained endogenously in a crude whole cell extract could initiate transcription accurately on exogenously added DNA, we decided to clone a restriction fragment of Ad2 DNA which we believed would contain the promoter and start-site for the late transcription unit. An ideal fragment for this purpose was the Bal I E fragment (14.7 - 21.5 map units). This DNA extends approximately 700 bp upstream from the site which encodes the capped 5' end of the *in vivo* RNA, and downstream about 1750 bp. It seemed reasonable that all the information required to initiate transcription would be contained within this

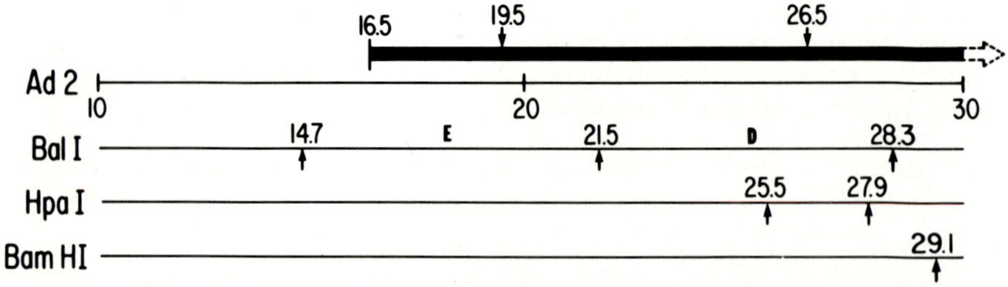

Fig. 1. The 5' end of the Ad2 late transcription unit. Transcription begins at approximately map coordinate 16.5 and continues rightward to almost the end of the Ad2 genome (coordinate 100). Since one map unit is roughly 350 bp, the length of the entire transcription unit is over 29,000 bp. Although at least 13 mRNAs are encoded in this transcription unit, none are located within the region shown in the figure. However, the approximately 200 nucleotides which ultimately form the leader segment common to all mRNAs from this transcription unit are located here. They are encoded in the genome in three segments, which are ultimately joined by RNA splicing. The coordinates of these segments are indicated by the numbers above the solid arrow. The sites at which the restriction endonuclease Bal I, as well as two other restriction enzymes, cleave this region of the Ad2 genome are indicated.

region. Also, if transcription initiates correctly *in vitro*, and continues to the end of the DNA molecule, the "run-off transcript" produced would be 1750 nucleotides in length. This size RNA species should be readily detectable by agarose gel electrophoresis. In addition, this RNA will contain both the 5'-donor and 3'-acceptor splice junctions which are required to make the first splice in the pathway leading to the production of mature mRNA. If accurate splicing of the *in vitro* synthesized RNA should occur, the products would be a 750 nucleotide spliced molecule and, if not degraded, an approximately 1,000 nucleotide long RNA which would be the excised intervening sequence

We therefore decided to clone the Bal I E fragment of Ad2 into the bacterial plasmid pBR322. At the same time, we felt it would be useful, for a control, to clone the Bal I D fragment. This fragment is adjacent to the Bal I E fragment on the viral chromosome, is nearly identical in size, but contains no known viral promoter (see Fig. 1). The recombinant plasmids were formed by standard procedures; i.e., Bam HI "linkers" were blunt-end ligated onto the previously purified restriction fragments, the mixture digested with Bam HI, and the restriction fragments were separated from the excess, restricted linkers by agarose gel electrophoresis. The two DNAs were eluted from the gel, and mixed, at a 3:1 molar ratio, with Bam HI digested, alkaline phospatase treated pBR322 DNA. Following ligation, the recombinant DNAs were transfected into the *E. coli* strain HB101. DNA was prepared from 12 colonies which were both ampicillin resistant and tetracycline sensitive. From these, strains containing either the Bal I E or Bal I D fragment inserted in pBR322 were identified (pBal E and pBal D). Plasmid DNAs were isolated, digested with the restriction endonuclease Bam HI, and the purified nucleic acids used as templates for *in vitro* transcription.

Sizing of *In Vitro* Run-Off Transcripts

We have previously described in detail the method used to prepare the whole cell extract (12). Briefly, HeLa cells were lysed by hypotonic swelling and homogenization, after which the nuclei were broken by addition of $(NH_4)_2SO_4$ to 10% of saturation. The lysate was centrifuged at 170,000 x g for 3 hr, and the supernatant concentrated by addition of $(NH_4)_2SO_4$ to 80% of saturation. The precipitate was collected, redissolved and dialyzed against a buffer which contained the appropriate ions, etc. for transcription.

If we assume that transcription initiates on the Bal E DNA templates at the site which encodes the 5' cap structure, and continues to the end of the restriction fragment, a run-off RNA fragment of 1750 nucleotides should be produced, as discussed above. However, if the DNA is restricted prior to its use as a transcription template with an enzyme which cuts within this coding region, then the 1750 nucleotide long transcript should not be synthesized. Instead, a shorter RNA species should now become apparent, its size dependent on the location of the restriction site. When several DNA templates are prepared with different restriction enzymes, analysis of the RNA products synthesized should indicate whether or not accurate transcription has occurred *in vitro*. An example of this type of analysis is shown in Figure 2. pBal E was digested with Hind III (190), SmaI (530) or Bam HI (1750), and the purified DNAs were used as templates for transcription. (The numbers in parentheses are the sizes expected, in nucleotides, of the appropriate run-off RNAs.) *In vitro* RNA synthesis reactions were carried out under standard

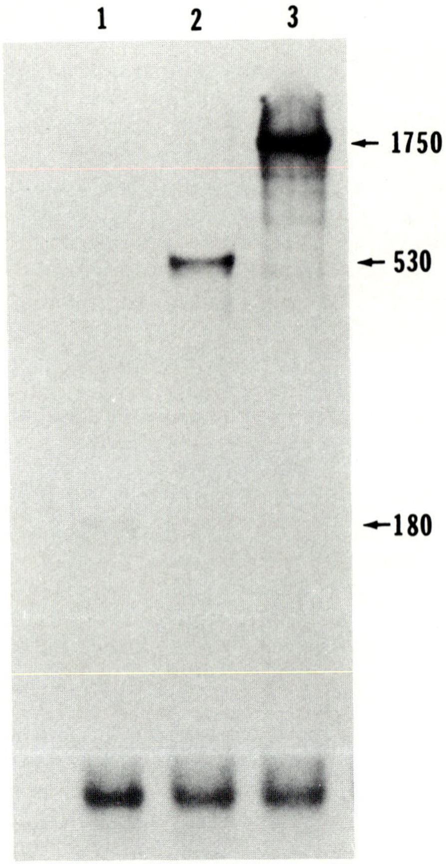

Fig. 2. Size analysis of transcription products obtained from truncated pBal E templates. pBal E restricted with either Hind III (lane 1), SmaI (lane 2) or Bam HI (lane 3) was used as template for *in vitro* transcription reactions in standard reaction mixtures. RNA was purified, denatured by treatment with glyoxal, and resolved by electrophoresis through an acrylamide gel as described in the text. The sizes indicated (in nucleotides) were determined by comparison with a denatured DNA restriction digest.

conditions (see ref. 12; the important concentrations are 50 μg/ml DNA, 60% extract and 10 μCi α-^{32}P UTP in a total volume of 50 μl). After 60 min of incubation at 30°C, RNA was purified, denatured by glyoxalation (16), and an aliquot of each sample resolved by electrophoresis in a 3.5% polyacrylamide-7M urea gel which contained 50 mM Tris borate, pH 8.3, 1 mM EDTA. The sizes of the RNA species shown in

Figure 2, which were determined by comparison with a Hind III digest of ^{32}P Ad2 DNA, which had also been glyoxalated prior to electrophoresis (not shown), agree well with the predicted values indicated above. This result is consistent with transcription initiating at the site utilized *in vivo*. Further results in support of this will be presented below. Note that the intensities of the bands produced by the run-off transcripts are proportional to their size. Thus, the 1750 nucleotide RNA gives rise to a very intense band, while the 180 nucleotide species is barely detectable. We have observed this pattern reproducibly. The simplest explanation is that RNA polymerase molecules which initiate accurately and elongate to the end of the restriction fragment are not able to reinitiate transcription. This would imply that some factor, present in limiting amounts, can not be recycled during *in vitro* incubations.

Analysis of *In Vitro* RNA by the S1 Gel Technique

We have also analyzed the products of *in vitro* transcription by the hybridization-nuclease S1-gel technique (17,18). In the variation of the method used in these experiments, ^{32}P-labeled RNA, synthesized *in vitro*, is hybridized to unlabeled DNA, under conditions of DNA excess. Unhybridized nucleic acid is digested with S1 nuclease, and the S1-resistant hybrids resolved by agarose gel electrophoresis. In Figure 3, we have used this method to examine the effect of varying the DNA concentration in the *in vitro* reaction mixtures. The samples shown in lanes 1-7 were derived from reaction mixtures which had contained varying amounts of pBal E which had been digested by Bam HI (pBal E_{Bam}), while the samples in lanes 8-12 were from reaction mixtures which had contained pBal D_{Bam}. Following purification of RNA, an aliquot of each sample was hybridized to pBal E_{Bam}, treated with nuclease S1, and electrophoresed through an agarose gel. At DNA concentrations of 25 μg/ml or less, with either DNA template, no bands were observed. When the concentration of DNA was increased to 50 μg/ml, a 1750 nucleotide RNA-DNA duplex could be detected with the RNA from the reaction mixture which had contained pBal E (lane 4), but not the one that had contained pBal D (lane 10). This is the result expected if RNA synthesis had initiated correctly on the Bal E fragment, and is consistent with the results in Figure 2. As the DNA concentration was further increased, the amount of this species decreased, and several other RNA-DNA duplexes appeared (lanes 5-7). The two larger species arise from

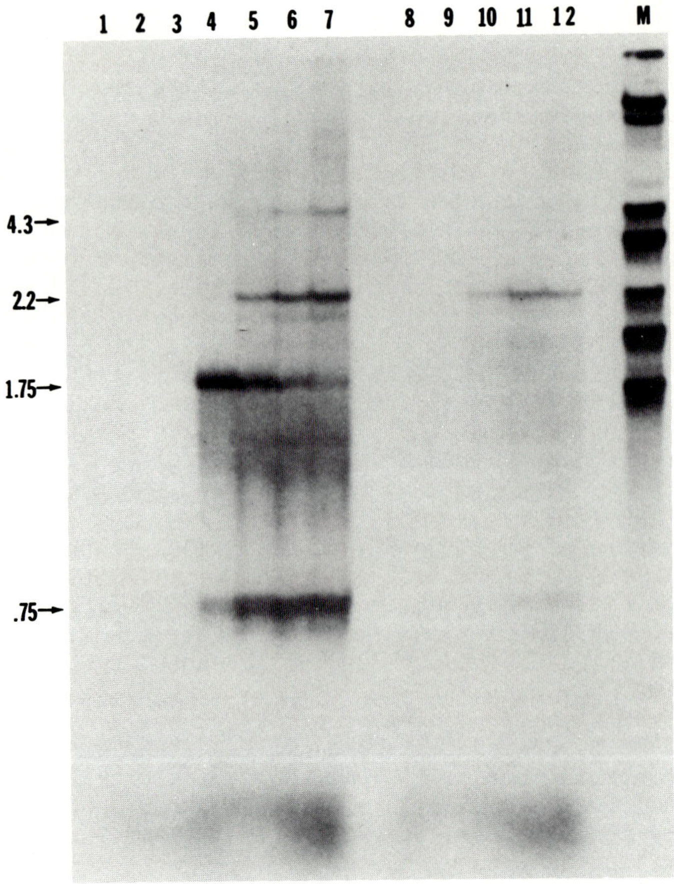

Fig. 3. Hybridization-S1 nuclease analysis of transcription products complementary to pBal E. Standard reaction mixtures contained either pBal E_{Bam} (lanes 2-7) or pBal D_{Bam} DNA (lanes 8-12). The sample shown in lane 1 was from a reaction mixture which contained no exogenously added DNA; lane 2, 12.5 μg/ml DNA; lane 3 and 8, 25 μg/ml; lanes 4 and 9, 50μg/ml; lanes 5 and 10, 75 μg/ml; lane 5 and 11, 100 μg/ml; and lanes 7 and 12, 125 μg/ml. After in vitro synthesis and purification of the RNA, 25% of each sample was hybridized to 0.5 μg pBal E_{Bam} DNA in 15 μl hybridization mixtures (18). After S1 nuclease treatment, RNA-DNA duplexes were resolved by electrophoresis in a 1.4% agarose gel and bands visualized by autoradiography of the dried gel. The sizes indicated are expressed in kilobases, and were determined by comparison with an Eco RI digest of ^{32}P-labeled Ad2 DNA (M).

hybridization of "end-to-end" transcripts, i.e., the 2.2 kb band results from transcription from one end of the Bal E molecule to the other, while the 4.3 kb species is transcribed from the pBR322 molecule in a similar manner. A large amount of 750 bp RNA-DNA duplex can also be detected. While we have not confirmed the identity of this species, it is interesting to note that this is the size expected of a molecule which had undergone the splicing reaction which joins the first leader segment (42 nucleotides) to the second, encoded at position 19.5. Direct nucleotide sequence analysis of this transcript will be required to confirm its identity. If this is indeed a spliced molecule, it is of interest to ask why its production is enhanced at high DNA concentrations. One possibility is that some sequence(s) in the DNA can interact with the RNA primary transcript to stabilize some intermediate in the splicing reaction. A similar role in splicing has been proposed for small nuclear RNAs (19), and it is not inconceivable that, under *in vitro* conditions, a DNA molecule might substitute in this role. In lanes 8-12, which display samples derived from reaction mixtures which had contained pBal D DNA, the most intense band corresponded in size to a hybrid consisting, in part, of an end-to-end transcript of the Bal D DNA molecule. However, since the hybridization probe was pBal E, this result was unexpected. At least two explanations are apparent. First, the band may result from an RNA-RNA duplex, since end-to-end transcription of either strand should be equally likely. Alternatively, if a small amount of the template pBal D DNA is left intact after the DNAse digestion, it could be sufficient to give rise to the band observed.

Fingerprint Analysis of the *In Vitro* Synthesized RNA

Recently we showed that the 5' end structure of the RNA synthesized *in vitro* with pBal E DNA as template was the same as that detected *in vivo* from the major late transcription unit (12). In Figure 4, we analyze the structure of RNA synthesized *in vitro* with pBal E DNA as template, with either a low (1 mg/ml of protein) or high (8 mg/ml protein) concentration of extract in the reaction mixtures. (The high concentration is the amount present in standard reaction mixtures.) We previously showed that when a low concentration of extract was used, approximately 2-3 fold more RNA was synthesized than at the high concentration, as measured by incorporation of radioactivity into RNA. However, when the RNA synthesized at the low extract concentration was analyzed by gel electrophoresis, only a heterogeneous smear of RNA

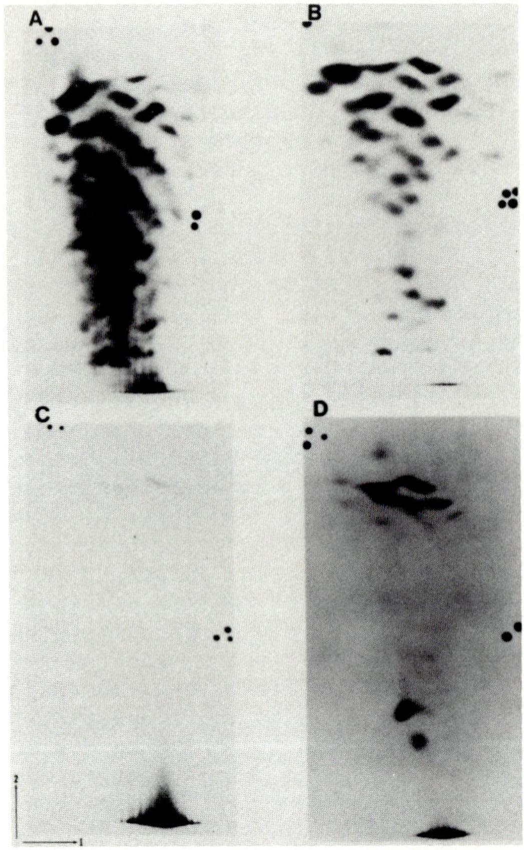

Fig. 4. Fingerprint analysis of in vitro synthesized RNA. RNA was synthesized in standard reaction mixtures which contained pBal E DNA as template, except that each contained 40 μCi α-^{32}P UTP. The samples shown in panels A and C were obtained from reaction mixtures which contained a low concentration of extract (approximately 1 mg/ml protein), while those in panels B and D contained the standard amount of extract (8 mg/ml protein). Purified RNA was digested with RNAse T1, and 5% of each sample was fingerprinted by standard procedures (electrophoresis on cellulose acetate strips at pH 3.5, followed by homochromatography on DEAE cellulose in the 2nd dimension, ref. 21). The remainder of each sample was bound to columns of dihydroxylboryl cellulose, washed, and eluted prior to fingerprinting.

could be detected,with molecules ranging in size from several hundred to several thousand nucleotides. No discrete species were detected. Figure 4A shows an RNAse T1 fingerprint of the RNA synthesized in vitro at the low extract concentration, while Figure 4B is a fingerprint of the RNA synthesized at the high concentration. The fingerprint in Figure 4A is extremely complex; this suggests that the RNA synthesized at the low extract concentration results from a large number of non-specific initiations on the pBal E DNA. The fingerprint in Figure 4B is simpler, consistent with the observation that when RNA from such a reaction mixture was analyzed by gel electrophoresis, virtually the only species detectable was the 1750 nucleotide transcript discussed above. Figure 4D shows that when an aliquot of the T1 digest displayed in 4B was selected on a column of dihydroxylboryl cellulose prior to fingerprinting, two large oligonucleotides (slow mobility in the second dimension) were detected. We previously showed that these oligonucleotides were identical to the capped undecanucleotide found by similar analyses at the 5' end of in vivo RNA (12). When an equal aliquot of the RNA synthesized at the low extract concentration was analyzed after dihydroxylboryl cellulose selection, no oligonucleotides could be detected (Figure 4C). This indicates that the large number of non-specific initiations which occur under these conditions are not capped at their 5' ends. A substantial amount of radioactivity could be detected at the origin of the second dimension. This material is resistant to further digestion with a variety of ribonucleases; it has not been further analyzed.

In Vitro Transcripts of SV40 DNA

We have begun to analyze the transcription products obtained from several other DNA templates to determine if accurate transcription initiation by the RNA polymerase II contained in our crude HeLa cell extracts can occur on genes other than those from adenovirus 2. Here we present evidence that both the SV40 early and late genes are accurately transcribed in vitro.

SV40 DNA was restricted with Pst I, which cleaves the DNA at positions 4 and 28 on the viral map (see Figure 5). The purified DNA was then used as template for in vitro transcription with the same concentrations of DNA and lysate as determined to be optimal for the Ad2 late genes. RNA was purified, denatured with glyoxal, and resolved by electrophoresis through an agarose gel (Figure 5, lane 5). Four major bands were detected, ranging in size from approximately 1650 to 2050 nucleotides. We have mapped the end-points of

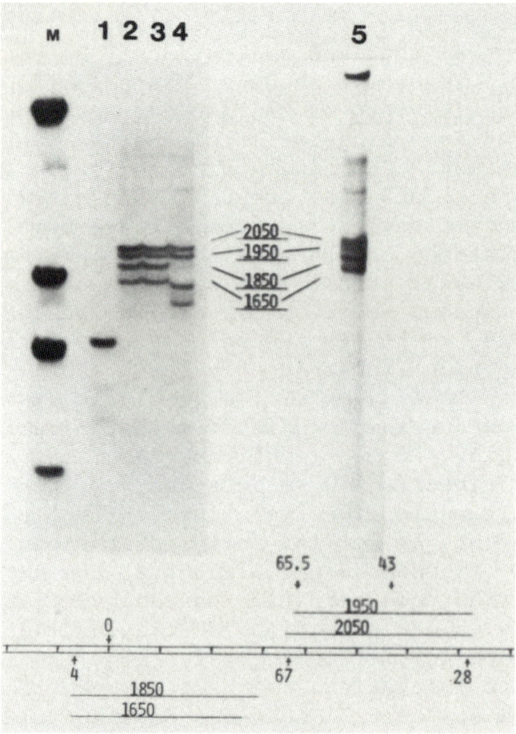

Fig. 5. Analysis of SV40 transcripts by gel electrophoresis. RNA was synthesized *in vitro* in standard reaction mixtures except that Pst I cut SV40 DNA was used as template. (Pst I cuts SV40 DNA at map coordinates 4 and 28.) An aliquot of RNA was denatured with glyoxal and electrophoresed through an agarose gel (Lane 5). The remainder was hybridized to different samples of SV40 DNA which had been digested with one of several restriction enzymes. Hybridization, S1 nuclease digestion, and gel electrophoresis were performed as described (17). The DNA probes used in each case were: Lane 1, the purified Hind III B fragment (65.5 to 43 m.u.); Lane 2, Pst I cut DNA; Lane 3, Bgl II cut DNA (Bgl II cleaves SV40 DNA at position 67); Lane 4, Eco RI cut DNA (Eco RI cleaves SV40 DNA at position o). The numbers indicate the sizes of the RNA transcripts (in bases), and were calculated by comparison with an Hpd I digest of ^{32}P-SV40 DNA. The diagram indicates the deduced coordinates of the *in vitro* synthesized transcripts. The two transcripts drawn above the double line are transcribed left to right, while the two below the line are transcribed right to left.

these transcripts by the S1-gel technique referred to above. Here we have hybridized aliquots of the RNA shown in lane 5 to various restriction fragments of SV40 DNA. An example of the results obtained are shown in lanes 1-4. Specifically, when an aliquot of RNA was hybridized to a purified restriction fragment which spans the region from 43 - 65.5, only one intense RNA-DNA duplex, 1,100 bp long, was detected (lane 1). When another aliquot of the RNA was hybridized to Pst I cut DNA, four RNA-DNA duplexes of sizes corresponding to the denatured transcripts were apparent (compare lanes 2 and 5). If the RNA was instead hybridized to SV40 DNA which had been cut at position 67, the same four RNA-DNA duplexes were observed (lane 3). This indicates that none of the transcripts span coordinate 67. Finally, lane 4 shows that if this RNA was hybridized to DNA restricted at position 0, the two smaller duplexes were each reduced in size approximately 200 nucleotides. These results are consistent with additional experiments (Handa et al., manuscript in preparation) which map the 5' ends of two of these transcripts at positions 67 and 65.5. These transcripts extend in the rightward direction on the map shown in Figure 5. Two leftward reading transcripts have 5' end-points at positions 68 and 72. All four species are run-off products.

It is interesting to compare both the 5' end points and polarities of the in vitro transcripts with the same properties of in vivo SV40 mRNAs. The replicative cycle of SV40 in vivo is divided into an early and late phase. During the early phase, transcription which leads to the production of mRNA is in the rightward-reading direction, and the 5' ends of the mRNAs map at position 67, when assayed by the S1 gel technique (17). One of the rightward reading transcripts synthesized in vitro has this same 5' end, and we believe it corresponds to the in vivo transcript. We are currently investigating the nature of the other rightward reading transcript for which no in vivo counterpart has been described. The 5' ends of the in vivo late mRNAs, which are leftward reading, have also been mapped by the S1 technique (20). In contrast to the early RNAs, these 5' ends map at several positions on the viral genome, although they are clustered in the region around position 70. The most abundant species has a 5' end at position 72, and one of the others has a 5' end at position 68. Thus both the leftward reading species appear to correspond to transcripts synthesized in vivo.

Both the early and late genes of SV40 are transcribed with roughly equal efficiencies in the uninfected HeLa cell extract. In fact, all four transcripts we have detected are synthesized with roughly equal efficiencies. In vivo, to a first approximation, early mRNAs are made only during the early phase of infection, and late mRNAs only during the late phase. We consider two explanations for the apparent contradiction between our *in vitro* results and the *in vivo* situation. First, transcription of both early and late viral genes may initiate with near equal efficiencies *in vivo*. This would then require that production of mature mRNAs be regulated by some other process, such as splicing and/or stability. A second possibility which we favor is that additional factors are required for the regulation which occurs *in vivo*. Possibilities include the minichromosome structure of the viral DNA and/or some repressor or activator protein, such as the viral T antigen. We are currently investigating which, if either, of these models is correct, as well as studying in more detail the structure of the *in vitro* synthesized RNAs.

IV. SUMMARY

We have developed a crude whole cell extract transcription system which accurately transcribes genes of human adenovirus and the simian virus SV40. All the enzymes and factors required to obtain readily detectable levels of specific transcription initiation are contained in a concentrated and dialyzed whole cell extract prepared from uninfected HeLa cells. We have shown here that the adenovirus major late transcription unit is initiated accurately *in vitro* when DNA containing this site is added to *in vitro* reaction mixtures. Furthermore, this transcript is accurately capped and methylated at its 5' end. Transcription initiation of SV40 early and late genes also appears to occur accurately, since we can detect transcripts with 5' ends which map at the same sites on the viral genome as do known viral mRNAs.

We feel this system will be useful not only to study the enzymes, factors and nucleotide sequences needed to bring about accurate initiation of transcription, but also to study the other biochemical reactions such as splicing, termination, etc. required to complete the synthesis of a mature mRNA.

REFERENCES

1. Roeder, R.G., *in* "RNA Polymerase" (R. Losick and M. Chamberlin, eds.), p 285. Cold Spring Harbor Labs, New York, (1976).
2. Zylbert, E., and Penman, S., *Proc. Nat. Acad. Sci. USA* *71*, 2337 (1971).
3. Schwartz, L. and Roeder, R.G., *J. Biol. Chem.* *250*, 3221 (1975).

4. Shatkin, A.J., Cell 9, 645 (1976).
5. Perry, R.P. and Kelley, D.E., Cell 1, 37 (1974).
6. Manley, J.L., Sharp, P.A., and Gefter, M.L., manuscript in preparation (1980).
7. Winters, M.A. and Edmonds, M., J. Biol. Chem. 248, 4763 (1973).
8. Chow, L.T., Gelinas, R.E., Broker, T.R., and Roberts, R.J., Cell 12, 1 (1977).
9. Berget, S.M., Moore, C., and Sharp, P.A., Proc. Nat. Acad. Sci. USA 74, 3171 (1977).
10. Manley, J.L., Sharp, P.A., and Gefter, M.L., Proc. Nat. Acad. Sci. USA 76, 160 (1979).
11. Weil, P.A., Luse, D.S., Segall, J., and Roeder, R.G., Cell 18, 469 (1979).
12. Manley, J.L., Fire, A., Cano, A., Sharp, P.A., and Gefter, M.L., Proc. Nat. Acad. Sci. USA, in press (1980).
13. Flint, S.J., Cell 10, 153 (1977).
14. Darnell, J.E., Prog. Nucl. Acid Res. Mol. Biol. 22, 327 (1978).
15. Ziff, E. and Evans, R., Cell 15, 1463 (1978).
16. McMaster, G.K. and Carmichael, G.C., Proc. Nat. Acad. Sci. USA 74, 4835 (1977).
17. Berk, A.J. and Sharp, P.A., Proc. Nat. Acad. Sci. USA 75, 1274 (1978).
18. Manley, J.L., Sharp, P.A., and Gefter, M.L., J. Mol. Biol. 135, 171 (1979).
19. Lerner, M.R., Boyle, J.A., Mount, S.M., Wolin, S.L., and Steitz, J.A., Nature 283, 220 (1980).
20. Lai, C-J., Dhar, R., and Khoury, G., Cell 14, 971 (1978).
21. Barrell, B., in "Proc. in Nucleic Acid Research" (G.L. Cantini and D.R. Davies, eds.), p. 751. Harper & Row, New York, (1971).

DISCUSSION

M. BESSMAN: Is the threshold phenomenon that you see with 50 micrograms/ml DNA in order to get transcription related in any way to the particular type of DNA that you use for transcription?

J.L. MANLEY: Yes, it is. We haven't done this in detail, but the human B globin gene doesn't show quite as sharp a DNA dependence as does the adenovirus late promoter containing DNA. There is a possibility that this phenomenon could have a physiological significance in that the adenovirus late promoter doesn't seem to be utilized *in vivo* until there is a very huge number of DNA templates present in the cell. I should also point out that when we add a non-promoter containing DNA as carrier to reaction mixes so that the total mass of DNA in each tube is the same, we find that the threshold phenomenon for accurate transcription disappears, and the amount of transcription is proportional to the promoter concentration in the reaction mixtures, up to 50 g/ml.

C. PRIVES: Do you see any differences in the abundance of early and late SV40 transcript?

J.L. MANLEY: There is a bit more late transcription, but it's only 10% to 20%. In fact all four transcripts that we have identified, the two earlys and the two lates, are roughly equal in their intensities on the gels.

G. WU: Can you tell me if you put Ad 2 DNA as template, what is the site of RNA transcript you made in your system?

J.L. MANLEY: If you use total Ad 2 DNA, the most prominent transcript we see is VA RNA - you perhaps saw it has a smear at the bottom of that gel. If you use total uncut Ad 2 DNA as a template and run the RNA on a gel, we see, in addition to VA RNA, a light smear of RNA extending up about 5,000 nucleotides in length.

H. AVIV: I noticed on your last slide that when you used human beta globin DNA as a template, there was an addition to the 1.4, then the 1.8. Is that a potential addition on promoter?

J.L. MANLEY: We don't know what that band is yet. However, we don't think it is related to B globin expression because we see it with DNAs digested by at least 3 different enzymes that we've used to cut the gene. There are several possibilities: one is that it's labeling of 185 ribosomal RNAs. I mentioned that this sometimes occurs in these reaction mixtures. We haven't tested its sensitivity to actinomycin or L-amanitin yet. Another possibility is that it arises from a pBR322 promoter which usually are not expressed efficiently. Sometimes we see bands apparently coming from pBR322 although they are usually not that intense.

H. AVIV: Another small technical thing. You used S1 for detection, I can't understand why S1 and not RNAse A.

J.L. MANLEY: You are talking about the early experiments with the nuclei.

H. AVIV: Yes. I suppose you used S1 all along to detect those bands.

J.L. MANLEY: Most of these gels are just RNA without any nuclease treatment from the soluble system - the RNA was just denatured with glyoxal. In the earlier experiments we were analyzing RNA-DNA hybrids on the gels, so we required digestion of unhybridized DNA sequences as well as RNA sequences to give a discrete band.

J. ABELSON: Jim, is there any evidence for splicing occurring in the reaction mixture?

J.L. MANLEY: All of the RNAs that we have looked at are spliced, all of the DNA templates have splice points, and we know the size of the expected RNAs from the Ad 2 promoter, from SV40 and from globin. And in fact we can see very faint bands at those positions in all cases, although we see a number of other faint bands, too. We are just now in the process of looking whether any of those faint bands might be splicing products. In fact something is funny about their synthesis, at least in the case of the adeno late transcripts, in that these RNA species come up very strongly under conditions suboptimal, for total transcription. However, it's very premature to say anything about splicing.

A. LANDY: Did you mention whether or not you used any protease or nuclease inhibitors in the extract?

J.L. MANLEY: I didn't mention that. We don't use any inhibitors.

J. McPherson: You said your nuclei were ineffective in splicing. Did you find then that none of the RNA was secreted from the nuclei?

J.L. MANLEY: We've looked for secretion of RNA from the nuclei *in vitro*. About 10% of the RNA synthesized *in vitro* is apparently secreted. However, we've looked at its structure and it doesn't seem to be any different than the

RNA that's retained inside the nucleus. In fact, we can't really see if that's secretion or perhaps just 10% of the nuclei lyse during *in vitro* incubation.

H. AVIV: You have used only HeLa cell extract. Is there any difference in extracts from different cells?

J.L. MANLEY: So far we've only used HeLa cell extracts. We have some other extract preparations that we are about to use but we haven't touched them yet.

YEAST MITOCHONDRIAL INTRON PRODUCTS REQUIRED INTRONS FOR RNA SPLICING

George M. Church and Walter Gilbert

Biological Laboratories, Harvard University
Cambridge, Massachusetts

With the goal of investigating the functions of introns we have been studying the cytochrome b (or *box*) and *oxi-3* genes of yeast mitochondria. They are encoded in the 68 to 76 kb mitochondrial DNA (1) along with only five other known proteins and all of the ribosomal and transfer RNA's required for mitochondrial protein synthesis (see Figure 1). Although several mitochondrial genes (oli-1, oli-2, oxi-2, and nine tRNA's) have been shown by sequencing to be continuous throughout their coding regions (2-8) three genes (*box*, lg rRNA, *oxi-3*) are thought to be split. The cytochrome b and lg rRNA introns have been located by R-loop electron microscopy, blot hybridization and genetic mapping (8-13). The arguments for *oxi-3* being split are more indirect (11,14).

For the investigation of split genes this system has a number of impressive advantages over most others at present:

1) Most mitochondrially encoded RNA's and proteins have been mapped and alterations in these macromolecules are particularly easy to detect.

2) Petite deletions provide a stable way to inhibit mitochondrial protein synthesis.

3) Mutations are easily isolated and characterized by complementation and fine structure petite deletion mapping.

4) Of about 200 mutations in cytochrome b characterized to date about half affect expression of another mitochondrial gene *oxi-3*. All of these pleiotropic cytochrome b mutations affect trans-acting elements (defined by complementation properties) and many map to within the introns of the cytochrome b gene.

ISBN 0-12-633360-2

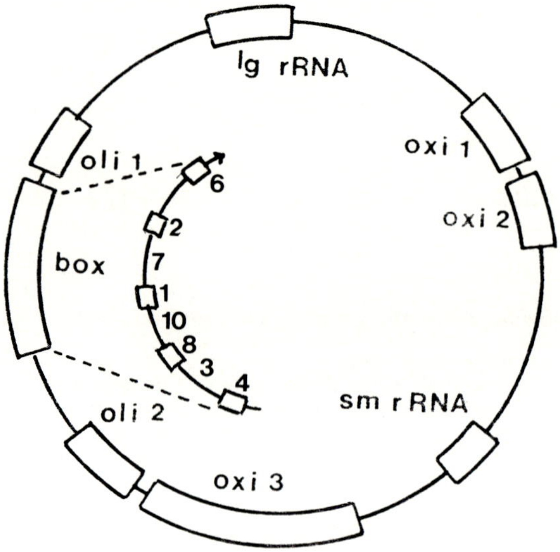

Figure 1. Genetic and physical map of the known mitochondrial genes and the cytochrome b (box) loci associated with five exons (rectangles) and three of the four introns (inner circle) .

figure 1 shows a detailed schematic of the cytochrome b (or box) gene derived from R-loop and petite deletion mapping. The coding region of 900 nucleotides is split into at least five exons (depicted as rectangles) by four large introns; the whole gene spans 8 to 9 kbp. Henceforth each locus will be refered to by its box locus number (defined by recombinational linkage in order of discovery) as well as its physical exon or intron number (in order from 5' to 3'). Thus box 7 is the third intron.

Functions within Introns

The first surprise in the study of this cytochrome b gene came from complementation between box mutants (15). Analysis of complementation was done by following the kinetics of oxygen utilization after mating two cytochrome b mutants (respiration deficient). The mutant mitochondria presumeably fuse in the zygotes soon after mixing and at 9 to 11 hours one sees respiration, if the mutants are in separate genes, due to complementation or at 17 to 19 hours due to recombination if in the same gene. A priori one expects all of the mutations within a single cytochrome b gene to be in the same

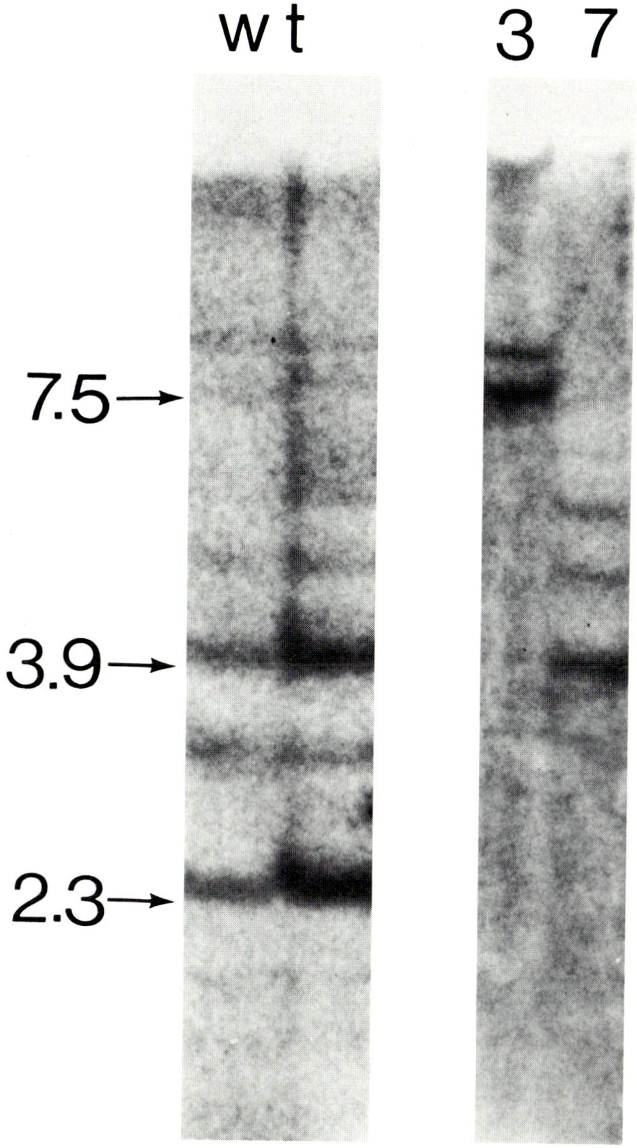

Figure 2. Analysis of mitochondrial RNA transfered to diazo paper (36) and hybridized to a 200 bp DNA fragment from the box-6 region. Wild type, box 3-2 and box 7-1 mutants are indicated. RNA sizes are in kb.

complementation group. Indeed all exon mutants (many of which terminate the cytochrome b chain) fail to complement each other in all combinations tested; intron mutants fail to complement other intron mutants alleles at the same locus. However, all intron mutants complement each other and most exon mutants. Thus box 3, box 10, and box 7 introns were shown to define three independent trans-acting elements involved in some unknown step in cytochrome b gene expression.

The next important finding came when the RNA's from the intron mutants were analyzed (12,14). figure 2 shows total mitochondrial RNA from wild type, box-3 (first intron) and box-7 (second intron) mutants separated by electrophoresis in denaturing methyl mercury agarose, blotted and hybridized with nick-translated DNA from the box-6 (fifth exon) region. The wild type pattern shows a prominent 2.3 kb RNA. Normally the 3.9 kb band is quite weak in the wild type. This preparation shows weak putative precursor RNA bands all the way up to 8.5 kb. The box-3 (first intron) mutants accumulate vast quantities of the 7.5 and 8.5 kb RNA's barely visible in wild type. The 8.5 Kb RNA may contain all introns intact. The 7.5 kb RNA probably has lost a piece of the 2kb first intron in the form of a 1 kb circular RNA (12). The box -7 (third intron) mutants accumulate the 3.9 kb RNA (which has lost the first two introns) and a subset of the larger RNA's. No 2.3 kb cytochrome b mRNA is visible in either mutant. We conclude that the trans-acting elements defined by the complementation data are required for RNA splicing.

How might the intron encoded trans-acting elements function? The RNA from these introns could act in trans either directly as base-pairing guides for splicing (figure 3) or as mRNA's for RNA splicing enzyme subunits (figure 4). We note at the top of figure 3 that in RNA guide models wild type RNA splicing normally need only use intramolecular RNA guidance (not trans acting RNA). The purely schematic bimolecular RNA complex shown at the bottom of figure 3, represents a variety of possible interactions of introns either free or as part of pre-mRNA's with sequences involved in aligning the splice junctions. Other alternative mechanisms involving recombination between mutant RNA's during splicing or otherwise has been shown to be an unlikely explanation for complementation. This was done by tagging exons with drug resistance and chain-termination markers in complementation tests (18,19). These experiments showed that only the exons from a single precursor RNA were brought together by splicing.

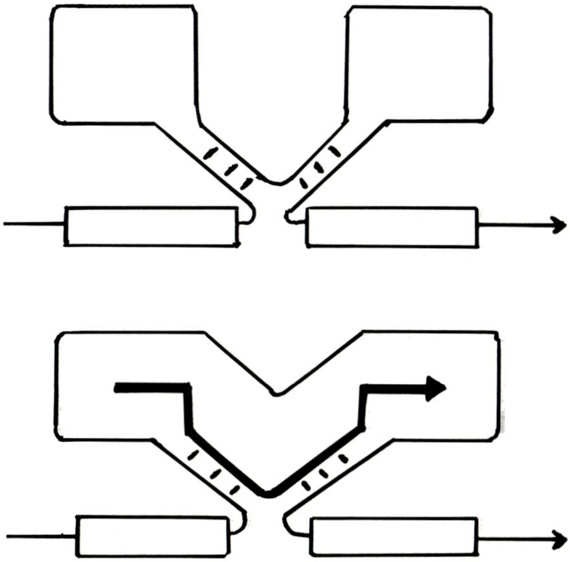

Figure 3. A typical guide RNA hypothesis for (A) wild type intramolecular RNA folding and (B) complementation of intron mutant RNA (thin outer line with rectangles as exons) by exon mutant RNA (thick inner arrow). The short regions of basepairing by intron sequences serve to align the junctions for splicing. The functional intron sequences found in the exon mutant replace the defective guide.

Cytochrome b RNA's Found in Petite Deletion Mutants

A basic prediction of the guide RNA hypotheses is that only nuclear encoded RNA splicing enzymes need be required for cytochrome b RNA processing. In contrast, intron encoded protein hypotheses require mitochondrial protein synthesis. How might one test this critical prediction? The use of mitochondrial protein synthesis inhibitors for enough cell divisions to allow dilution of mitochondrially encoded proteins results in accumulation of drug resistant and petite deletion mutants (20). We chose to take advantage of a unique property of the petite deletion mutants. These deletions lack tRNA's or rRNA's required for mitochondrial protein synthesis and can be propagated in this state indefinitely.

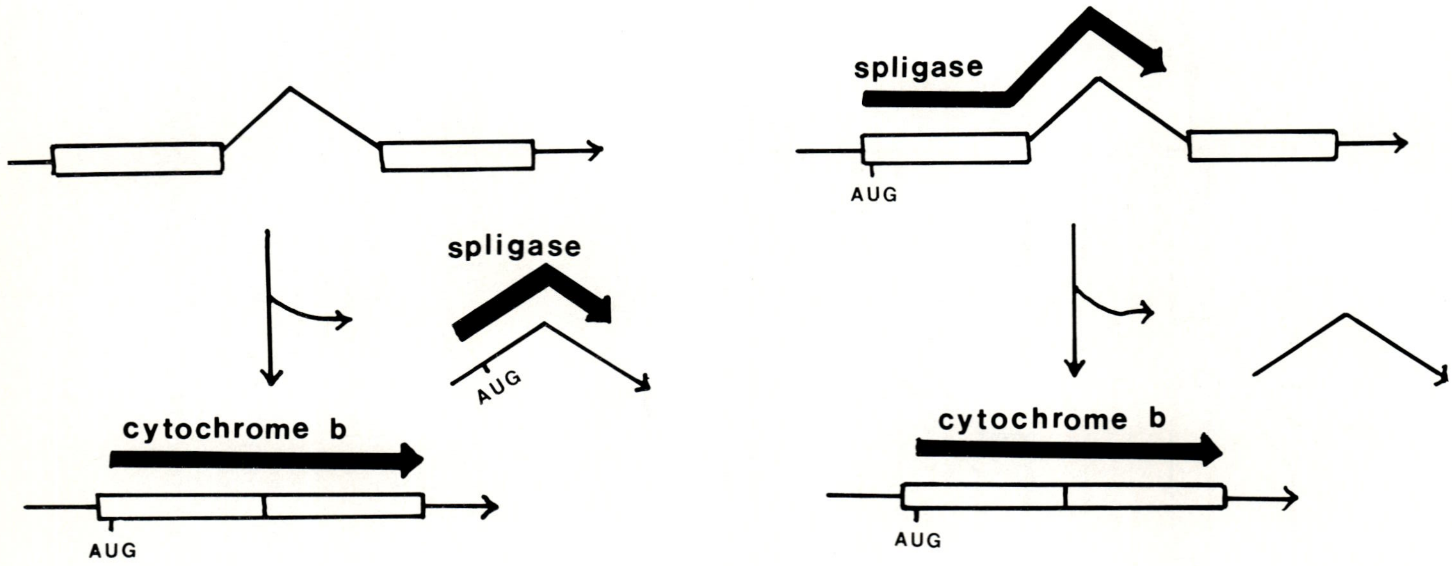

Figure 4. Intron encoded spligase hypotheses. Two representatives of a variety of translational start sites and phases are shown. Solid arrows indicate proteins encoded by exons (rectangles), introns (thin lines), or both.

Two of the petites that we have studied are depicted in figure 5; one has retained the large rRNA and the other the small rRNA gene. Both have an intact cytochrome b gene since they can restore respiratory function by recombination with a deletion of the entire box region. figure 6 shows that rRNA's in the two petites are transcribed and processed normally in the absence of mitochondrial protein synthesis (more than 60 generations after the deletion of the rRNA genes). The wild type RNA enriched for mitochondrial RNA displays, in lane C, four prominent bands: in order from the top, large cytoplasmic, large mitochodrial, small cytoplasmic, and the small mitochondrial rRNA's. The petite OIP2 has only the small mt rRNA. The petite IS2509 has normal sized large mt rRNA. Furthermore, hybridization of the IS2509 RNA with pure intron DNA (21) also indicates that splicing of the lg rRNA is normal in this petite (not shown). We conclude that all of the protein components required for the

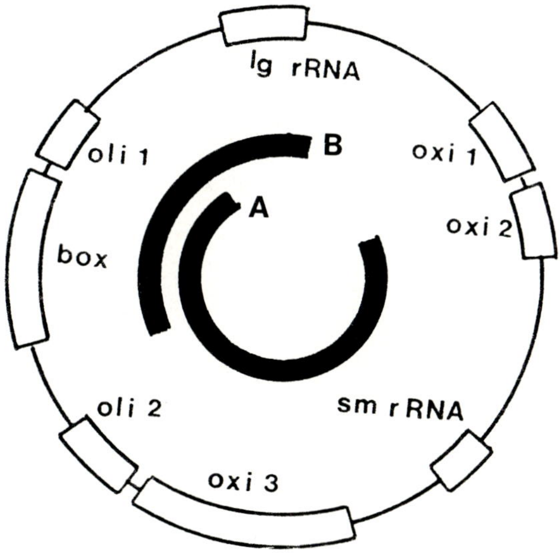

Figure 5. DNA sequences retained in petites used for analysis of rRNA and cytb RNA are indicated by the black inner curved bars. A) Petite OIP2 (37). B) petites K114-4A/I21/IS2509, 3319 (Dujon and Church unpublished).

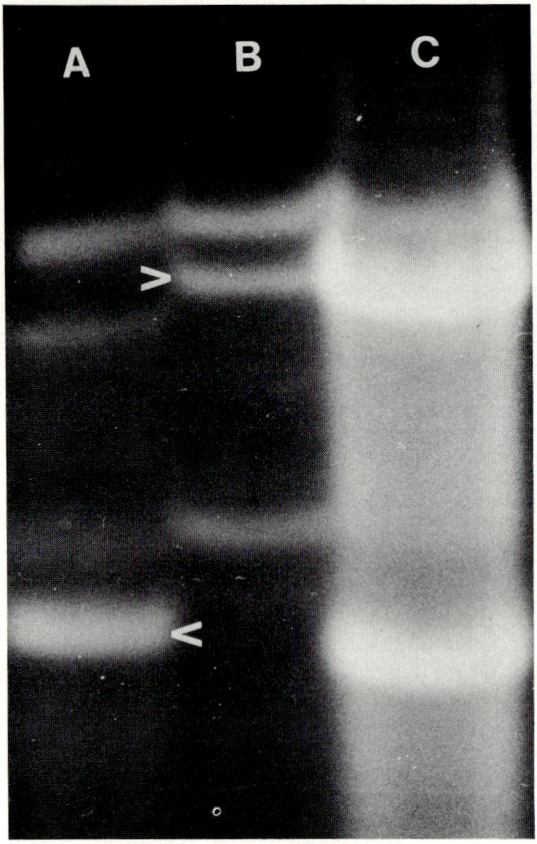

Figure 6. Petite deletion strains (A=OIP2, B= IS2509) and wild type (C) rRNA's visualized by ethidium. Normal sized small (A) and large (B) mt rRNA's are evident in the petites (white arrows).

processing for the large rRNA must be encoded in the yeast nucleus. The lanes labeled petite s in figure 7 show the pattern arising after hybridization of the petite RNA's with the box-6 (fifth exon) probe. There is a major 7.5 kb partially spliced RNA and a minor 8.5 kb RNA. No fully processed cytochrome b RNA appears at 2.3 kb (the pattern is identical to that found in box-3 mutants). So both petites fail to process cytochrome b RNA past the initial initial splicing step (which is also unaffected by all intron mutants). Although the mitochondrial ribosome or RNAs not found in either petite may be required as a cofactors for the splicing, the simplest interpretation of this result is that a mitochondrially encoded protein is required for cytochrome b RNA splicing.

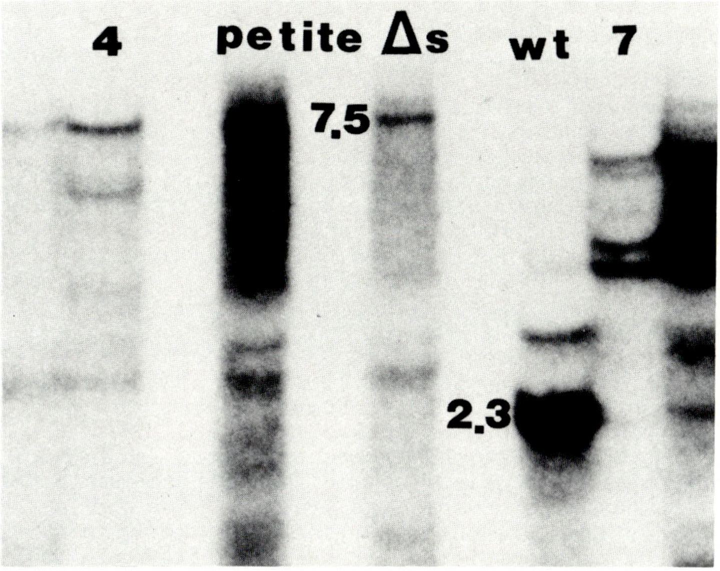

Figure 7. Cytochrome b RNA's found in the petite deletions lacking mitochondrial protein synthesis wild type and two pleiotropic box point mutants. The lanes displayed are: 4 = box 4-2 (G1334); Δs = IS2509,3319; wt = 7773A wild type mitochondrial genes; 7 = box 7-1 (G1659).

Further Predictions of the Hypothesis that Splicing Enzymes are Encoded by the Introns

These observations force us to look more closely at those models in which the trans-acting functions encoded in the cytochrome b introns would be splicing enzymes. As shown in figure 4 an initiation site for protein synthesis within the intron might be recognized in the pre-mRNA or on the free intron. Alternatively translation across a splice junction could occur either in phase or out of phase with the cytochrome b gene. The hypothesis that the active splicing enzymes are initiated in phase with and near to the cytochrome b N-terminus and then read through the splice junction, seems most consistent with a variety of additional observations on cytochrome b mutants. figure 8 briefly summarizes data on the molecular phenotypes of several exon, intron, and putative junction mutants that have been analyzed, including the sizes of the cytochrome b RNA's that accumulate (14,12), the sizes of the new proteins that appear (22-27) and the pleiotropic effects on other mitochondrial gene products: oxi-3 protein and RNA (14,22,23,25).

The single assumption that protein synthesis across splice junctions results in splicing enzyme (hereafter called spligase) C-terminal domains attached to cytochrome b N-terminal domains makes the following predictions (which are not made by any of the other intron protein or RNA models):

1) Long protein coding regions must exist within the introns in phase with the adjacent upstream cytochrome b exon (in contrast, for random DNA 18% G+C, the probability of avoiding an UAG or UAA stop for 35 codons is 0.05 and for 250 is 4×10^{-10}). A 250 amino acid reading frame in the box-3 intron is implied by data discussed below.
2) Since the complementation data is consistent with each spligase specifically cleaving its own intron, each protein would be autoregulatory: destroying its own mRNA by splicing out. In such a way only trace catalytic amounts of each protein should exist in wild type mitochondria, but overproduction to the high levels of the cytochrome b protein itself should be evident in the intron mutants due to the accumulation of the partially spliced RNA's. Exactly such levels of new proteins larger than cytochrome b are generally seen (23,24,25,27).
3) If these new proteins are read through from the cytochrome b start into the intron, they must have N-terminal homology to cytochrome b. Preliminary immunoprecipitation and peptide mapping results are consistent with this prediction (24,27).
4) The intron mutants, which have been chosen to be non leaky, must be missense or nonsense mutants in some mitochondrial protein. If the mutations are terminators, the sizes of the new overproduced proteins must be colinear with the genetic map position of the alleles. The genetic order (figure 8) of the box-3 alleles (5' to 3'): 3-1, (3-2,3-3,3-4), 3-5 fits well with a 42000 dalton spligase terminated at 30,34, and 38000 daltons (15). In this interpretation, box3-2 and 3-4 must be missense mutations.
5) Double mutants with an exon mutation upstream from an intron mutation should have detectable levels of only the short protein (terminated within the exon), as was found in all of the five such double mutants inspected (28).
6) The box-3 (first intron) mutants should accumulate the largest cytochrome b RNA's since the mRNA's for the spligases specific for subsequent steps are not matured (see figure 9). Box-10 (second intron) mutant RNA's should be one intron smaller. Box-7 (third intron) mutant RNA's still one smaller. True: the sizes are 7.5, 6.2, 3.9 kb, respectively (12,14).

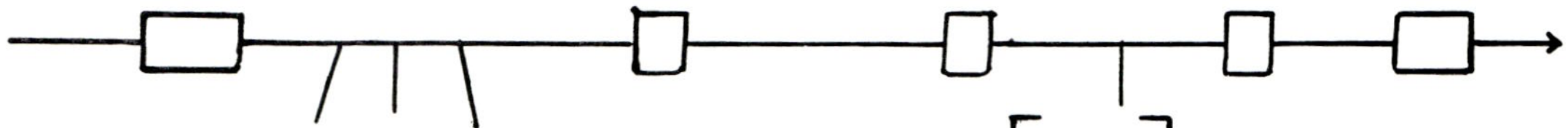

box loci:	4	3	8	10	1	7	2	6
New mt proteins (kdaltons)	13	30,34,38 42	42		18	[26,70 36,42]	22-26	27
Major cyt b RNAs (kb)	2.3 to 7.5	7.5	8.5	6.2	2.3 to 7.5	3.9	2.3	2.3
oxi-3 gene expression	++ or 0	0	0	0	++ or ±	0	++	++

Figure 8. Summary of the sizes of the mutant cytochrome b RNA's and mitochondrial proteins and the pleiotropic effects on oxi-3 gene expression. Each column is headed by the box locus with the phenotypes below. The protein data for 14 exon (or junction) and 8 intron mutants are summarized as are the RNA data for 8 exon and 5 intron mutants. The genetic order of five alleles at three box-3 subloci (13) is indicated. Different box-4 and box-1 alleles give different RNA distributions and degree of oxi-3 expression (see text). Oxi-3 phenotype refers to disappearance of spectral band, enzymatic activity, 40000 dalton protein, and 2.3 kb oxi-3 RNA (with concomittant increase in 3.3 kb oxi-3 RNA).

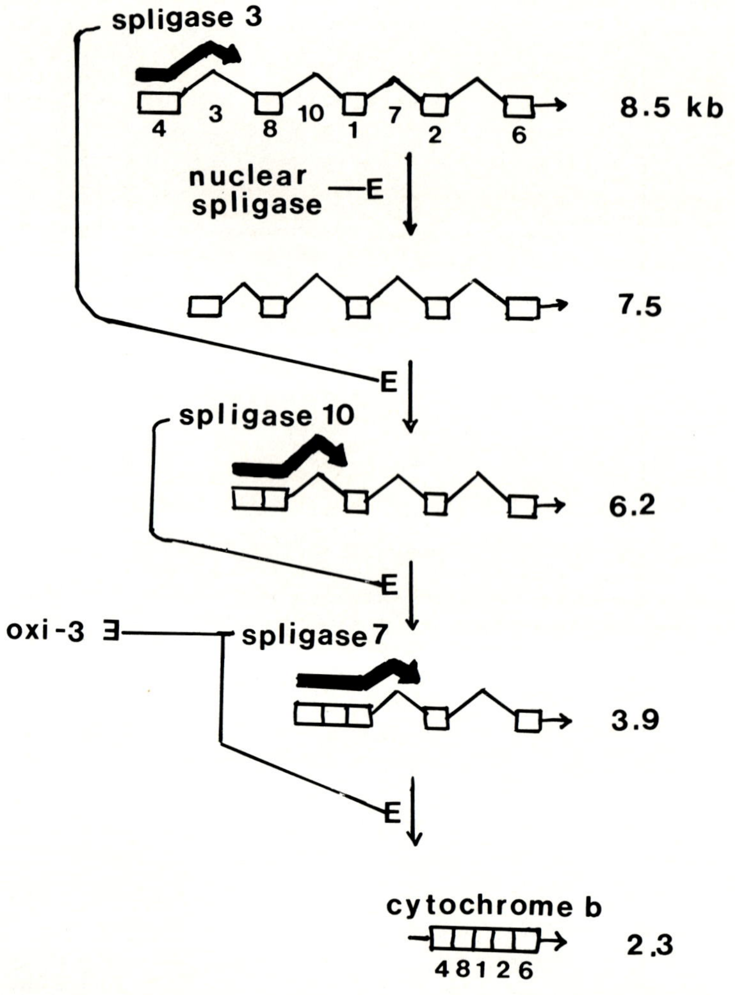

Figure 9. Detailed model for cytochrome b RNA splicing showing sequence of spligase mRNA generation and autoregulation. The RNA structures and sizes in kb are hypothetical and refer to RNA's accumulating in the mutants (figure 8).

7) Since box-7 mutants fail in some step of oxi-3 RNA processing (accumulating a 3.3 kb RNA in place of the normal 2.3 kb species), this intron must encode a product required for this process. This may be the same protein which is required for the splicing out the last two cytochrome b introns (figure 9).

8) Exon termination mutants in box-4,-8, and -1 (first,second, and third exons) but not necessarily missense mutants and not box-2 and box-6 mutants should:

8a) accumulate cytochrome b RNA characteristic of adjacent downstream intron mutants. Box 4-2 resembles 3-2 and 3-4; 4-1 and 1-2 show abnormal levels of cytochrome b RNA precursors but detectable levels of 2.3kb mRNA (12,14, figure 7).

8b) complement only downstream intron mutants poorly. Box 4-1,4-2 fail to complement box 3-1; box 8-1 complements 3-5 better than 10-1 and 7-1; box 1-1 complements 3-1 better than 7-1; however, box 1-2 complements box 7-1 fine (15,29).

8c) fail in the same step of oxi-3 RNA processing as box-7 mutants. For alleles box 4-1,4-2,1-1, and 1-2 this seems to be the case (14,15, Church unpublished).

This hypothesis is an example of the general suggestion that alternate readings and splicings play a role in providing alternate proteins, that one protein's introns can be another protein's exons.

Questions Raised by the Specific Model

A few observations not obviously explained by this hypothesis deserve mention. At first glance, the chain termination mutations within the exons should fail to complement downstrean intron mutations, since they should also terminate the corresponding spligase. However, the overproduction then of the spligase messenger could permit enough read-through of a slightly leaky nonsense codon (which is often observed in other systems (30-33)) to generate a functional, enzymatic amount of spligase. In fact, in these mutants the splicing pattern is aberrant, only a reduced amount of mature mRNA but high levels of precursors.

Box 2-4 has temperature sensitive effects on oxi-3 but may lie at the junction of box-7 and box-2 (Claisse et al., 1980) or affect some function in the last intron. Far more mutations have been found within the cytochrome b exons and box-3 locus than within the box-10 or box-7 loci. This may reflect a smaller functional unit or active site (e.g. an oligonucleotide guide RNA) or peculiarities of the mutagenic or screening procedures.

Several novel large proteins found in some pleiotropic mutants are still unexplained. These may represent post-translational modification, new reading frames, or oxi-3 encoded proteins (whose mRNA's have accumulated for reasons outlined above). We would suggest that these other proteins are not translated from random non-coding regions but like the hypothetical fused splicing enzymes in figure 9 have functions in wild type, some possibly unrelated to RNA processing.

Some intriguing directions for inquiry are suggested by this hypothesis of fused splocing enzymes encoded in introns: Are the intron protein functions sufficient for splicing or are nuclear proteins and/or mitochondrial guide RNA's also required? Are the three intron spligases homologous? If so, was a special transposition mechanism used to generate a domain triplication (34)? Of what use is an N-terminal cytochrome b domain for a splicing enzyme? Perhaps membrane binding? How many proteins in this and other systems translate across splice junctions and read through terminator codons normally? So far papovavirus early genes (35) represent examples of the former, TMV and MuLV polymerase genes the latter (30-32). How many other uncommon proteins lie hidden in the "extra" DNA of the yeast mitochondrion whose genome is five times larger than that of other mitochondria? Finally, if the cytochrome b mRNA processing really requires four separate RNA splicing recognition subunits how does one interpret the apparent universality of nuclear splicing activities?

ACKNOWLEDGMENTS

We thank B.Dujon and H.Fukuhara for gifts of petite strains and advice; P.P. Slonimski, A. Halbreich, W. Herr, P. Lomedico, R. Morimoto, and D. Wirth for provoking discussions. This work was supported by National Institutes of Health grant GM09541 to W.G. G.M.C. is supported by Training Grant GM07598-02. W.G. is an American Cancer Society Professor of Molecular Biology.

Claude Jacq, Jaga Lazowska, and Piotr P. Slonimski have determined the DNA sequence of part of the intron containing box-3 mutations and find a continuous reading frame of at least 170 amino acids. Box-3-5 is a terminator in this frame, while box 3-2 is a double missense mutation. They have suggested that box-3 might encode a protein involved in RNA splicing. (Comptes rendus Acad. Sci. Paris 290).

REFERENCES

1. Borst, P. and Grivell, L.A. (1978) Cell 15, 705.
2. Fox, T.D. (1979) Proc. Nat. Acad. Sci. USA 76, 6534.
3. Corruzi,G. and Tzagoloff, A. (1979) J. Biol. Chem. 254, 9324.
4. Li,M. and Tzagoloff, A. (1979) Cell 18, 47.
5. Hensgens, L.A.M., Grivell, L.A., Borst, P. and Bos, J.L. (1979) Proc. Nat. Acad. Sci. USA 76, 1663.
6. Macino, G. and Tzagoloff, A. (1979) J. Biol. Chem. 254,4617.
7. Miller, D.L., Martin, N.C., Pham, H.D., and Donelson, J.E. (1979) J. Biol. Chem. 254, 11735.
8. Grivell, L.A., Arnberg, A.C., Boer, P.H., Borst, P., Bos, J.L., Van Bruggen, E.F.J., Groot, G.S.P., Hecht, N.B., Hensgens, L.A.M., Van Ommen, G.J.B., and Tabak, H.F. (1979) in Extrachromosomal DNA: ICN-UCLA Symposia on Molecular and Cellular Biology, vol 15 ,Cummings, D., Borst, P., Dawid, I., Weissman, S., and Fox, C.F., eds. (New York:Academic Press), in press.
9. Bos, J.L., Heyting, C., Borst, P., Arnberg, A.C., and Van Brugger, E.F.J. (1978) Nature 275, 336.
10. Faye, G., Dennebouy, N., Kujawa, C., and Jacq, C. (1979) Mol. Gen. Genet. 168, 101.
11. Van Ommen, G.-J.B., Groot, G.S.P., and Grivell, L.A. (1979) Cell 18, 511.
12. Halbreich, A., Pajot, P., Foucher, M., Grandchamp, C., and Slonimski, P. (1980) Cell in press.
13. Slonimski, P.P., Pajot, P., Jacq, C., Foucher, M., Perrodin, G., Kochko, A., and Lamouroux, A. (1978b) in Biochemistry and Genetics of Yeast, Bacila, M., Horecker, B.L., and Stoppani, A.W.N., eds. (New York: Academic Press), pp. 339.
14. Church, G.M., Slonimski, P., and Gilbert, W. (1979) Cell 18,1209.
15. Slonimski, P.P., Claisse, M.L., Foucher,M., Jacq, C., Kochko, A., Lamouroux, A., Pajot, P., Perrodin, G., Spyridakis, A., and Wambier-Kuppel, M.L. (1978a) in Biochemistry and Genetics of Yeast, Bacila, M., Horecker, B.L., and Stoppani, A.W.N., eds. (New York: Academic Press), pp. 391.
16. Lerner, M.R., Boyle, J.A., Mount, S.M., Wolin, S.L., Steitz, J.A. (1980) Nature 283, 220.
17. Murray, V. and Holliday, R. (1979) FEBS Lett. 106, 5.
18. Kochko, A., Colson, A.M., Lamouroux, A., Pajot, P., and Slonimski, P.P. in preparation.
19. Dujon, B. (1979) Nature 282, 777.
20. Williamson, D.H., Maroudas, N.G., and Wilkie, D. (1971) Molec. Gen. Genet. 111, 209.

21. Morimoto, R. and Dujon, B. (1980) in preparation.
22. Alexander, N.J., Vincent, R.D., Perlman, P.S., Miller, D.H., Hanson, D.K., and Mahler, H.R. (1979) J. Biol. Chem. 254, 2471.
23. Claisse, M.L., Spyridakis, A., Wambier-Kuppel, M.L., Pajot, P., and Slonimski, P.P. (1978) in Biochemistry and Genetics of Yeast, Bacila, M., Horecker, B.L., and Stoppani, A.W.N., eds. (New York: Academic Press), pp. 369.
24. Claisse, M., Slonimski, P.P., Johnson,J., and Mahler, H. (1980) Molec. Gen. Genet. in press.
25. Haid, A., Schweyen, R.J., Bechmann, H., Kaudewitz, F., Solioz,M., and Schatz,G. (1979) Eur. J. Biochem. 94, 451.
26. Hanson, D.K., Miller, D.H., Mahler, H.R., Alexander, N.J., and Perlman, P.S. (1979) J. Biol. Chem. 254, 2480.
27. Solioz, M. and Schatz, G. (1979) J. Biol. Chem. 254, 9331.
28. Mahler, H.R. and Perlman, P.S. (1979) in Extrachromosomal DNA: ICN-UCLA Symposia on Molecular and Cellular Biology, vol 15 ,Cummings, D., Borst, P., Dawid, I., Weissman, S., and Fox, C.F., eds. (New York:Academic Press), in press.
29. Lamouroux, A., Pajot, P., Kochko, A., Colson, A.M., and Slonimski, P.P. in preperation.
30. Pelham, H.R. (1978) Nature 272, 469.
31. Phillipson, L., Andersson, P., Olshevsky, U., Weinberg, R., and Baltimore, D. (1978) Cell 13,189.
32. Murphy, E.C., Kopchick, J.J., Watson, K.F., and Arlinghaus,R.B. (1978) Cell 13, 359.
33. Geller, A.I. and Rich, A. (1980) Nature 283, 41.
34. Gilbert, W. (1978) Nature 271, 501.
35. Rigby, P. (1979) Nature 282, 781.
36. Alwine, J.C., Kemp, D.J. and Stark, G.R. (1977) Proc. Nat. Acad. Sci. USA 74, 5350.
37. Lewin, A., Morimoto, R., Rabinowitz, M., and Fukuhara, H. (1978) Molec. Gen. Genet. 163, 257.

DISCUSSION

I. HERSKOWITZ: Is it possible that some of your complementation results with different <u>oxi-3</u> mutants are due to recombination rather than complementation?

G.M. CHURCH: I have not done these experiments myself so I can't give an expert opinion on this, but this complementation has been done with a number of genes. For example, when you pair cytochrome B with <u>oxi-1</u>, <u>2</u> and <u>3</u> mutants you see complementation. For all pairs of exon mutants that have been looked at, you see only the recombinations not complementation. Those are the only internal controls that I know of. In every case, time points are taken every half hour and screened for recombinants.

R. BUTOW: Do you know if the pattern of splicing that you see in a given petite mutant is independent of the nuclear background?

G.M. CHURCH: I have looked at diploids where the nuclear background has been changed to introduce the OP-1+ allele for the box mutants and those look the same, but I haven't looked at effects on petites yet.

A. HALBREICH: There are a few things I wanted to correct here. First of all, there is a distinct difference between complementation and recombination in the sense that any pairwise combination of two mutants will eventually restore wild type phenotype by recombination, but only the four complementation groups have been seen. Furthermore, by using the petite mutants to complement box mutants, one can make the distinction between those mutants which are complemented and those that can complement. In this way it can be seen that no exon mutant can be complemented. They can only complement intron mutants. The same conclusion can be derived from the experiments on the combinations of diuron resistance with exon and intron mutations, namely that in all cases of complementation, all the exons from one parental gene with the mutated intron are expressed. In other words, the only thing that you can do in complementation is to overcome an intron mutation and allow splicing to occur. In this way, one can say with certainty that box-8 is not complemented; it complements box-3 mutations but does not complement box-10 and box-7 mutations.

G.M. CHURCH: I agree.

A. HALBREICH: The other point of criticism, on your proposed spligase concerns the fact that box-4 mutants have a stop codon inside that exon; consequently they cannot translate beyond that exon into the next intron, nevertheless, most of them are processed normally. Similarly, the box-1 mutant, cannot be translated beyond _box-1_ and therefore cannot possibly make the last spligase that you eluded to. In conclusion, there is no possibility for this kind of a model.

G.M. CHURCH: I'll basically just address the _box-1_ situation. As I've mentioned, certain _box-1_ alleles, in fact two out of the original three, are pleiotropic in terms of gene expression, which indicates that they are, in fact, affecting a downstream function. In addition, one of them complements _box-7-1_ poorly. I agree in the sense that this is a weak point in the theory.

FREE COMMUNICATIONS

FUNCTIONAL VIRAL POLY(A)+ AND POLY(A)-MESSAGE IN ADENOVIRUS 5-INFECTED CELLS

Catherine Baird and James M. Wilhelm Department of Microbiology, University of Rochester, Rochester, New York

In the processing of late adenovirus mRNA, the long primary transcript is polyadenylated and spliced to form a mature mRNA species which is transported to the cytoplasm. It is thought that most, if not all, viral mRNAs enter the cytoplasm as poly(A)+ RNAs. To investigate whether viral mRNA can be metabolized to a poly(A)- RNA and still retain its function as a message, we have isolated RNA from adenovirus 5-infected KB cells 18 hours P.I. and assayed for the presence of translatable poly(A)+ and poly(A)- viral mRNA.

Total cytoplasmic RNA from a post-mitochondrial extract was fractionated on oligo(dT)cellulose into poly(A)+ and poly(A)- fractions which were used to program cell-free protein synthesis in wheat germ extracts. Poly(A)+ RNA codes for all of the major viral proteins. Poly(A)- RNA also codes for viral proteins, although the mRNA for several proteins, notably proteins IV(fiber) and IX, seems to be present in very low amounts. Poly(A)+ and (A)- RNA was isolated from the polysomal and nonpolysomal regions of the cytoplasm to determine whether the poly(A)- mRNA is found in a translatable state or as free mRNP in the infected cell. Both regions were found to contain translatable poly(A)+ and poly(A)- viral mRNA, suggesting that viral poly(A)- RNA is present as a functional message and is utilized in the infected cell.

Shortening of the poly(A) segment is a normal metabolic process and it is not unexpected that functional viral poly (A)- mRNA exists. Tryptic mapping studies show that proteins synthesized by poly(A)+ and (A)- mRNAs are identical, indicating that although the structure of the RNA has been changed, the coding region has not been altered. The very low amounts of proteins IV and IX in the poly(A)- translation products may be explained by the close proximity of the C-terminal ends of the protein-coding regions to the poly(A) segment in the two messenger species. Any nucleolytic activity at the 3' end of these mRNAs may cleave the poly(A) and extend into the coding region, resulting in non-functional mRNAs.

Supported by PHS grant A1-11427 from the NIH and by USPHS grant GM-07102-04

ISBN 0-12-633360-2

PURIFICATION AND CHARACTERIZATION OF DNA-DEPENDENT RNA POLYMERASE FROM VACCINIA VIRIONS

B. M. Baroudy and B. Moss. National Institute of Allergy and Infectious Diseases, N.I.H., Bethesda, Maryland 20205

A DNA-dependent RNA polymerase was extracted from vaccinia virus particles. Maximal soluble activity was obtained by incubating purified virions in 0.25 M NaCl, 10 mM dithiothreitol, 0.2% sodium deoxycholate, 0.05 M Tris-HCl (pH 8.5) at 0°C. Insoluble structural proteins were sedimented and the supernatant was passed through a DEAE-cellulose column to remove nucleic acids. The RNA polymerase was purified about 800-fold and was separated from poly(A) polymerase and other viral enzymes by successive chromatography on DEAE-Biogel, phosphocellulose, and aminopentyl-agarose. A single protein peak containing RNA polymerase activity was detected by glycerol gradient centrifugation. The enzyme which had a M.W. of 519,800 daltons was judged to be homogeneous upon electrophoresis on a non-denaturing polyacrylamide gel. Seven polypeptides ranging in M.W. from 140,000-17,000 daltons were resolved by polyacrylamide gel electrophoresis in the presence of sodium dodecyl sulfate. RNA polymerase activity was dependent on the addition of a suitable DNA template, all four ribonucleoside triphosphates and Mn^{2+}. Maximal activity was obtained at pH 7.9 with 100 mM KCl. The Km for GTP, ATP, UTP, and CTP were determined to be 333, 80, 22, and 12 μM, respectively. Double-stranded DNA from vaccinia virus, adenovirus, T4 phage, lambda phage, and calf thymus were poor templates as were supercoiled SV40 and plasmid pBR 322 DNAs. All of the above DNAs were good templates, however, after denaturation.

ISBN 0-12-633360-2

INDUCTION OF THE TRANSFORMED PHENOTYPE IN RAT FIBROBLASTS BY TRANSFECTION OF CHROMOSOMAL DNA EXTRACTED FROM POLYOMA TRANSFORMED RAT CELLS.

S. Gattoni, J.D. Laskin and I.B. Weinstein, Columbia University/College of Physicians and Surgeons, Cancer Center/Institute of Cancer Research, New York, New York 10032

Rat fibroblasts transformed by the ts-a mutant of polyoma produce free viral molecules only at a temperature which is permissive for the A gene (33°), but not at the non-permissive temperature (40°). H6, a rat cell line transformed by the polyoma ts-a mutant, contains about 3 copies of a full size viral DNA integrated, in a head to tail tandem arrangement, in the host genome (Basilico *et al.*, Cell 17, 645-659, 1979). We have extracted the chromosomal DNA from H6 cells grown at 40°, so that only viral integrated sequences were present in our preparation. We then used this DNA to transfect Fischer rat fibroblasts at 33°, using the calcium phosphate precipitation method. Two days after transfection, the cells were trypsinized and plated in 0.3% agar. Four weeks later, several clones of growing cells appeared in the agar plates of the transfected cells but none were seen in control cultures. These clones were picked and tested for a) the production of polyoma T antigen by immunofluorescence, b) the presence of viral DNA integrated into the host cell genome and c) the induction of free viral DNA. Some of these clones gave positive results for all three parameters. One clone (G5) contains tandem repeats of a full polyoma genome integrated into the host DNA, as in the case of the parental line, although the arrangement appears to be different. The extent of expression of phenotypic properties of transformation in the transfected clones did not correlate with the number of integrated viral equivalents, although the latter did correlate with the production of free viral DNA.

Our results appear to represent the first successful transfection of polyoma DNA using as donor DNA viral sequences integrated into cellular DNA. (Supported by NCI Grant CA-21111-03).

ISBN 0-12-633360-2

WHAT IS THE CHEMICAL NATURE OF CYTOPLASMIC REGULATORY FACTOR FOR PHENYLALANINE HYDROXYLASE GENE?

T.V. Gopalakrishnan. Department of Pediatrics, The Johns Hopkins Hospital, Baltimore, MD 21205.

We have previously shown that the fusion of rat hepatoma cytoplasts with mouse erythroleukemia (MEL) cells results in cybrids expressing phenylalanine hydroxylase (PH) constitutively due to the activation of the normally silent mouse PH gene (1). Activation of the PH gene allows the cells to grow in tyrosine free medium. The activation could be due to the action of self-perpetuating, positive cytoplasmic regulatory factor (CRF).

To identify the chemical nature of the CRF, fusions were carried out between MEL cells and rat hepatoma cells, treated before fusion as follows: 1) with a lethal does of mitomycin (10 μg/ml for 3 hr), which binds covalently to DNA causing depolymerization of DNA, or 2) irradiated with UV ($4x10^3$ $erg/sec/cm^2$ for 2 min), a dose that results in >98% inhibition of protein synthesis, possibly due to the inactivation of cellular mRNAs. Fusion products were selected in appropriate tyrosine free media for the expression of phenylalanine hydroxylase. Both mitomycin treated and UV irradiated hepatoma cells when fused with MEL cells gave rise to a growing population of cells with the same chromosome complement as MEL cells, but expressing phenylalanine hydroxylase, thus resembling the original cybrids (1). These results suggest that the CRF for phenylalanine hydroxylase gene is probably neither DNA nor RNA in nature. Experiments with reagents that inactivate proteins are in progress.

Supported by USPHS Grant # HD 13316.01.

1. Gopalakrishnan, T.V., and Anderson, W.F. Proc. Natl. Acad. Sci. USA 76:3932-3936 (1979).

ISBN 0-12-633360-2

SEQUENCES THAT FLANK THE INTERVENING SEQUENCE IN rDNA ARE HIGHLY CONSERVED IN EVOLUTION.

Richard L. Gourse and Susan A. Gerbi, Division of Biology and Medicine, Brown University, Providence, Rhode Island, 02912

By hybridization studies, we have located regions conserved between the rDNAs of distantly related eukaryotes and a prokaryote to defined restriction fragments of Xenopus laevis rDNA on the order of 50-300 base pairs in length. (Gourse & Gerbi, manuscript submitted). Some of these conserved fragments are located about 2/3 of the way along the 28S coding region. In various eukaryote rDNAs, intervening sequences (IVS) have been found at a position about 2/3 of the way along the 28S coding region. In Tetrahymena pigmentosa, some strains contain an IVS while others do not. It was therefore possible to identify a fragment containing the point where an IVS interrupts the Tetrahymena rDNA (Wild & Gall, [1979] Cell 16: 565-573). One of the evolutionarily conserved Xenopus restriction fragments hybridized strongly to this Tetrahymena fragment. We have sequenced the rDNA from this Xenopus region (which does not contain an IVS) and compared it with the Tetrahymena sequence (M. Wild, personal communication). The sequences revealed that there is almost perfect homology extending some 100 base pairs on either side of the point where the IVS interrupts the Tetrahymena rDNA. This IVS flanking region is the most highly conserved section of rDNA we have yet identified.

Supported by PHS-GM20261; S. Gerbi is recipient of an RCDA: PHS-GM00036.

ISBN 0-12-633360-2

A PATHWAY OF SPECIFIC SPLICING STEPS IN CYTOCHROME b mRNA PROCESSING REVEALED IN YEAST MITOCHONDRIA BY MUTATIONAL BLOCKS WITHIN THE INTRONS AND CHARACTERIZATION OF A CIRCULAR RNA DERIVED FROM A COMPLEMENTABLE INTRON.

A. Halbreich, P.Pajot, M. Foucher, C. Grandchamp, P. Slonimski

Centre de Génétique Moléculaire du C.N.R.S.,91190 Gif-sur-Yvette,France and Université Pierre et Marie Curie (Paris VI)

We have studied the expression of the mosaic mitochondrial gene, cob-box, which codes for the cytochrome b in Saccharomyces cerevisiae. Mitochondrial RNAs were analyzed by the technique of Alwaine et al.(1977) using DNA fragments containing parts of the cytochrome b gene and by electron microscopy. Only high MW, not fully processed transcripts are seen in intron mutants, when a carboxy-terminal exon and an amino-proximal intron probes are used, whereas exon mutants contain 2.7 kb mRNA which does not hybridize to the intron probe. The 8.6, 8.1 and 7.1 kb species accumulate in box8, box3 and box10 mutants respectively. The processing of the first two species above is also marked by the sequential appearance of 0.8 and 0.9 kb stable intron RNAs respectively ("1 kb RNAs"). The interruption of mRNA processing by intron mutations, thus allowed the deduction of a pathway for the maturation of cytochrome b mRNA in which introns are sequentially spliced with more than one step for at least some of them.

The "1 kb RNAs" were also observed in three rho$^-$ petites which complement box3 mutations. They were not found in two non-complementing petites even though these also contain the wild type box3 allele and produce mtRNA transcripts. The "1 kb RNAs" map inside the amino-proximal intron which extends from box4 to box1, but outside the segment encompassing box3 mutations. The "1 kb RNA" fraction comprises RNA circular molecules 0.275 $\pm$ 0.017 μm long and linear molecules 0.238 $\pm$ 0.028 μm long. Larger transcripts were also seen in these petites indicating that petites are capable of transcript processing. These results point out that the normal sequence of nucleotides in an intron and its flanking sites is necessary but insufficient for the correct splice to occur. Other sequences, quite remote from it are also important. The onset and progression of splicing three-dimensional structures which can be recognized by the enzymic machinery. Mutations which deform these structures and corrections by RNA complementation may be a powerful tool for a better understanding of the phenomenon.

ISBN 0-12-633360-2

TYPE C VIRUS EXPRESSION IN HUMAN PLACENTA

E. Hefti, and S. Panem. Department of Pathology, The University of Chicago, Chicago, Illinois 60637

There is extensive evidence from subhuman primate and human studies that placenta is a privileged site for type C virus expression in contrast to adult tissue. The failure to isolate type C viruses from human placentas to date most probably reflects inappropriate culture conditions. Other possible reasons are: (1) the low level of release of mature virions, (2) the production of defective particles, (3) a restricted virus-host relationship inhibiting the production of infectious virions. In view of these problems, we have analyzed type C virus expression using techniques which do not require isolation of virus particles. Simian sarcoma virus (SiSV) has been used as probe to detect viral expression, since all so far known human type C viral isolates have been shown to be related to SiSV. The information obtained from the intracellular studies may at a later time be helpful in the isolation and cultivation of a human type C virus.

We have established *in vitro* cell cultures from 15 term human placentas. Each cell strain has a limited life span *in vitro* lasting between 2.5 - 5 months. These cells have been monitored throughout their *in vitro* life span for virus particle production by electronmicroscopy and reverse transcriptase assay of culture fluids. So far, no cell strain has shown spontaneous virus release. However, 5 of 5 cell strains examined synthesized type C viral antigens detected by competition RIA using anti-SiSV sera and purified SiSV antigens. The data indicate that placenta cells express a subset, but not all, of SiSV antigenic determinants.

Parallel to the immunological studies, fresh human placenta and *in vitro* cell cultures established from human placenta have been analyzed for type C viral related sequences by molecular hybridizations. The analysis of cellular DNA and RNA from human placenta tissue and *in vitro* cell cultures has provided information in regard to the following questions:
(1) Does *in vitro* culture of human placenta allow selective growth of cells containing viral related nucleic acid sequences?
(2) Does the extent of viral expression alter as a consequence of *in vitro* cell passage?
(3) Is viral RNA expressed in all cells containing proviral DNA. The implications of these data for understanding a possible endogenous human type C virus and the frequency of exogenous virus infection and spread in human placenta will be discussed.

ISBN 0-12-633360-2

ORGANIZATION OF EARLY HISTONE REPEATS IN THE SEA URCHIN *STRONGULOCENTROTUS PURPURATUS*

D.S. Holmes. Department of Biological Sciences, State University of New York at Albany, 1400 Washington Avenue, Albany, New York 12222.

DNA prepared from the sperm of several sea urchins was sheared to 20-50 kilobases and enriched for histone genes by several cycles of centrifugation in actionomycin D- CsCl. Hybridization measurements using radioactively labeled cloned early repeats indicate that approximately 20% of this DNA is histone DNA. The enriched DNA was R-looped with crude early embryonic histone mRNA. Plasmids containing the five early histone genes were included in the R-loop reaction as internal length markers and to determine the efficiency of R-looping. A full complement of 5 R-loops was observed on these plasmids in 70-80% of all cases. 20% of the enriched DNA formed R-loops with a pattern recognizable of the early repeat type as described in the literature. Typically these repeats were formed in blocks of 2-3 even though many molecules were sufficiently long to contain up to 8 contiguous repeats. The junction of the histone repeat with the flanking non-histone DNA was most frequently in the spacer between H2B and H3. Thus the several hundred early histone repeats of *S. purpuratus* are not arranged as a single contiguous array but are more likely to be grouped in blocks of a few repeats separated by other DNA. Variants of the basic early repeat were observed. Most common was a variation in the length of the spacer between H4 and H2B, but variation in the H1-H2A spacer was also common. Length variations of spacers of less than 100 bps would not be detected in our experiments. Generally contiguous repeats had the same R-loop topology but occasionally molecules containing more than one type of early repeat were observed, hence repeat heterogenity is found at the individual as well as the population level. Histone mRNA was also hybridized to single stranded enriched DNA and visualized in the electron microscope by the gene-32 spreading method. Hybrids were found on one strand consistent with a model where contiguous repeats are transcribed in the same direction and no structures that might be interpreted as intervening sequences were observed.

ISBN 0-12-633360-2

ORGANIZATION OF LATE HISTONE GENE REPEATS IN THE SEA URCHIN *STRONGULOCENTROTUS PURPURATUS*

D.S. Holmes and R. Orris. Department of Biological Sciences State University of New York at Albany, 1400 Washington Avenue, Albany, New York 12222.

Total RNA was prepared from late sea urchin embryos (36 hour stage) and 7-14S RNA selected by centrifugation in a non-denaturing sucrose gradient. This RNA was R-looped to plasmids containing the early histone gene repeat. R-loops were formed at high criterion with very low efficiency to all genes except H2B even at extreme Rot. Hence there is little residual early histone mRNA in this preparation and little hybridization of late histone mRNA to the early genes at the chosen criterion. There is no detectable hybridization to the spacers. However when the RNA is R-looped to the early genes at relatively low criterion (as described in Holmes *et al*. (1977) Biochemistry 16, 1504) R-loops are formed with high efficiency with all 5 genes but none of the spacers. The R-loops of the H4 and H1 were frequently complicated and could be interpreted by assuming that regions of homology and non-homology exist between late histone mRNA and early histone genes. When the RNA is R-looped with total unfractionated, unsheared genomic DNA at high criterion, approximately 1/500 molecule forms R-loops which have detectable topological homology with the early histone genes but with a larger (approx. 6.9 kb) repeat. However since the late RNA forms R-loops inefficiently with the early genes we interpret these R-loops to represent late histone genes. Most importantly R-loops are fomed consistently in the spacer DNA, with the exception of short regions of the spacers flanking H4 which never form R-loops. We conclude that the late histone genes are organized into repeats with the same gene order and with "spacers" of approximately the same size as the early repeats, but that the spacers are transcribed and have no detectable homology with the early spacers. Generally only 3-4 contiguous repeats were observed even though some molecules have enough DNA to contain more than 10 contiguous repeats. Hence the late histone genes, like the early histone genes, are probably organized into blocks and not as a single contiguous array. By quantitating the frequency of late histone gene R-loops, we estimate roughly that 0.1% of the genomic DNA consists of late histone repeats.

ISBN 0-12-633360-2

INTERFERENCE OF AN AZOCARCINOGEN WITH RNA PROCESSING. A POSSIBLE FUNCTION OF REPEAT RNA SEQUENCES IN THE MAINTENANCE OF CONTINUOUS PROCESSING OF NUCLEAR RNA.

Viktor Holoubek and Nutan T. Patel. Department of Human Biological Chemistry and Genetics, University of Texas Medical Branch, Galveston, Texas 77550

We isolated from the liver chromatin 13,000 daltons RNA which contain nucleotide sequences restricted to the cell nucleus. This RNA is transcribed from middle repetitive DNA as a part of metabolically highly active large molecular weight RNA. During the processing of this RNA the repeat RNA sequences are split from the large RNA molecules and associated with chromatin. After the feeding of 3'-methyl-4-dimethylaminoazobenzene the processing of the newly synthesized RNA molecules is disrupted. The nucleotide sequences of the chromatin-associated RNA are released into the cytoplasm while they are still parts of large RNA molecules. The sequence complexity of the chromatin-associated RNA which is released into the cytoplasm increases continuously with the time of the feeding of the azocarcinogen. The appearance of this nuclear RNA in the cytoplasm is concomitant with the depletion of the nucleus of the same nucleotide sequences in the form of chromatin-associated small molecular weight RNA. This by the azocarcinogen induced release of the sequences of nuclear repeat RNA into the cytoplasm will continue even after the feeding of the carcinogen has been stopped. According to the Davidson and Britten model of the regulation of gene expression on the level of RNA processing (1), repetitive RNA sequences which are split from the RNA during its processing are required to maintain this processing. Any temporary removal of such sequences from chromatin would permanently switch off the processing of a given species of RNA. Our findings support the Davidson and Britten model. At the same time they explain how a temporary action of azocarcinogens could permanently change the gene expression in the affected cells resulting in a hepatoma with restricted gene expression.

Supported by a grant No. CA 22559 awarded by the National Cancer Institute, DHEW and by a grant No. H-393 from the R. A. Welch Foundation.

(1) Davidson, E. H. and Britten, R. J., Science 204, 1052-1059 (1979).

ISBN 0-12-633360-2

INTEGRATION OF PLASMID SEQUENCES INTO MITOCHONDRIAL CHROMOSOMAL DNA UPON REVERSION OF S-TYPE CYTOPLASMIC MALE STERILE PLANTS TO FERTILITY IN MAIZE.

B.D. Kim[1], M.F. Conde[2], S.J. Gabay-Laughnan[4], J.R. Laughnan[4] C.S. Levings, III[3], R.J. Mans[1], D.R. Pring[2]. [1]Dept. Biochemistry & Molecular Biology, [2]Dept. Plant Pathology, Univ. of Florida, Gainesville FL 32610; [3]Genetics Dept., North Carolina State Univ., Raleigh, NC 27650; [4]Depts. Agronomy and Genetics & Development, Univ. of Illinois, Urbana, IL 61801.

Nuclease restriction fragment profiles of mitochondrial DNAs (mtDNAs) isolated from maize plants exhibiting cytoplasmic male sterility (CMS) are distinct from normal (fertile) mtDNA (1). Electrophoresis of mtDNAs on agarose gels showed the presence of small, discrete DNAs in addition to the large mtDNAs (2). In S-type CMS, two of these plasmid-like DNAs are often lost from the complement of mtDNAs when S-type lines revert to fertility. Concomitant with the loss of the plasmid-like DNAs, unique restriction fragments (XhoI or Bam HI) were observed in the large mtDNAs isolated from revertant S-type maize. We asked if the plasmid-like DNA sequences were represented among the sequences in the unique restriction fragments.

Plasmid-like DNAs S-1 and S-2 (4.10 and 3.42 megadaltons, respectively) were isolated and labeled with [α-^{32}P] dGTP by nick-translation. Large mtDNAs from 3 fertile revertants (S-733, Vg-369 and Vg-296) and from sterile, parental Vg-CMS and S-CMS lines were isolated and restricted (XhoI or Bam HI). Using Southern blotting of the resolved fragments we sought sequence homology with each of the labeled probes, *i.e.* S-1 and S-2 and fragments of large mtDNAs.

Several sources of homology among mtDNA components were revealed: *i.* S-1 and S-2 share some homologous sequences but S-2 is not a sub-set of S-1; *ii.* portions of S-1 and S-2 are homologous with specific restriction fragments of large mtDNA from both S-CMS and Vg-revertant lines; *iii.* Hybridization with [^{32}P]dGMP-labeled S-2 probe gave rise to unique bands in the revertants that superimposed with the new restriction fragments of revertant mtDNAs. The data suggest the integration of plasmid-like sequences into large mtDNA upon reversion of S-type CMS maize to fertility.

(1) Levings, C.S. III and Pring, D.R., Science 193 158-160 (1976).
(2) Pring, D.R., Levings, C.S. III, Hu, W.W.L. and Timothy, D.H. Proc. Nat. Acad. Sci. 74 2904-2908 (1977).

ISBN 0-12-633360-2

mRNA(NUCLEOSIDE-2'-)METHYLTRANSFERASES FROM HELA CELLS

STEVEN LANGBERG and BERNARD MOSS. Laboratory of Biology of Viruses, National Institute of Allergy and Infectious Diseases, National Institutes of Health, Bethesda, Maryland.

HeLa cell enzymes which convert 5' terminal cap 0 ($m^7GpppNpN$-) of mRNA to cap I (m^7GpppN^mpN-) and cap I to Cap II ($m^7GpppN^mpN^m$-) by methylation of the 2'-OH groups of the riboses of the two nucleotides nearest to the 7-methylguanosine residue were isolated and characterized. A sensitive assay for these enzymes, which exploited the inability of RNAse T2 to cleave phosphodiester bonds after methylation of the 2'-OH group of the ribose esterified at its 3' position, was developed. To use as substrates, ^{32}P-labeled $m^7GpppA(pA)n$ and $m^7GpppA^m(pA)n$ were enzymatically synthesized from pppAp(pA)n which had been prepared with E. coli RNA polymerase. For the first substrate, purified RNA guanylyltransferase-RNA (guanine-7-)methyltransferase complex from vaccinia virus (1), [α-^{32}P]GTP and unlabeled S-adenosylmethionine were used. For the second, a mixture of the latter enzyme complex and mRNA (nucleoside-2'-)methyltransferase from vaccinia virus (2) was used. Following incubation of the labeled substrates with HeLa cell fractions and S-adenosylmethionine, the products were characterized by RNase T2 and alkaline phosphatase digestion and electrophoresis on DEAE paper at pH 3.5.

Using these assays we could demonstrate that activities which sequentially convert cap 0 to cap I (cap I methyltransferase) and cap I to cap II (cap II methyltransferase) are present in the postribosomal supernatant prepared from the cytoplasm of HeLa cells. That two separate enzymes are involved was determined by their separation on ion exchange chromatography columns. The two enzymes also differ in their salt optima and other properties. Thus, with the isolation of two mRNA (nucleoside-2'-)methyltransferases, the number of enzymes involved in cap formation that have been purified from HeLa cells has been increased to five.

(1) Barbosa, E. and Moss, B., J. Biol. Chem. 253 7692-7697 (1978).

(2) Martin, S. A., Paoletti, E. and Moss, B., J. Biol. Chem. 250 9322-9329 (1975).

ISBN 0-12-633360-2

The intracisternal A-particle genome: An extensively reiterated interspersed genetic element of Mus musculus. Kira K. Lueders and Edward L. Kuff, Laboratory of Biochemistry, NCI, NIH, Bethesda, Md. 20205

Mouse intracisternal type-A particles contain polyadenylated RNA molecules of three size classes: 35S, 32S and 29S. These RNAs share extensive sequence homology and at least two of them (35S and 29S) have been shown to code for the 73,000 dalton A-particle structural protein. The normal genome of Mus musculus contains multiple copies (500-1000) of DNA sequences related to those found in the A-particle RNAs. Using labeled 35S RNA as probe in blot-hybridization experiments, we have now identified a characteristic electrophoretic pattern of A-particle-related fragments generated by restriction endonuclease cleavage of mouse DNA. Discrete, strongly hybridizing bands were seen after digestion with a number of enzymes, confirming the presence of multiple gene copies with similar restriction sites. Hybridizing fragments with heterogeneous size distributions were also observed, perhaps reflecting diverse integration sites of the A-particle genes. There was no evidence for regular tandem arrangement of these genes. Similar patterns were obtained with DNA from non-producer cells (liver and F9 teratoma) and cells rich in A-particles (myeloma and neuroblastoma).

A major 5.3 kb Hind III/EcoR1 genomic fragment containing A-particle sequences was cloned by insertion into plasmid pBR322 and transfection of E.coli x1776. This insert contained internal restriction fragments corresponding to the predominant reactive bands seen after digestion of the whole mouse genome. The cloned fragment was then used to scan a mouse embryo gene library (average 16 kb inserts into Charon 4A; obtained from Dr. Jon Seidman, NIH). About 1% of the total recombinants in the library reacted with this probe, providing independent evidence for the multiplicity of A-particle genes. Half of 34 independent plaque-pufified isolates contained the complete 5.3 kb Hind III/EcoR1 fragment together with flanking mouse sequences. More detailed restriction analysis of six of the isolates, using probes of labeled A-particle RNA, showed that most of the internal structure of these individual genes has been highly conserved over approximately 7 kb, but some diversity does occur. Flanking sequences appeared to be different in each case, confirming our observation of multiple integration sites within genomic DNA.

ISBN 0-12-633360-2

mRNAs AND PROTEINS CODED BY AGROBACTERIUM TUMEFACIENS T-DNA OF CROWN GALL TUMORS

J.C. McPherson*, E.W. Nester†, and M.P. Gordon*. Departments of Biochemistry* and Microbiology†, University of Washington, Seattle, WA.

Crown gall tumors contain sequences of DNA (T-DNA) derived from large tumor inducing (Ti) plasmids present in virulent strains of Agrobacterium tumefaciens (1, 2). Several studies in this and other laboratories have detected RNA from tumor tissues which hybridizes with T-DNA sequences of Ti plasmids. We have now demonstrated that transcripts derived from T-DNA of tumor tissues give discrete and reproducible protein products in eucaryotic cell-free translation systems.

Messenger RNA has been isolated from cultured tumor and normal tissues and translated in cell-free systems derived from rabbit reticulocytes and wheat germ. Analyses of the translation products by SDS-polyacrylamide gel electrophoresis reveals a large number of proteins ranging in size from above 90,000 to less than 15,000. Slight differences in the protein profiles obtained from different tumor lines and from normal callus were observed. In order to characterize products which may arise in the tumor as a result of expression of the T-DNA we have hybridized RNA from tumor tissue with cloned fragments of T-DNA immobilized on cellulose nitrate filters. The RNA which specifically hybridized to the filters was released and translated to give protein products which were analyzed by SDS gel electrophoresis. Individual proteins have been obtained which are characteristic of a particular tumor line. In addition, a common protein derived from sequences of T-DNA which are conserved in crown gall tumors has been found in each tumor line studied.

Supported by NIH grant #CA 13015.

(1) Van Larebeke, N., Engler, Holsters, M., Van der Elsaker, S., Zaenen, I., Schilperoort, R. A., and Schell, J. (1974) Nature 252, 169-170.

(2) Chilton, M.-D., Drummond, M. H., Merlo, D. J., Sciaky, D., Montoya, A. L., Gordon, M. P., and Nester, E. W. (1977) Cell 11, 263-271.

ISBN 0-12-633360-2

CHROMOSOMAL LOCALIZATION IN ISOGENIC LINES OF A CLONED DROSOPHILA DNA SEQUENCE: ANALYSIS BY IN SITU HYBRIDIZATION

Dorothy A. Pierce and John C. Lucchesi. Department of Zoology and the Genetics Curriculum, University of North Carolina at Chapel Hill, Chapel Hill, North Carolina

The location of sequences homologous to a cloned *D. melanogaster* DNA segment, Dm 25, has been examined in polytene chromosomes by means of *in situ* hybridization. Dm 25 shows properties similar to other dispersed repetitive Drosophila sequences (1) considered to be transposable: (a) localization at multiple sites including the chromocenter; (b) variation in patterns between two wild-type laboratory strains, Canton S and Oregon R; and (c) variation in patterns among individuals within each of these strains.

To determine whether this individual variation is transmitted through the germ line or is the product of somatic transpositions, Dm 25 hybridizations were performed on 11 different isogenic lines, each isogenic for all major chromosomes. Each line showed a unique pattern of sites which is identical in each individual of a particular line. Larvae of an F1 generation obtained from reciprocal crosses between two isogenic lines demonstrated a hybrid pattern consistent with strict inheritance of each parental pattern.

A total of 38 different Dm 25 sites has been detected in these stocks. In general there is wide variation in Dm 25 patterns among different lines, but a significant number of sites are common to several distinct lines. Dm 25 hybridized effectively to the sibling species *D. simulans*, but not to members of five other species subgroups.

The location of Dm 25 sites within the genome in these isogenic lines has shown considerable stability over numerous generations. We are currently analyzing other specially constructed genetic lines to determine the actual rate of transposition of Dm 25 within fly populations.

Supported by NIH Research Grant GM-15691 and Training Grant GM-07121.

(1) Strobel, E., Dunsmuir, P. and Rubin, G.M., *Cell* *17* 429-439 (1979).

ISBN 0-12-633360-2

BASE SEQUENCE OF THE INTERSPERSED 300 nt REPEATED SEQUENCES IN HUMAN DNA.

C.M. Rubin, C.M. Houck and C.W. Schmid, Department of Chemistry, University of California, Davis, California.

Most of the interspersed 300 nt repeated sequences in human DNA share identical restriction sites and belong to a single family of sequences (the "Alu" family)[1]. We have determined an overall average base sequence of this family by sequencing restriction fragments using the Maxam-Gilbert procedure. This composite sequence is compared to the sequence of individual cloned members of the Alu family.

Members of this family include all of the long oligonucleotides which are found in repetitive double stranded hnRNA[2]. The agreement between these sequences shows this family of interspersed repeats to be transcriptionally active and supports the previous conclusion that most repetitive hnRNA sequences are transcribed from a single family of sequences[2].

The sequence of this interspersed repeated family is similar to: a) regions of cloned DNA fragments containing mRNA coding sequences from both humans and rodents; b) regions of cloned DNA fragments that can serve as templates for *in vitro* RNA polymerase III transcription; c) regions in a low molecular weight RNA found hydrogen bonded to rodent hnRNA; d) inverted repeated hnRNA from cultured human cells; e) regions at or near the origins of replication of BK, SV40, and polyoma viruses and also a nucleotide sequence in the human hepatitis B virus; f) a sequence located at one of the junctions between the large intervening sequence and the coding sequence in the human beta globin gene[3].

Sequence comparisons indicate that these repeats might function as origins of DNA replication, as initiation sites for RNA polymerase III transcription or in the processing of hnRNA into mRNA.

This work was supported by USPHS Grant GM 21346 and a AAUW predoctoral fellowship award to C.H.

1. Houck, C.M., Rinehart, F.P. and Schmid, C.W. J. Mol. Biol. 131, 289-306 (1979).
2. Robertson, H.D., Dickson, E. and Jelinek, W. J. Mol. Biol. 115, 571-589, 1977.
3. Jelinek, W.R., Toomey, T.P., Leinwand, L., Duncan, C.H., Biro, P.A., Choudary, P.V., Weissman, S.M., Rubin, C.M., Houck, C.H., Deininger, P.L. and Schmid, C.W. submitted for publication.

ISBN 0-12-633360-2

INTEGRATION AND EXPRESSION OF ROUS SARCOMA VIRUS DNA IN TRANSFORMED CHICK EMBRYO FIBROBLASTS

Robert A. Schulz, Panagiotis Pantazis, Riccardo DellaFavera+ and Takis S. Papas Laboratories of Tumor Virus Genetics and +Tumor Cell Biology, National Cancer Institute, National Institutes of Health, Bethesda, Maryland 20014.

Viral transformation of cells by avian retroviruses depends on the conversion of the RNA genome into proviral DNA and its subsequent integration into the host DNA. This is a vital part of the virus life cycle since it establishes the viral genes in the host genetic makeup, which then provides the template for synthesis of progeny genomes and viral mRNAs. The presence of viral specific molecules within the infected cell leads to complex molecular changes seen at the DNA, RNA, and protein levels in the transformed state.

We have been interested in understanding the changes seen after infection of chick embryo fibroblasts (CEF) by Rous Sarcoma Virus (SR-RSV-D). At the DNA level, sheared chromatin prepared from normal and RSV infected CEF was separated by rate sedimentation on glycerol gradients into two components. These two components were shown by a number of criteria to represent active and inactive chromatin regions. DNA extracted from these fractions were assayed by molecular hybridization for proviral sequences utilizing a DNA probe complementary to AMV RNA. These studies indicated that the exogenous proviral DNA was preferentially integrated in active chromatin regions in the infected cells. By contrast, endogenous RAV 0 sequences were found equally distributed between active and inactive fractions in control cells.

Since virus specific mRNAs constitute such a small fraction of total cellular RNA, the sensitive Northern hybridization technique was used to verify expression of subgenomic viral RNAs in the transformed cells. Using ^{32}P-AMV-cDNA as a probe, two polyadenylated mRNAs of 38S and 28S were detected in cytoplasmic RNA populations of infected cells, which were not seen in normal CEF. These are presumed to be messengers for the viral "gag-pol" and "env" genes, respectively (1). A homologous probe should allow the detection of the 21S "src" mRNA. In addition, the methodology used for chromatin fragmentation and fractionation provides fragments suitable for a number of studies. Ongoing experiments attempt to prepare viral RNA transcripts from the chromatin fractions using eukaryotic RNA Polymerase II. This may allow us to correlate their synthesis with the aspects of viral integration and expression described above.

(1) Hayward, W.S. (1977) J. Virol. 24, 47-63.

ISBN 0-12-633360-2

TRANSPOSON INDUCED MUTATIONS IN THE GENES CODING FOR THE ACETOHYDROXY ACID SYNTHASE (AHAS) ISOZYMES IN *Salmonella typhimurium*.

Karen J. Shaw and Claire M. Berg

U-131 Univ. of Connecticut, Storrs, Ct. 06268

Tn*10*-induced mutants of *Salmonella typhimurium* which require isoleucine or α-ketobutyrate or threonine for growth were found to lack AHAS II (*ilvG*). Polarity studies on these mutants indicate that *ilvG* is the first structural gene in the *ilv* operon, *ilvOGEDA*. These mutants do not require the other branched-chain amino acids because an internal promoter before *ilvE* permits low level constitutive expression of the distal *ilv* structural genes.

A Tn*5*-induced mutant lacking AHAS I (ilvB) was isolated in an *ilvG*::Tn*10* mutant strain following Tn*5* mutagenesis and penicillin enrichment, selecting for a strain which required valine for growth (in the presence of isoleucine). This double mutant requires both amino acids and has ~1% of wild type AHAS activity, indicating the existence of a third (minor) AHAS activity. The *ilvB* single mutant was found to be a prototroph, although the AHAS level is only ~30% of wild type activity.

A drug resistance marker is associated with each of the AHAS mutations. This is particularly important in the case of the *ilvB*::Tn*5* single mutant which is Ilv$^+$. Mutant isolation and manipulation has therefore been greatly facilitated by the readily detectable antibiotic resistance phenotype.

ISBN 0-12-633360-2

CHARACTERIZATION OF HUMAN $^{G}\gamma$ AND $^{A}\gamma$ FETAL GLOBIN GENES

J. L. Slightom, A. E. Blechl and O. Smithies
Laboratory of Genetics, University of Wisconsin, Madison, Wisconsin 53706

In order to study the switch between fetal and adult globin synthesis, we cloned the $^{A}\gamma$ and $^{G}\gamma$ fetal globin genes into Charon 4A. In particular, we have obtained clone 165.24 which contains both the $^{A}\gamma$ and $^{G}\gamma$ globin genes and clone 51.1 (1) which contains the $^{A}\gamma$ gene.

We have sequenced both the $^{G}\gamma$ and $^{A}\gamma$ genes from clone 165.24 (both fetal genes from one chromosome) and the $^{A}\gamma$ gene from clone 51.1. Both $^{G}\gamma$ and $^{A}\gamma$ genes are split into three segments by two intervening sequences (IVS); IVS 1 contains 122 bp, and IVS 2 contains about 880 bp. IVS 1 is located between residues 30 and 31, and IVS 2 is located between residues 104 and 105. Sequence homologies surrounding these splice sites will be discussed.

The DNA sequence for IVS 1 from both $^{G}\gamma$ and $^{A}\gamma$ genes in clone 165.24 and the $^{A}\gamma$ gene in clone 51.1 are identical, indicating the importance of this intervening sequence. Similar comparison of the IVS 2 sequences reveals the existance of both constant and variable DNA sequence regions. These regions vary in size and variability depending on whether the comparison is between $^{A}\gamma$ and $^{G}\gamma$ genes located on the same or opposite chromosome. The possible significance of these variabilities will be discussed.

Supported by NIH Grants AM 20120 and GM 20069.

(1) Smithies, O., Blechl, A. E., Denniston-Thompson, K., Newell, N., Richards, J. E., Slightom, J. L., Tucker, P. W., and Blattner, F. R., Science 202:1284-1289 (1978).

ISBN 0-12-633360-2

ISOLATION AND CHARACTERIZATION OF GENOMIC DNA SEQUENCES CODING FOR THE PRO α2 CHAIN OF SHEEP TYPE I PROCOLLAGEN

Paul Tolstoshev, Charles D. Boyd, Millie P. Schafer, Bruce C. Trapnell, Helen C. Coon, Peter J. Kretschmer, Arthur W. Nienhuis, and Ronald G. Crystal. National Heart, Lung, and Blood Institute, Bethesda, Maryland 20205.

A sheep genomic library contained in the bacteriophage vector, Charon 4A, was screened with [^{32}P]cDNA complementary to sheep type I procollagen mRNA. Two recombinant phage (SpC3 and SpC7) were identified as containing pro α2 procollagen mRNA coding sequences by the technique of hybrid selected cell-free translation. A restriction endonuclease map of the two genomic DNA fragments (13.8 and 15.1 kb in length) was established, and the orientation of the coding sequences was determined by hybridization of restriction fragments of the genomic DNAs to characterized fragments of a cloned pro α2 cDNA probe. Electron microscopy of R-loops generated by annealing the recombinant bacteriophage DNA with type I procollagen mRNA revealed that 12 to 16 intervening sequences interrupt the pro α2 coding sequence contained in SpC3. Thus, approximately half of the 5 kb mRNA coding sequence of pro α2 procollagen is distributed throughout a 15 kb region of the genome, making this gene one of the most complex studied to date. Restriction mapping revealed that the two cloned pro α2 procollagen sequences were closely related, with the inserted DNA of SpC7 overlapping that of SpC3 by 1.8 kb at the 5' end and being 3.3 kb shorter at the 3' end. The 1.8 kb 5' overlapping fragment has been used as a [^{32}P]-labeled probe to screen the library, and recombinant bacteriophage with genomic sequences containing regions of the 3' half to the pro α2 procollagen gene have been identified. Direct sequencing of restriction fragments from the 5' end of SpC3 has yielded a sequence which may define part of the amino acid sequence of the C-terminal propeptide of the pro α2 procollagen chain.

ISBN 0-12-633360-2

FORMATION OF P1*argF* SPECIALIZED TRANSDUCING BACTERIOPHAGE: INVOLVEMENT OF IS1.

Mary K. York and Marvin Stodolsky
Department of Microbiology, Stritch School of Medicine
Loyola University, Maywood, Illinois 60153

Bacteriophage and plasmid P1 chromosomes encode a single IS1 sequence (1). Integration of the *Escherichia coli* chromosomal *argF* gene segment into the P1 chromosome occurs with frequencies as high as one per 10^5 pfu of the low frequency transducing lysate. Fourteen P1*argF* transductants selected from separate lysates have been examined. They are structurally indistinguishable by cleavage analysis. Further characterization of one such P1*argF* chromosome by heteroduplex analysis against P1 and P1CmO (2), has identified a 10.7kb insertion of DNA precisely at the naturally occurring IS1 locus of P1. To further define the *argF* insertion, a P1Cm89*argF* was constructed, having the IS1 sequences of Cm89 (1) and *argF* in opposite orientation. Intrastrand annealing of P1Cm89*argF* DNA establishes that the *argF* segment is flanked by directly repeated IS1 sequences. These results suggest that as P1 acquires foreign genes, it converts them to transposons with flanking IS1 sequences.

Clonal populations of 10^9 *E. coli* lysogenic for P1*argF* contain about 0.3% *argF*$^-$ bacteria. Loss of the *argF* phenotype commonly results from excision of the *argF*-IS1 insertion, since most P1*argF*$^-$ are indistinguishable from the parent P1 chromosome by cleavage analysis.

The union of the *argF* gene segment and the P1 genome occurs during the lytic cycle of P1 infected bacteria. Comparison of lysates produced by lytic infection and induction of established lysogens suggests that *immI* function represses IS1 action during the lytic cycle.

Insertion of the Tn9 transposon flanked by IS1 at sites on the chromosome other than the naturally occuring IS1 reduces by 10-fold the number of detectable P1*argF* transductants. P7, a close relative of P1 having no known IS1 sequence (1), rarely forms P7*argF*. This further implies a role for IS1 in the formation of specialized transducing derivatives of P1.

Supported by USPHS Grant #GM 21435.

(1) Iida, S., Meyer J. and Arber W., *Plasmid 1* 357-365 (1978).

(2) Kondo, E. and Mitsuhashi, S., *J. Bacteriol. 88* 1266-1276 (1964).

ISBN 0-12-633360-2

NEW SYNTHETIC OLIGONUCLEOTIDES FOR SPECIFIC TRIMMING OF DNA AND FOR SEQUENCING DNA

Ray Wu, Lester Lau, Hansen Hsiung*, Wing Sung*, Roland Brousseau*, and Saran A. Narang*
Section of Biochemistry, Cornell University, Ithaca, NY, and
*Division of Biological Science, NRC of Canada, Ottawa, Canada

We have cloned several genes (1) after blunt-end ligation to EcoRI (or BamI) adaptors. After cloning and enzymic digestion, the additional 8 nucleotides contributed by the adaptor can be removed by using a synthetic retrieving adaptor designed according to the knowledge that MboII and HphI cleave DNA 8 nucleotides away from the recognition sites. For example:

```
        MboII                1      8
5'  TCTTCGAAGA          5'  AATTCATG ┌────────┐ CATGAATT
3'  AGAAGCTTCT     +    3'  TTAAGTAC │ cloned │ GTACTTAA
                                     │  gene  │
                                     └────────┘
retrieving adaptor      cloned gene with cohesive ends repaired

                                          1      8
 Blunt-end    cleave with    5'  TCTTCGAAGAAATTCATG       ┌────────┐C
 ligation  →    MboII     →  3'  AGAAGCTTCTTTAAGTA    +  C│ cloned │
                                                          │  gene  │
                                                          └────────┘
```

The cloned gene can then be cleaved with the S_1 nuclease to remove the single protruding residue at each 3' end. In principle, retrieving adaptors of this type containing 1 to 7 additional nucleotides beyond the MboII or HphI recogntion sites can be used to trim off 7 to 1 base pairs from the ends of any DNA duplex.

We have also synthesized a 19-long oligonucleotide primer, 5' TTGTAAAACGACGGCCAGT-OH, to increase the speed of DNA sequence analysis of genes cloned into the EcoRI site (2) of M13mp2 vector (3), the HindIII site of mWJ 22 vector (4) or the BamI site of mWJ 43 vector (4). Hybridization of this primer to single-stranded M13 DNA and extension of the primer toward the EcoRI site (or the HindIII site, or the BamI site adjacent to the EcoRI site) will allow sequence analysis of the gene cloned at this site.

1. Lau, L., Narang, S.A., and Wu, R. XI Int'l. Congress Biochem. 29 (1979)
2. Schreier, P.H. and Cortese, R., JMB, 129, 169 (1979)
3. Gronenborn, B. and Messing, J., Nature 272, 375 (1978)
4. Rothstein, R.J., Lau, L.F., Bahl, C.P., Narang, S.A. and Wu, R. Methods Enzymol. 68, (1979) in press.

ISBN 0-12-633360-2

A Cloning Vehicle with Poly (dA) and Poly (dT) Stretches on Complementary Strands; Applications for Strand Separation

Kenshi Hayashi and Norman Davidson

Department of Chemistry, California Institute of Technology, Pasadena, California, 91125

A cloning vehicle, the strands of which are readily separable, is useful for: (1) sequencing of inserted DNA by the dideoxy method[1]; (2) restriction mapping by the pulse-chase method[2]; (3) mapping of transcribed regions by the S1 protection method[3]; and (4) directed site-specific mutation.[4]

We describe one approach to this problem, involving the insertion of dA and dT stretches into opposite strands of pBR322. pBR322 was cleaved with Pvu II which cuts at a non-essential site. By use of terminal transferase, poly (dA) was added to both ends of one cleaved sample and poly (dT) to both ends of another sample. A mixture of the two samples was annealed, the DNA cleaved with Hind III, diluted, and cyclized with T4 DNA ligase. Tc^r transformants were selected and amplified in E. coli HB101. A resulting plasmid, pKH 4-7, was shown by appropriate restriction digests to contain an approximately 100 base pair insert at the former Pvu II site. Strand separation was achieved by denaturation of the restriction endonuclease linearized plasmid and affinity chromatography on oligo (dT) and oligo (dA) cellulose. The separation is efficient and rapid, giving each strand in high purity and in greater than 90% yield.

The modified plasmid is useful as a cloning vehicle, by ligating inserts to the usual cleavage sites of pBR322. Preliminary sequence studies of a cloned insert containing a Drosophila tRNA gene cluster will be reported.

---

[1]Sanger, F., Nicklen, S. and Coulson, A.R., Proc. Nat. Acad. Sci. 74, 5463-5467 (1977).

[2]Jeppesen, P.G., Sanders, L. and Slocombe, P.M., Nuc. Acid. Res. 3, 1323-1339 (1976).

[3]Berk, A.J. and Sharp, P.A., Cell 12, 721-732 (1977).

[4]Razin, A., Hirose, T., Ikakura, K. and Riggs, A.D., Proc. Nat. Acad. Sci. 75, 4268-4270 (1978).

ISBN 0-12-633360-2

JUNCTION SEQUENCES OF AN ERYTHROMYCIN TRANSPOSON (Tn551) FROM S. aureus

Saleem Khan and Richard Novick

Department of Plasmid Biology, The Public Health Research Institute of the City of New York, New York

The erythromycin resistance determinant of Staphylococcus aureus plasmid pI258 has previously been found to reside on a 5.3 kilobase transposon with short flanking inverted repeat DNA sequences (1, 2). This transposon is similar to E. coli transposons in causing insertional inactivation and other rearrangements; it differs from Tn3 and other E. coli transposons in frequently undergoing what appears to be precise excision (on the basis of restriction analysis). We report here the DNA sequences surrounding the junctions of the transposon with the flanking DNA in the wild type plasmid, in several insertions and in one of the precise excision-deletions. The ends of the transposon consist of an inverted repeat of 39 base pairs flanked by a direct repeat of 5, thus placing the transposon in the same class as Tn3, IS2 and Tn501. As shown below, 14 of the first 17 base pairs of Tn551 are identical with the corresponding segment of Tn3 (3). This match is obviously much too close to be coincidental; in fact it is much better than the match between Tn3 and Tn501 (J. Grinsted, personal communication), two transposons that are evidently closely related. This result strongly suggests common ancestry for Tn3 and Tn551 and raises the interesting questions of how the common sequence is connected evolutionarily and of how an apparently common transposon system carries erythromycin resistance in one organism and penicillin resistance in another.

```
Tn551  5'   GGGTCCGA^G CGCAC^G AGAAATTTGTATCGATAAGGGGTA
Tn3    5'  _G GGGTCTGA CGCTC AGTGGAACGAAAACTCACGTTAAT

Tn501  5'  GGGGGAACCGCAGAATTCGGAAAAAATCGTACGCTAAG
```

Supported by Grant #VC 229 From NSF.

(1) Pattee, P.A., Thompson, N.E., Haubrich, D. and Novick, R.P., Plasmid 1 38-51 (1977).

(2) Novick, R.P., Edelman, I.,Schwesinger, M.D.,Gruss, A.D., Swanson, E.C.and Pattee, P.A., Proc. Nat. Acad. Sci. 76 400-404 (1979).

(3) Ohtsubo, H., Ohmori, H. and Ohtsubo, E. Cold Spring Harbor Symp. Quant. Biol. 43 (1978).

ISBN 0-12-633360-2

SUBUNIT STRUCTURES OF DIFFERENT ELECTROPHORETIC FORMS OF NUCLEOSOMES

W.T. Garrard, S.C. Albright, J.M. Wiseman, and R.A. Lange
Department of Biochemistry, University of Texas Health Science Center at Dallas, Dallas, Texas 75235

We have reported previously that five different electrophoretic forms of mononucleosomes (MI, MII,....MV) are produced upon treatment of mammalian nuclear chromatin with micrococcal nuclease (1,2). We report here the reconstitution of each of these mononucleosome classes. We demonstrate that HMG-17 is the most abundant nonhistone protein associated with native bovine thymus nucleosomes, and that this protein in combination with histone H1 can reassociate with nucleosomes and generate the same five major electrophoretic forms of nucleosomes found in native chromatin. From the results of one- and two-dimensional electrophoretic analyses on reconstituted samples, we conclude that different mononucleosome classes possess the following subunit structures: MI, core histone octamer (8-mer); MII, 8-mer plus 1 copy of HMG-17; MIIIA, 8-mer plus 1 copy of histone H1; MIIIB, 8-mer plus 2 copies of HMG-17; MIV, 8-mer plus 1 copy each of histone H1 and HMG-17; and MV, 8-mer plus 1 copy of histone H1 and 2 copies of HMG-17. Equal numbers of HMG-14 molecules can substitute for HMG-17 and generate the same nucleosome components. Thus, mononucleosomes possess independent specific binding sites for at least 1 histone H1 molecule and 2 nonhistone chromosomal protein molecules.

Supported by USPHS Grant GM 22201 and the National Foundation, March of Dimes.

(1) Todd, R.D., and Garrard, W.T. (1977) J. Biol. Chem. 252, 4729-4738.

(2) Todd, R.D., and Garrard, W.T. (1979) J. Biol. Chem. 254, 3074-3083.

ISBN 0-12-633360-2

PROMOTOR SITES IN BOVINE MITOCHONDRIAL DNA

V. L. Cole, P. C. Crawford, *P. J. Laipis and W. W. Hauswirth, Departments of Immunology and Medical Microbiology and *Biochemistry and Molecular Biology, University of Florida, Gainesville, Florida

Mitochondrial DNA constitutes the only example of a functional non-viral eukaryotic genetic element which can be easily isolated in its entirety from animal tissue. Because of this, the mitochondrial genome has been recognized as a model system for understanding gene organization, control, and evolution in eukaryotic organisms. We have determined the location and orientation of the mitochondrial ribosomal RNA genes on the bovine mitochondrial genome. These two genes are adjacent and transcribed from the same strand of mitochondrial DNA. It is not clear, however, that a single promotor site controls this transcriptional event, although nearly genome length *in vivo* RNA transcripts have been isolated here and in other mammalian systems. In order to determine the number and polarity of promotor sites in the mitochondrial genome, we have determined the strand polarity and 5' terminus location of several of these large transcripts. We have also carried out *in vitro* RNA polymerase binding studies similar to those reported for animal virus DNAs. We have mapped the binding positions by electron microscopy using the benzylalkonium chloride spreading technique. In addition, these results have been confirmed by "missing fragment" analysis of restriction enzyme digests of mitochondrial DNA.

The results of these independent experiments suggest that there are no more than four RNA polymerase binding sites. None of these binding sites map within the mitochondrial ribosomal RNA genes. Fine structure restriction mapping and DNA sequencing studies will more accurately locate these control regions.

Supported by National Science Foundation grant PCM 78-23221 and Biomedical Research Support Grant RR-05788.

ISBN 0-12-633360-2

EUKARYOTIC JUMPING GENES WITH LARGE INVERTED REPEATS.

Steven Potter, Martha Truett, Mark Schiro, Andrew Maher, and Mark Phillips. Department of Biology, Wesleyan University, Middletown, CT. 06457.

A family of DNA segments with sequence homology was purified from *Drosophila melanogaster* by molecular cloning. Several of these DNA segments contain large inverted repeats detectable by electron microscopy. Restriction maps and cross hybridization studies suggest a subunit construction. Southern blot experiments using the moderately repetitive components as probes showed distinct DNA arrangements in the genomes of the five different *Drosophila melanogaster* strains tested. Southern blot experiments using single copy flanking sequences as probes allowed the analysis of sequence arrangements at specific sites in the genomes. The results indicated that three of the moderately repetitive sequences with large inverted repeats are transposable genetic elements. The corresponding DNA sequences "before" and "after" the insertion events have been cloned.

These transposable DNA sequences are distinct from *copia* and others previously characterized in two ways. First they contain inverted terminal repeats instead of direct terminal repeats. Second, they have a more complex construction.

ISBN 0-12-633360-2

ENDODEOXYRIBONUCLEASE MUTANTS OF *SALMONELLA TYPHIMURIUM* LT2

Michele Trucksis, Donna J. Zabel, and Richard E. Depew. Program in Microbiology/Immunology, Northeastern Ohio Universities College of Medicine, Rootstown, Ohio 44272

Chemically induced and Tn10 insertion mutants of *Salmonella typhimurium* LT2 which have reduced levels of endodeoxyribonucleolytic activity have been isolated and characterized. The chemically induced mutations define a locus, *endA*, which is cotransduced with *serA* at a low frequency in P22-mediated crosses. Three factor crosses demonstrate a *serA*, *metK*, *endA* gene order for this region. The tetracycline resistance of an insertion mutation of the *endA* locus, *endA*41::tn10, permits selection of *this* otherwise unselectable marker during the construction of endonuclease deficient strains. Two other tn10 insertions which reduce endonucleolytic activity do not map in the *endA* locus.

The major endodeoxyribonucleolytic activity of crude extracts of *Salmonella* cells harvested in mid-log phase is similar to the activity of endonuclease I of *Escherichia coli*. The *Salmonella* activity in fresh extracts is masked, but autolysis or RNase A treatment results in a dramatic increase in activity. This activity is the predominant activity which cleaves native DNA, requires magnesium ion, and can be inhibited by tRNA.

Of the various alleles studied, *endA*41::tn10 results in the most severe loss of endonucleolytic activity. Two of the chemically derived *endA* alleles result in considerable residual endonucleolytic activities which are significantly more thermolabile than the wild-type activity.

These results suggest that the *endA* locus contains the structural gene for *Salmonella* endonuclease I.

This work was supported by PHS Research Grant GM25500.

ISBN 0-12-633360-2

ORGANIZATION OF THE MOUSE HISTONE GENES

Don Sittman, Charles Card, Ing-Ming Chiu, Chi-Jiunn Pan and William Marzluff. Department of Chemistry, Florida State University, Gainesville, Florida

Histone gene expression is regulated during the cell cycle. Histone mRNA only accumulates in the cytoplasm during early S phase. In order to understand this regulation we have undertaken to determine the histone gene organization in the mouse and to isolate the mouse histone genes.

To map the gene organization we have used the approach Crawford et al. (1) used for the chicken histone genes. Mouse myeloma cells were synchronized by isoleucine starvation and polyribosomes prepared from S phase cells. RNA sedimenting from 7-14S was purified over oligo dT-cellulose to remove poly A-RNA and subjected to repeated sucrose gradient centrifugation. The vast majority (>90%) of the polypeptides coded for by this RNA in the wheat germ cell-free translation system were the five histones.

A cDNA copy was made from this partially purified histone mRNA using a random primer (2). This cDNA was hybridized to ribosomal RNA and rRNA-DNA hybrids were removed by centrifugation. This mouse cDNA hybridized to at least three of the sea urchin histone genes contained in the plasmids PSP102 and PSP117 (3). We show that a significant portion (30-40%) of this cDNA is derived from contaminating mitochondrial RNA. This cDNA will lead to artifactual mapping of the mitochondrial ribosomal genes if mitochondrial DNA is present in the DNA restriction cleaved for Southern (4) analysis.

After removing this mitochondrial cDNA contaminate or using nuclear DNA a complex pattern of histone gene organization is observed, possibly due to the presence of several non-allelic histone variants expressed in these mouse cells.

(1) Crawford, R.J., Krieg, P., Harvey, R.P., Hewish, D.A. and Wells, J.R.E. (1979) Nature 279: 132-136.
(2) Taylor, J.M., Illmensee, R. and Summers, J. (1976) Biochem. Biophys. Acta 442: 324-330.
(3) Sures, I., Lowry, J. and Kedes, L.H. (1978) Cell 15: 1033-1044.
(4) Southern, E.M. (1975) J. Mol. Biol. 78: 503.

ISBN 0-12-633360-2

IMPROVED METHODS FOR RECOMBINANT cDNA CONSTRUCTION

Robert Frankis, Jim Gaubatz, Fu-Kuen Lin and Gary V. Paddock
Department of Basic and Clinical Immunology and Microbiology, Medical Univ. of South Carolina, Charleston, S.C. 29403

Nucleotide sequences containing 5' ends of mRNA are important to regulation of gene expression because they contain signals for the initiation of transcription or processing of nuclear pre-mRNA, capping of mRNA, and the subsequent translation of mRNA. Methods have been developed for sequencing these regions (1-3) in abundant mRNAs such as globin, but none is applicable to mRNAs obtainable only as minor components. The current recombinant cDNA cloning method (4) developed by several laboratories requires S1 nuclease to open the hairpin loop of *in vitro* synthesized double-stranded cDNA. This results in the destruction of sequences representing the important 5'-end region.

To overcome this difficulty we have developed cloning methods which do not use S1 nuclease. The first method uses a ribosubstitution step after cDNA synthesis. The ribonucleotides become a linker segment which primes the synthesis of a second strand by DNA polymerase. The ribonucleotides are hydrolyzed by alkali or RNase to open the hairpin. This double stranded molecule is then tailed with poly(dC) for annealing with poly(dG)-tailed, Pst-cleaved pBR322. The second technique involves construction of symmetrical cDNA molecules via primed synthesis of a "third" DNA strand after extension of double-stranded hairpin cDNA at the 3' end with a homopolymer tail. The resulting molecules should have a center of symmetry at sequences representing the 5' end of mRNA. This should facilitate sequence analysis of this region by the Maxam and Gilbert technique (5).

We have obtained clones using both these methods for duck and rabbit globin mRNAs. Since we did not size select our cDNA, we have obtained a variety of sizes as determined by restriction enzyme analysis. A few of the larger recombinant cDNAs may contain 5' regions; we are analyzing their sequences. We are also constructing new duck globin recombinant cDNAs, using cDNA size selected in acrylamide gels, which should provide full length recombinant cDNAs.

Supported by USPHS Grant GM-24783 and a Biomedical Research Grant from the State of South Carolina.

1. Lockard RE, Raj Bhandary UL, *Cell* *9* 747-760, 1976.
2. Baralle FE, *Cell* *10* 549-558, 1977.
3. Chang JC, et al, *PNAS* *74* 5145-5149, 1977.
4. Higuchi R, Paddock GV, Wall R, Salser W, *PNAS* *73* 3146-3150, 1976.
5. Maxam AM, Gilbert W, *PNAS* *74* 560-564, 1977.

ISBN 0-12-633360-2

GENETIC STUDIES ON TRANS-ACTING REGULATORY EFFECTS IN HIGHER EUKARYOTES AS REVEALED BY CHROMOSOMAL IMBALANCE

James A. Birchler
Univ. of Tenn.-Oak Ridge Grad. School of Biomed. Sci. and Biol. Div., Oak Ridge Nat'l. Lab.,Oak Ridge, TN.

In maize, certain enzyme levels have been found to be negatively correlated with several segmental chromosomal dosage series(1-2-3 doses). The magnitude of deviation is dependent upon the individual enzyme monitored and the particular chromosomal region varied, but the extreme limits observed, appear to be the inverse of the dosage relative to the diploid. SDS polyacrylamide gel electrophoretic analysis revealed characteristic alterations of specific protein levels in each dosage series and also revealed a number of positive correlations with dosage. The structural loci encoding alcohol dehydrogenase and another protein species are present in the long arm of chromosome one, but neither exhibits the customary gene dosage effect in whole arm comparisons. By subdividing the chromosome arm, this compensation was demonstrated to be due to a cancellation of a structural gene dosage effect by a negative inverse effect simultaneously produced by the varied chromosome segment.

In DROSOPHILA, the question of whether dosage compensation of X-linked genes was an evolutionary specialization of the type of compensation demonstrated in maize was examined. The X chromosome was divided into a number of single or combined trisomic regions with six enzymes being studied. Multiple duplications in females specifically reduce one or more encoded by X or autosomal genes. The addition of the same duplications to males does not in general result in reductions of greater magnitude. Longer duplications in females are cumulative only to the inverse level relative to normal. Structural genes included in segmental trisomics no longer than a third to a half of the X show compensation. Deficiencies in females produce elevations or reductions. Although numerous dosage sensitive regions alter the level of any one enzyme, there is a positive correlation between enzyme level/cell and ploidy. These observations will be related to genetic balance phenomena found in higher eukaryotes. Research sponsored by Department of Energy under contract with Union Carbide Corporation.

ISBN 0-12-633360-2

Y CHROMOSOME CONTROL OF SPERMATOGENESIS IN *DROSOPHILA MELANOGASTER*

J. A. Lovett, A. P. Mahowald, and T. C. Kaufman, Dept. of Biology, Indiana University, Bloomington, IN

In *Drosophila melanogaster* the Y chromosome does not determine the sex of the fly but is necessary for the production of viable sperm. An XO fly which is phenotypically male shows defective spermatogenesis at elongation of the spermatids. The Y chromosome has been shown to be genetically active only in the germ line and except in primary spermatocytes is heterochromatic. Deletion-complementation studies have defined six fertility factors on the Y chromosome, Although its DNA content is similar to that of the X chromosome which has an estimated 1000 genes. The bulk of the Y chromosome DNA appears to be present as large blocks of simple sequence satellite DNA. We have attempted to understand the effect of the Y chromosome on the control of spermatogenesis.

A library of *Drosophila melanogaster* DNA cloned in lambda was screened using ^{32}P labelled cDNA made to poly (A) containing RNA isolated from testes of either XO or XY larvae. Among those clones isolated were a class which gave a strong positive signal with the XO probe, but only a very weak signal with the XY probe. This suggests that there is a gene either on the X chromosome or autosomes which is not expressed in the larval testes in the presence of the Y chromosome.

To determine whether any one fertility factor is responsible for this effect, RNA is being isolated from testes of larvae containing individual Y chromosomes deleted for each fertility factor. A cDNA probe made from each of the six RNA types will be hybridized to the selected clones. Deletion of the factor responsible for the suppression may be expected to mimic an XO condition.

The selected clones are also being characterized by restriction mapping to determine their relationship to each other; and *in situ* hybridization to polytene chromosomes is being done to determine their location and possible correlation with existing male sterile mutations.

Supported by Training Grants #82 and #7757 from the National Institute of General Medical Science and NSF Grant PCM-25427.

ISBN 0-12-633360-2

LOCALIZATION OF *E. COLI* RNA POLYMERASE INITIATION SITES ON SV40 DNA.

C. Lavialle, Y. Reuveni and N. P. Salzman. National Institute of Allergy and Infectious Diseases, National Institutes of Health, Bethesda, Maryland

We have used transcription of SV40 DNA by *E. coli* RNA polymerase to develop a method allowing for precise localization of promoters. Stable RNA polymerase-SV40 DNA complexes are cleaved with single cut restriction endonucleases (Bam HI, Hpa II or Eco RI). These cleavages generate termination sites for transcription at precisely defined locations. The RNA synthesized on these linearized complexes is denatured with glyoxal and analyzed by electrophoresis on agarose and acrylamide-agarose gels. Discrete RNA species can be demonstrated as defined bands on the gels, the migration of which allows the determination of their size, using denatured SV40 DNA fragments as markers. As expected, the electrophoretic pattern varies according to the restriction endonuclease used to cleave the complexes. The strand of DNA on which each RNA species is synthesized is analyzed by transfer of the RNA to DBM paper and hybridization with appropriate DNA probes. The determination of the size and direction of transcription of each RNA species makes it possible to position the 5' ends on the SV40 DNA map. The first nucleotide incorporated at the 5' end of each RNA species has been assigned by labeling with γ [^{32}P]-ATP or -GTP. A comparison of promoters recognized on superhelical or linear SV40 DNA templates have also been determined by using these procedures.

This method should be useful for the study of the specificity of initiation by monkey cell RNA polymerase B, whose purification has been undertaken in our laboratory. The *in vitro* synthesis of well characterized primary transcripts may also provide appropriate substrates for the analysis of the enzymes involved in RNA processing.

ISBN 0-12-633360-2

MOLECULAR RECOMBINATION BETWEEN SIMIAN VIRUS 40 AND MONKEY CELL DNAs

Bharti Dhruva and Kiranur N. Subramanian
Department of Microbiology and Immunology, University of Illinois Medical Center, Chicago, Illinois 60612

A viable mutant of SV40 has been isolated and provided by to us by Dr. T. Shenk of the University of Connecticut. We have fully characterized the DNA of this mutant by restriction analysis and DNA sequencing. We have found an insertion of 157 nucleotides at map position 0.650 of SV40 DNA in this mutant and determined its complete nucleotide sequence. This insert is not derived from any part of SV40 DNA and is presumably of host cell origin.

The two viral/cell DNA junctions at either end of the insert do not have any sequence-similarity. However, an octanucleotide contributed by the viral DNA part of the first junction point is exactly repeated 15 nucleotides away within the insert; likewise, another octanucleotide present at the cell DNA part of the second junction point is exactly repeated 21 nucleotides away within the viral DNA. These findings suggest that molecular recombination between viral and cellular DNA has no sequence specificity, but is dependent on the recognition of homologous sequences in the two DNA components. The complete nucleotide sequence of the cell DNA insert and flanking viral DNA will be presented.

This mutant differs from other previously characterized mutants of SV40 carrying host sequences in the following respects: 1) the mutant is viable; 2) the insert is within the 5' untranslated part of the early region of SV40; 3) the sequence of the insert differs from others determined previously; 4) the insertion is not due to infection of the cells by high multiplicties of the virus.

Supported in full by NIH Contract # N01 CP 71061.

ISBN 0-12-633360-2

STUDIES ON CLONED HUMAN GLOBIN GENES.

V. Prabhakara Choudary*, Paul A. Biro*, J. T. Elder*, Craig H. Duncan*, P. Jagadeeswaran*, Jon Kim de Riel*, Richard A. Spritz*, Bernard G. Forget** and Sherman M. Weissman*. Department of Human Genetics*, Department of Internal Medicine**, Yale University School of Medicine, New Haven, Ct. 06510.

We have cloned EcoRI and partial EcoRI restriction fragments of DNA from patients exhibiting either normal or thalassemic globin gene expression. By screening hybrid bacteriophage libraries of unfractionated and of size-selected DNA, we have isolated clones containing regions of the globin genes from normal and thalassemic subjects, and other sequences possessing structural homology to globin cDNA. Our findings on the nature of these "shadow" genes and the structure of the globin genes will be presented.

To investigate the expression of these genes, the cloned DNA fragments were incubated with cell extracts containing RNA polymerase III. DNA sequences preceding the delta-beta pair of linked genes and preceding the gamma genes directed the transcription of discrete RNA species over 500 nucleotides in length.The sequence of the template for the transcript preceding the gamma globin genes has been determined in its entirety and shows that the transcript includes a sequence through 12 consecutive adenylic acids. In parallel work by J. Pan and S. Weissman (unpublished data), unrelated clones of human DNA were selected for their ability to hybridize the low molecular weight RNA from human HeLa cells. Clones were found which contained polymerase III transcription units producing RNAs of over 200 nucleotides in chain length. The DNA templates for these RNAs included long stretches of adenylic acid. A substantial amount of the oligo-adenylic acids encoded in human DNA may be included within potential polymerase III transcription units and RNA exists in cells showing complementarity to at least some of these transcription units.

ISBN 0-12-633360-2

TRANSPOSITION IN THE DIMORPHIC BACTERIUM *CAULOBACTER CRESCENTUS*

B. Ely. Department of Biology, and Microbiology and Immunology, University of South Carolina, Columbia, S.C. 29208

Promiscuous drug resistance factors such as RP4 can be transformed to *Caulobacter crescentus* at a high frequency. The plasmids are stably maintained, confer drug resistance upon their host, and can be serially transferred to other drug sensitive recipients. RP4 can also be used to mobilize small plasmids such as ColE1::Tn5 to *Caulobacter*. The ColE1::Tn5 plasmid is stably maintained, confers Kan^R, and can be serially transferred. Thus, transposons can be introduced to *Caulobacter* on either conjugative or non-conjugative plasmids.

When RP4 was introduced into a *C. crescentus* strain harboring RP4 Kan::Tn7, simultaneous selection Kan^R and Str^R resulted in the isolation of strains which had lost the resident plasmid and but retained Tn7 in a chromosomal location. Therefore, transposition occurs in *C. crescentus*. Several thousand isolates were examined and no nutritional mutants were observed, suggesting that Tn7 has a high specificity of insertion in *C. crescentus* similar to that found in *E. coli*. Under the appropriate conditions, Tn7 inserted in the *Caulobacter* chromosome could be transferred to *E. coli* on an RP4 plasmid, demonstrating that serial transposition can occur in *Caulobacter*.

A "suicide" plasmid also could be used to select for transposition in *Caulobacter*. When an *E. coli* strain containing the "suicide" plasmid pPH1JI::Mu::Tn5 was mated with a *C. crescentus* recipient, the "suicide" plasmid was transferred but not maintained. Selection for Kan^R resulted in *C. crescentus* clones containing Tn5 inserted in the chromosome. Transposition of Tn5 resulted in nutritional mutants among the survivors at a frequency of 4% and motility mutants at a frequency of 2%. Thus, as in *E. coli*, Tn5 has a low site specificity for transposition in *C. crescentus* and transfer of Tn5 on a "suicide" plasmid is an effective means of obtaining chromosomal insertions of Tn5.

ISBN 0-12-633360-2

TWO-DIMENSIONAL NATIVE-SDS PAGE ANALYSIS OF MAIZE MALATE DEHYDROGENASE ISOZYMES: EVIDENCE FOR "SIZE ALLELES"

Kathleen J. Newton and Drew Schwartz
Biology Department, Indiana University, Bloomington, Ind.

Extensive genetic analyses on a large number of maize stocks variant for MDH isozymes recently led to the elucidation of the genetic basis of the major bands observed on starch gel electrophoretograms (1). Total MDH has been isolated from several variant lines and analysed on a native-SDS two-dimensional PAGE system. Some electrophoretic variants, identified on starch and native polyacrylamide gels, also exhibit different migration rates on SDS gels, indicating apparent size differences.

Duplicate, independently inherited genes, *Mdh4* (on chromosome 1) and *Mdh5* specify the soluble MDH isozymes. In most maize stocks, the products of these genes comigrate on starch and on the two-dimensional PAGE. However, a full activity allele of *Mdh4* has been found whose product exhibits altered migration in both the native and SDS dimensions.

Many electrophoretic and null variants for the mitochondrial MDH isozymes occur naturally. Three nuclear genes, *Mdh1* (on chr. 8), *Mdh2* (chr. 6) and *Mdh3* (chr. 3) code for the major mitochondrial MDH isozymes. The different gene products exhibit apparent size differences. However, we also find fully active, naturally occurring *allelic* variants at two of these three loci which differ in apparent M.W., as defined by their altered migrations in SDS gels. On the other hand, a modifier (*Mmm*, on chr. 1) which affects the native electrophoretic mobilities of all the mitochondrial MDH isozymes, does not cause detectable size differences in the SDS dimension under the conditions used in this study.

Supported by training grants #82 and #7757 from NIGMS and NSF grant PCM 76-]]009.

(1) Goodman, M.M., Stuber, C., Lee, C., Johnson, F., *Genetics*. (In press)

ISBN 0-12-633360-2

ISOGENIC RESCUE, A METHOD FOR STUDYING NONCONDITIONAL LETHAL MUTANTS OF VIRULENT PHAGES: APPLICATION TO ϕX 174

M. Zafri HUMAYUN and Katherine KALTY-MORIN
Department of Microbiology
CMDNJ - New Jersey Medical School, Newark, NJ 07103

Isolation and propagation of non-conditional lethal mutants such as missense mutants, frameshifts, deletions or rearrangements in virulent phages has been a long-insurmountable technical problem. To overcome this problem, we have devised a simple strategy, termed _isogenic rescue_ or _isogenic complementation_. Isogenic rescue is defined as an experimental strategy for propagation of both conditional _and_ nonconditional lethal mutants of a virus on host strains in which constitutive, plasmid-mediated complementation for the affected viral gene has been set up prior to infection. This strategy involves the _in vitro_ construction of stable bacterial plasmids carrying selected, _functional_ viral genes and subsequent introduction of such plasmids into uninfected phage hosts. In principle, such transformed host strains are capable of providing a continuous supply of a given phage gene product, so that an infecting mutant (whether conditional or non-conditional lethal) will be able to replicate despite a defect in the target gene. Such host strains are expected to act as "gene-specific, universal suppressors," for phage mutants because they are capable of suppressing all types of mutations ("universal suppression") in a specific phage gene ("gene specific suppression").

The strategy of isogenic rescue has been successfully applied to an essential gene (gene G, a spike protein gene) of the virulent single-stranded DNA phage ϕX174. ϕX gene G has been incorporated in a functional state into the cloning vehicle pMB9 to obtain a hybrid plasmid termed pϕXG. pϕXG has been shown to confer gene G-specific universal-suppressor properties on various host strains. Use of this plasmid has permitted the isolation and characterization of a number of nonconditional lethal ϕX mutants to the nucleotide sequence level. Specifically, this strategy has permitted the characterization of deletion, frameshift and insertion mutants of ϕX, the first examples of such mutants in ϕX174.

Humayun, M. Z. and Chambers, R. (1979), _Nature 278_, 524-529.

ISBN 0-12-633360-2

AMPLIFICATION OF DIHYDROFOLATE REDUCTASE GENES IN ANTIFOLATE RESISTANT CHINESE HAMSTER LUNG FIBROBLASTS

P. W. Melera,* J. A. Lewis,* J. L. Biedler† and C. Hession.*
Laboratories of RNA Synthesis and Regulation* and Cellular and Biochemical Genetics,† Sloan-Kettering Institute for Cancer Research, Walker Laboratory, Rye, NY 10580

When challenged by the antifolate drugs methotrexate or methasquin Chinese hamster lung fibroblast cells acquire resistance via overproduction of the target enzyme dihydrofolate reductase (DHFR). We have recently shown that this overproduction is accompanied by a relative increase in both the rate of synthesis of DHFR and the intercellular amount of DHFR-specific mRNA in resistant cells (1,2). We have also shown that, whereas only one molecular weight form of DHFR is detectable in parental drug-sensitive cells, drug resistant cells can overproduce one of apparently two different molecular weight forms of DHFR, only one of which is parental-like (1,2).

Using a DHFR mRNA-specific cDNA probe kinetically purified as described by Alt *et al.* (1978), we now show that the overproduction of either parental DHFR or nonparental DHFR is accompanied by amplification of DHFR gene number.

Supported in part by NIH Grant CA-24635, and the Fairchild New Frontiers Fund.

(1) Melera, P.W., Wolgemuth, D., Biedler, J.L. and Hession, C. (1979), submitted for publication.

(2) Melera, P.W., Shanske, S., Scotto, K., Meyers, M. and Biedler, J.L. (1979), submitted for publication.

(3) Alt, F.W., Kellems, R.E., Bertino, J.R. and Schimke, R.T., *J. Biol. Chem.* 253 1357-1370 (1978).

ISBN 0-12-633360-2

TRANSFER OF A TEMPERATURE-SENSITIVE THYMIDINE KINASE GENE TO MOUSE CELLS

Larry McReynolds, Michele Manos, Scot Munroe, and Ewa Przybytkowska
Department of Biochemistry, College of Medicine, University of Arizona, Tucson, AZ 85724

Herpes simplex virus Type I (HSV-1) codes for its own thymidine kinase. A mutant strain of HSV-1, ts1117 (or ts-23b), produces a temperature-sensitive thymidine kinase (1). The temperature sensitivity of the enzyme was determined by preincubation of a lysate of infected cells for various times at elevated temperatures. The enzyme was then assayed at the permissive temperature of 34° C. The enzymatic activity decays with a half-life of 10-12 min at 40° C and 22-24 min when preincubated at 38° C. At 34° the enzyme is quite stable, having a $t_{1/2}$ of 300-400 min. The mutant enzyme is 3-4 times more sensitive than the thymidine kinase enzyme from the wild type virus.

The mutant enzyme's activity is independent of the host. Viral infection of Vero cells and L cells produces an enzyme with the same thermolability.

The inhibition of enzyme activity was shown not to be caused by a production of an inhibitor. The addition of heat treated extract to an unheated extract did not cause suppression of the enzyme activity.

A restriction digest of the DNA from the mutant virus with Hpa-I produces an 8.4 kb band. This is similar in size to the fragment of DNA which has been shown to contain the wild type gene (2). A thymidine kinase deficient mouse cell line, LMTK$^-$, was transformed with this fragment at the permissive temperature of 34° C. The TK$^+$ clones were selected for their growth in HAT media. The growth property of these clones is being investigated at the non-permissive temperature.

Supported by NIH grant No. GM 25692.

(1) Aron, G., Schaffer, P., Courtney, R., Benyesh-Melnick, M., and Kit, S. Intervirology 1, 99-109 (1973).

(2) Wigler, M., Silverstein, S., Lee, S., Pellicer, A., Cheng, G., and Axel, R. Cell 11, 223-232 (1977).

ISBN 0-12-633360-2

IMMUNOPRECIPITATION AND CHARACTERIZATION OF TOTAL RNA TRANSLATION PRODUCTS FROM A PATHOGENIC CESTODE: *TAENIA SOLIUM* LARVAE.

Díaz de León, L.*, Arcos, L. and Willms, K.
Departments of Developmental Biology* and Biotechnology. Instituto de Investigaciones Biomédicas. National University of Mexico. A. Postal 70228. México 19, D.F.

One of the most interesting features of helminth infection is their capacity to survive for prolonged periods in their hosts.(1). Several mechanisms have been proposed to explain the evasion of the host response by these complex parasites (2). It has been demonstrated that some helminths such as *Taenia solium* larvae bear host-like proteins on their surfaces (3). Thus, a possible explanation for these observations is that the parasite has the genetic information necessary to synthesize either proteins sharing common antigenic determinants with the host, or surface receptors which have affinity for specific host proteins such as immunoglobulins and histocompatibility antigens.

To explore this possibility *Taenia solium* larvae were dissected from infected swine. Total RNA was isolated and translated in a reticulocyte derived cell-free system. Cell-free products were analyzed by slab gel electrophoresis and also by immunoprecipitation with different antisera. The electrophoretic patterns revealed 13 main radioactive bands, ranging from 15,000 to 220,000 daltons. Two major proteins with approximate molecular weights of 30,000 and 130,000 daltons were immunoprecipitated by antibodies raised against porcine immunoglobulins. When antibodies to whole larvae extracts were used, four main proteins were immunoprecipitated from the cell-free products. One of these bands corresponded to the 30,000 daltons protein precipitated with anti-porcine immunoglobulin antibodies. It may not be coincidental that the molecular weight of this protein is similar to that of the immunoglobulin light chain. However, the evaluation of this possibility as well as its role as a specific receptor for host immunoglobulin molecules is currently under investigation.

(1) Gemmel, M.A. and Macnamara, F.N., in: Immunity to animal parasites, (E.J. Saulby Ed.), Academic Press, New York and London, (1972), pp. 235.
(2) Ogilvie, B.M. and Wilson, R.J. British Med. Bull., 32, 177 (1976).
(3) Willms, K. and Arcos, L. Exper. Parasitol., 43, 396 (1977).

ISBN 0-12-633360-2

Differential Hormone Responsiveness of the Ovalbumin Gene and its Pseudogenes in the Chick Oviduct.

Donald A. Colbert, Savio L.C. Woo, Ming-Jer Tsai, Myles L. Mace, and Bert W. O'Malley. Dept. of Cell Biology, Baylor College of Medicine, Houston, Texas, 77030

We describe the isolation of a recombinant phage from a chicken gene library which contains two genes, designated X and Y, which are linked to the ovalbumin gene in the order 5'-X-Y-ovalbumin-3'. Both the X and Y genes share limited sequence homology with the ovalbumin gene, but are expressed in oviduct cells at different levels. In order to study the hormone responsiveness of these three genes in the chick oviduct, X and Y specific hybridization probes were constructed. The sequence specificity of the probes was demonstrated by electron microscope heteroduplex mapping, Southern hybridization assays, and DNA-RNA hybrid competition studies. Using northern hybridization and RNA excess DNA-RNA (Rot) hybridization analysis we demonstrated that the Y gene, like ovalbumin, is stimulated by the steroid hormone estrogen, but to a lower level. In contrast, X gene transcript concentrations are not significantly affected by estrogen treatment. Studies of the rates of transcription of these three closely linked genes suggest that the different steady-state levels of the X, Y, and ovalbumin gene transcripts within estrogen stimulated oviduct cells maybe reflective of their differential transcription rates. Thus, the X, Y, ovalbumin gene complex presents an attractive model system for studies of the mechanisms operative in steroid hormone mediated control of differential gene expression.

ISBN 0-12-633360-2

EXPRESSION OF HUMAN α-GLOBIN GENES IN HYBRID MOUSE ERYTHROLEUKEMIA CELLS

A. Deisseroth, U. Bode, J. Fontana, R. Lebo, Y. Kan, A. Dozy and D. Hendrick, Experimental Hematology Section, NCI,NIH, Bethesda, Md., and the Howard Hughes Medical Institute Laboratory, UCSF, SanFrancisco, Ca.

In order to study the genetic mechanisms governing differentiated gene expression in human hematopoietic cells, we have used cell fusion to generate hybrid mouse erythroleukemia(MEL) cells which permanently retain the chromosome bearing the human alpha globin(HAG) genes from erythroid and non-erythroid as well as thalassemic human hematopoietic donor cells. Hybrid MEL cells with HAG genes from erythroid donor cells exhibit full expression of these genes through globin chain synthesis, while hybrid MEL cells containing a near diploid complement of MEL chromosomes as well as HAG genes from non-erythroid human hematopoietic donor cells contain very low levels of HAG mRNA and no detectable levels of HAG chains. Tetraploid MEL cellswhich retain HAG genes from non-erythroid human hematopoietic donor cells do contain detectable levels of HAG chains, suggesting that the expression of HAG genes in these hybrid cells is dependant upon factors present in the MEL cell as well as upon the differentiated state of the human donor cell. We have also isolated hybrid MEL cells which retain deletion or non-deletion mutations of the HAG genes from a patient with alpha thalassemia(hemoglobin H disease) in which genetic modification of at least three of the four alpha globin genes has taken place. These experiments have been used to differentiate between cis and trans acting defects in this thalassemic syndrome, and show that the abnormal pattern of expression of HAG genes found in the patient's erythroid cells is also observed after transfer of these genes to MEL cells. These studies suggest that the MEL hybrid system may permit correlation of disorders found in the patients with the structure and expression of their globin genes found after introduction into the MEL cells. This sytem may therefore be useful for characterization of genetic mechanisms governing hematopoietic differentiation in man as well as studying how mutations affecting expression of these differentiated genes lead to disease states in man.

ISBN 0-12-633360-2

Transitions of genomic DNA organisation

J.T. Epplen, Institut fur Humangenetik der Universitat Gottingen, Nikolausberger Weg 5a, D-3400, Gottingen, Germany

It has been proposed that transcription level control depends on the organization of repetitive and nonrepetitive DNA sequences (Britten and Davidson, 1969 and 1977; Davidson and Britten, 1979). Studies of genomes over a wide phylogenetic range indicate that repeated and unique elements are organized in highly ordered fashions. In most eukaryotic genomes these elements alternate in short period ("Xenopus-type"), and this pattern is thought to be of functional significance (lit.cit.). We investigated the mode of interspersion of repetitive and nonrepetitive DNA sequences in several different protochordate and chordate genomes including avian, reptilian, fish and acraniate species (Epplen et al., 1978, 1979, unpublished; Schmidtke, unpublished; Schmidtke et al., 1979). We have shown that five bird species lack the typical short period interspersion of repetitive and nonrepetitive elements, whereas reptile, fish species and Amphioxus are organized in the usual Xenopus-like arrangement. Among birds, we also observed a genomic organization pattern which possibly represents a transitory interspersion type. These results together with other findings reported in the literature show that alternative genome organization patterns are of relatively wide-spread occurrence. The apparent lack of intimate sequence interspersion in many genomes makes the general functional implications of short type genome organization questionable.

References

Britten, R.J., Davidson, E.H. (1969) Science 165: 349-357.

Davidson, E.H., Britten, R.J. (1979) Science 204: 1052-1059.

Epplen, J.T., Leipoldt, M., Engel, W., Schmidtke, J. (1978) Chromosoma 69: 307-321.

Epplen, J.T., Diedrich, U., Wagenmann, M., Schmidtke, J., Engel, W. (1979) Chromosoma (in press).

Schmidtke, J., Epplen, J.T., Engel, W. (1979) Comp. Biochem. Physiol. 638: 455-458.

ISBN 0-12-633360-2

NUCLEOTIDE SEQUENCE AND STRUCTURAL FEATURES OF THE COMMON REGION OF AVIAN SARCOMA VIRUS

Tadashi Yamamoto, Jaya Sivaswami, Gilbert Jay* and Ira Pastan. Laboratory of Molecular Biology and Molecular Virology*, NCI, NIH, Bethesda, Maryland 20205

We have constructed a recombinant plasmid, pSR1, containing a 700 bp cDNA copy of the common region present at the 3' end of SR-ASV 21S mRNA. The cDNA was inserted into plasmid pBR322 at the *Pst* I site by the G-C tailing method. A restriction map of the cloned insert indicates that it corresponds to the 3' end of the ASV genome. The nucleotide sequence of the insert was determined and shows that the insert contains extra sequences not found at the 3' end of ASV genomic RNA. Following the terminally redundant sequence of 20 bp that has been found at the extreme 3' end of genomic RNA, there is a sequence of 79 bp that is almost identical to that located immediately next to the 20 bp repeat at the 5' end of ASV genomic RNA. This is followed by 18 bp of unique sequence, possibly of host origin. The structure of the clone suggests that ASV mRNA may differ from genomic RNA at its 3' end, and that 21S mRNA is transcribed from integrated ASV DNA and contains at its 3' end sequences derived both from the 5' end of the ASV genome and from host DNA adjacent to the site of integration (see Figure). The results of "northern" hybridizations to viral RNA coding either *src* or *env* deletions using [^{32}P]-labeled restriction fragments of the insert show that the 5' end of the insert contains *src* gene sequences. This suggests that common region in ASV genomic RNA is no more than 550 bases in length. In the common region the presence of sequences that resemble probable promoter sites indicates that this region may be involved in the regulation of transcription (see Figure).

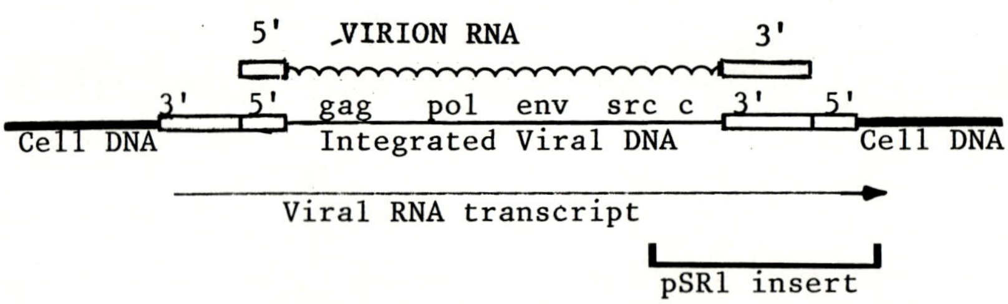

ISBN 0-12-633360-2

LOCALIZATION OF SINGLE COPY DNA SEQUENCES ON G-BANDED HUMAN CHROMOSOMES BY *IN SITU* HYBRIDIZATION

Mary E. Chandler, Michael B. Mann and Grady F. Saunders. Department of Biochemistry, University of Texas System Cancer Center, M.D. Anderson Hospital and Tumor Institute, Houston, TX 77030.

An 18 kb human DNA segment cloned in bacteriophage λCh4A was obtained from the genomic DNA library of T. Maniatis. The segment, comprised of >70% single copy DNA, was localized to band p36 of chromosome 1 according to two approaches. In the first, recombinant DNA, nick-translated with ^{3}H-NTPs to specific activity 1.2 x 10^6 cpm/μg, was hybridized in the presence of 10% dextran sulfate and a 1000-fold excess of unlabeled competitor total human repetitive DNA (C_ot<40) to a C_ot of either 0.17 or 0.8. After exposure for 37 days or 5 days, respectively, 71% and 53% of mitotic cells exhibited grains on band p36 of one or both chromosomes 1, which accounted for 18% and 24% of all chromosomal grains. In the absence of competitor, locus 1p36 was again significantly labeled (61% of cells), although the average number of grains per cell was increased (4.7 vs 2.2). In the second approach, chromosome preparations were hybridized in dextran sulfate with the intact unlabeled λCh4A-human DNA in the presence of competitor repetitive DNA to a C_ot of 1.2, then incubated with the ^{3}H-labeled λCh4A-human DNA to a C_ot of 0.2 and exposed for 5 days. Again a high number of cells (60%) exhibited label on band 1p36, with these grains comprising 23% of the total. Furthermore, amplification of the signal from single or double grains to clusters of 5-20 grains could be accomplished by increasing the time of incubation with the labeled probe to C_ot 0.8. In the absence of dextran sulfate, neither amplification of signal nor significant labeling of any chromosome were observed. When applied to the localization of specific gene sequences, this methodology now allows the rapid and refined mapping of any cloned single copy gene on mammalian chromosomes.

Supported by NCI grant CA 20124.

ISBN 0-12-633360-2

OVALBUMIN IS SYNTHESIZED IN MOUSE CELLS TRANFORMED WITH THE NATURAL CHICKEN OVALBUMIN GENE.

Eugene C. Lai*, Savio L.C. Woo+, Mary E. Bordelon-Riser*, Thomas H. Fraser#, and Bert W. O'Malley*. *Department of Cell Biology and +Howard Hughes Medical Institute Laboratory at Baylor College of Medicine, Houston, Texas, 77030, and #Research Laboratories, The Upjohn Company, Kalamazoo, Michigan, 49001.

The entire chicken ovalbumin gene, accompanied by genomic DNA sequences flanking both termini of the gene and three copies of the Herpes simplex virus thymidine kinase (TK) gene, has been cloned in plasmid pBR322. This recombinant plasmid was linearized and used to transform thymidine kinase-deficient ($LMTK^-$) mouse cells. Thymidine kinase positive (TK^+) transformants were selected by their ability to grow in the hypoxanthine-aminopterin-thymidine (HAT) medium. The entire ovalbumin gene has integrated into high molecular weight DNA within all the transformants and retained its original sequence organization. In all of the transformants examined, a protein identified as chicken ovalbumin by a solid-phase immunoassay was detected within the cells. It is estimated that between 1,000-100,000 molecules of chicken ovalbumin were produced per mouse cell in each of these transformants. Our results demonstrate that the mouse cellular machinery can be utilized to accurately express genetic information encoded in a cloned gene from a different eucaryotic organism into its specific protein product.

When three of the transformed cell lines were allowed to grow in medium without HAT for more than 10 generations and then transferred back to a HAT medium, no decrease in plating efficiency was obvserved and there was no loss of the acquired phenotype. Furthermore, when 30 micrograms per ml of BrdU was added to the HAT medium, more than 99% of the cells died within two days of culture. These results suggest that the thymidine kinase and ovalbumin genes are stably integrated into mouse cellular DNA.

ISBN 0-12-633360-2

SITE PREFERENCE BY RESTRICTION ENDONUCLEASES

Kamalendu Nath and Barbara Azzolina. Merck Institute for Therapeutic Research, Rahway, N.J. 07065

The action of restriction endonucleases on bacteriophage lambda DNA was followed by monitoring the appearance of restriction fragments at various intervals. The restriction fragments were resolved by electrophoresis in agarose or agarose-polyacrylamide gels. A relative quantitation was achieved via scanning of the photographic negative plate of the gel. Further, resolution and quantitation was performed by introducing [^{32}P] at the 5'-termini of lambda DNA (specific activity of about 150,000 pm per pmole) and by following the electrophoretically-separated gel profile via autoradiography.

In all cases, it was observed that the restriction sites situated close to the termini of lambda DNA were preferred over those situated towards the middle. The Bam HI site situated at 6.76 KbP from the right terminal of lambda DNA (fragment C) is cleaved at least 10 times faster than the other four sites. The three Hind III sites situated within 40 to 60% of lambda DNA (fragments E and F) are preferred about seven times less than the site situated at 4.26 KbP from the right terminal (fragment D). Similarly, Eco RI cleaves the site situated at 3.38 KbP from the right end of lambda DNA (fragment F) with a frequency of about 10 times greater than the two sites within 40 to 60% of the molecule (fragment E). The later observation supports previous report on non-random cleavage by Eco RI (1). This site preference of restriction endonucleases can be further modified or attenuated by the use of intercalators (2).

The restriction endonucleases prefer the cleavage of linear DNA molecules over the corresponding circular form. Thus, pBR312 plasmid DNA when linearized by the action of Hind III (single site) is digested by Hinc II (four sites) at least twice as fast as the covalently closed circles. Irrespective of the number of restriction sites present on a DNA, the restriction enzymes prefer a single site on circular DNA. However, this cleavage is not restricted to a particular site.

(1) Thomas, M. and Davis, R. W., J. Mol. Biol. 91: 315-328 (1975).

(2) Nath, K., XIth International Congress Biochem. (abstract) p. 45 (1979).

ISBN 0-12-633360-2

MAPPING OF THE TRANSFORMING REGION OF HARVEY SARCOMA VIRUS GENOME USING INSERTION-DELETION MUTANTS CONSTRUCTED *IN VITRO*

Cha-Mer Wei*, Douglas R. Lowy** and Edward M. Scolnick*
Tumor Virus Genetics Branch* and Dermatology Branch**,
National Cancer Institute, Bethesda, Md. 20014

Harvey murine sarcoma virus (HaMuSV) is a defective virus with a genome of 6.0 Kb. DNA intermediates, including linear and circular forms, accumulate during virus replication. The circular DNAs were cloned in λWES·λB and shown to be capable of transforming mouse NIH3T3 cells by focus formation assay (1).

In order to define the transforming region of HaMuSV, insertion-deletion mutants were constructed and mapped. The HaMuSV DNA insert (5.4 Kb) in the λ vector was excised from the recombinant clone by EcoRI digestion and circularized by T4 ligase reaction. The reconstructed circle was partially digested by HaeIII which has multiple cleavage sites in the HaMuSV genome. The SalI linker, GGTCGACC, was then ligated to the ends of linearized molecules. The modified HaMuSV DNA with SalI linker was inserted in a λWES·λB-Sal vector (2) to transfect E. coli LE392.

Fourteen insertion-deletion mutants (id#1 to 14) have been isolated as recombinant clones and sites of mutation characterized. The transforming ability of modified HaMuSV DNAs was assayed by focus formation on NIH3T3 cells. The mutants (id#1 to 8, 11 and 14) containing insertion-deletions within 3.0 Kb from the 3' end of the genome retain full ability to transform NIH3T3 cells. Among them two deletion mutants (id#11 and 14) mapped in the presumptive C region. These results indicate that the sequence in the 3' half of the genome including C region is not essential for transforming activity. However, the mutants (id#9,10,12 and 13) mapping within 1.5 Kb from the 5' end were unable or had a decreased capability of forming foci in the transfection assay. Thus we conclude that the region responsible for the expression of transformation by HaMuSV lies within 2.5 KB from the 5' end of the genome and at least 1.5 Kb from the 5' end are required for full expression.

(1) Hager, G.L. et. al.(1979) J. Virol. 31, 795-809.
(2) A modified λWES·λB vector constructed by Wei, C. et.al. for cloning 3.5 Kb to 18.5 Kb fragments generated by SalI enzyme.

ISBN 0-12-633360-2

SELECTION OF RAT HEPATOMA CELL VARIANTS WITH ALTERED GLUCOCORTICOID RESPONSE.

J. Russell Grove and Gordon M. Ringold
Department of Pharmacology, School of Medicine, Stanford University, Stanford, Ca. 94305.

A mouse mammary tumor virus (MMTV) infected rat hepatoma cell line has been used as a model system for studying glucocorticoid action. These cells induce the expression of tyrosine amino-transferase (TAT) and MMTV genes in response to dexamethasone. The major antigen, a glycoprotein of M_r 52,000 (gp52), appears on the surface of infected cells in amounts that reflect the cytoplasmic content of viral RNA.

Using an anti-gp52 antiserum and a fluorescence activated cell sorter, we have selected variants which display low levels of gp52 in the presence of hormone. Five cycles of enrichment for poorly labeled cells have produced a population which fails to produce a detectable hormone-inducible increase in cell surface gp52.

This population of nonresponders and a number of independent clones were analyzed for their ability to induce gp52 and TAT and for the presence of glucocorticoid receptors. In no case did hormone administration appear to alter the level of expression of gp52 or TAT. Preliminary receptor assays indicate that all the clones contain very low or undetectable levels of specific hormone binding.

These results provide genetic evidence, complementary to the pharmacological evidence, that glucocorticoid receptors are involved in the induction of TAT and MMTV in these cells. Further studies using this selection system are under way in attempts to identify additional classes of variants which would correspond to additional components in the pathway of hormone response.

ISBN 0-12-633360-2

A RESTRICTION MAP OF BACTERIOPHAGE T3

David R. Dembinsky, Jean N. Bailey and William T. McAllister. Department of Microbiology, CMDNJ-Rutgers Medical School, Piscataway, N. J. 08854.

T3 and T7 are closely related bacteriophages of *E. coli* which share many common features but also have interesting differences. Restriction maps of the DNA of both viruses are necessary for a detailed comparison of the genetic structure of the two viruses. Although a restriction map of T7 has been prepared (1), no comparable physical data for T3 has been reported. In this work the derivation of a restriction map of T3 is described. The restriction enzymes Ava I, XbaI, Hind III and Bgl II each cleave T3 DNA at one site. The enzymes Kpn I, Mbo I and Hpa I cleave the DNA at two, nine and seventeen sites respectively. The sizes of all fragments produced by cleavage with the above enzymes have been determined using T7 Hpa I fragments as molecular weight standards. All fragments produced by these enzymes have been ordered with respect to each other and to the genetic map (see below).

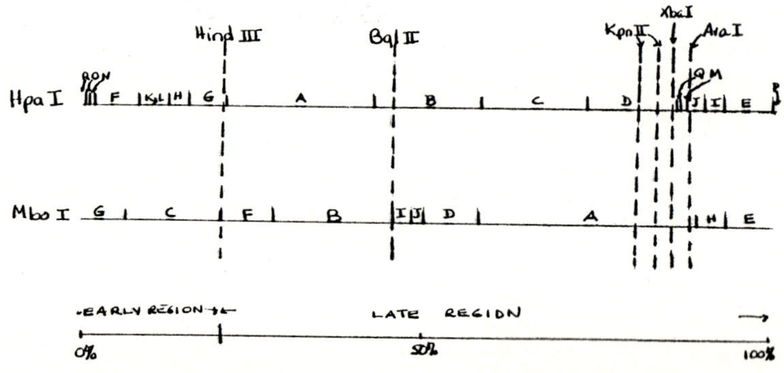

1. McDonnell, M.W., Simon, M.N., and Studier, F.W., J. Mol. Biol., 110, 119-146 (1977).

This research was supported by NIH Grant GM-21783 to William T. McAllister.

ISBN 0-12-633360-2

CHARACTERIZATION OF T7 PROMOTER SEQUENCES

Anthony D. Carter, Claire E. Morris, William T. McAllister, and F. W. Studier*. Department of Microbiology, CMDNJ-Rutgers Medical School, Piscataway, N. J. 08854.

*Brookhaven National Laboratory, Upton, New York.

In vivo the late genes of bacteriophage T7 are transcribed by a phage specified RNA polymerase in two temporal classes: class II mRNA from 4 until 16 minutes after infection; class III mRNA from 6-8 minutes after infection until lysis (at 30 minutes). *In vitro*, the purified enzyme discriminates between the class II and class III promoters as a function of changes in temperature and ionic conditions. Restriction fragments of T7 DNA containing promoters for either class II or class III RNAs have been inserted into the plasmid pBR322. In infected cells, T7 promoters located within the plasmid are utilized by the T7 RNA polymerase with the same kinetics as their counterparts in the intact phage genome. As promoter selection by this enzyme must reflect differences in class II and class III promoter structure, a comparison of the sequences surrounding the two types of promoters is of considerable interest. All of the class III promoters and several of the class II promoters have been sequenced by other investigators. We have sequenced two class II promoters at 33.3 and 34.8 percent on the T7 map (shown below). Comparison of the class II and class III promoters reveals differences that may be functionally important.

X = purine   Y = pyrimidine
+1 = site of initiation of transcription

```
                                    -1 +1
33.33  A A T A C G A C T C A C T A A A G G A X X C
       T T A T G C T G A G T G A T T T C C T Y Y G

                                    -1 +1
34.75  T A T T C G A C T C A C T A T A G G A G A T
       A T A A G C T G A G T G A T A T C C T C T A
```

This research was supported by NIH Grant GM-21783 to William T. McAllister. Anthony D. Carter is a predoctoral trainee supported by NCI Training Grant 27-1252.

ISBN 0-12-633360-2

THE STRUCTURE OF TRANSPOSABLE MATING TYPE GENES IN YEAST

J. Hicks, J. Strathern, A. Klar and J. Broach

Cold Spring Harbor Laboratory
Cold Spring Harbor, New York 11724

During the mitotic growth of homothallic Saccharomyces yeasts cell type changes occur that are the result of transposition of cryptic regulatory genes from unlinked "storage" loci to the MAT locus. Expression of these regulatory alleles, MATa and MATα determines the sexual type of the cell by controlling the expression of sex-specific functions coded elsewhere in the genome.

The transposons carrying the a and α structural genes reside at the HMR and HML loci, 120 kbp to the right and 130 kbp to the left of MAT on opposite arms of yeast chromosome III. We have cloned the components of this site-specific transposition system and determined that each element consists of two regions common to all 3 loci flanking a mating type specific region of approximately 0.7 kbp in MATa elements and 0.9 kbp in MATα elements (see Figure). During cell type switching information is copied alternately from HML and HMR and inserted into the MAT locus, replacing the resident MAT information. This site specific transposition, which is functionally equivalent to gene conversion, results in the loss of the genetic material originally present at MAT but leaves the information at HML and HMR intact.

The cryptic a and α information at HMR and HML is apparently under negative transcriptional control involving the products of at least three different, unlinked genetic loci. Subsequent transposition to the MAT locus results in the release of the structural genes from this negative control presumably by removing them from a cis-dominant regulatory site.

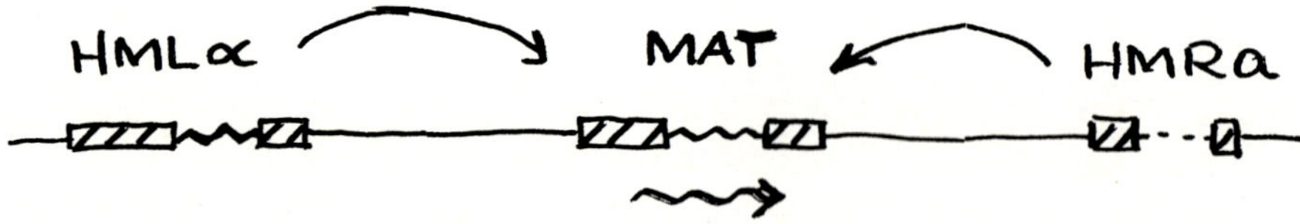

▨ - regions of homology as determined by E.M. heteroduplex analysis; ⁓ - α specific sequences; ---- - a specific sequences.

Arrangement of cryptic (HML and HMR) and active (MAT) loci containing mating type genes on chromosome III of yeast (not to scale).

ISBN 0-12-633360-2

COMPLEMENTATION OF *Saccharomyces cerevisiae leu2* AUXOTROPHS BY THE LEUCINE OPERON OF *Escherichia coli*

R.K. Storms, E.W. Holowachuk and J.D. Friesen. Department of Biology, York University, Downsview, Ontario, Canada.

A chimeric plasmid carrying the intact leucine operon of *Escherichia coli* has been constructed and found to complement *leu2* mutants of *Saccharomyces cerevisiae*. The parental plasmid pYF85, which contains an *Eco*R1 DNA fragment of the yeast plasmid 2 μm circle inserted into the *Eco*R1 site of the bacterial plasmid pBR322, is capable of transforming both yeast and *E. coli* at high frequencies. A DNA fragment of 4.6 mD in size, generated by *Bam*H1 cleavage of λd*leu*7 DNA, was cloned into the unique *Bam*H1 site of pYF85 in both orientations. Plasmids carrying the *Bam*H1 fragment in both orientations, pYF*leu*101 and pYF*leu*102, are capable of transforming *leuA*, *leuB*, *leuC* and *leuD* mutants of *E. coli* to the Leu$^+$ phenotype, indicating the presence and expression of the intact leucine operon. Only plasmid pYF*leu*101 was found to transform yeast *leu2* auxotrophs to prototrophy. This strongly suggests that the *E. coli* leucine operon on pYF*leu*101 is expressed in yeast from a control region, other than the *E. coli* leucine promoter, which lies outside of the *E. coli* *Bam*H1 DNA fragment bearing the leucine operon. This is supported by the lack of expression in yeast of the leucine operon on plasmid pYF*leu*102, which contains the leucine operon in the opposite orientation.

The yeast Leu$^+$ transformants obtained with pYF*leu*101 grow with a doubling time 3 to 4 times that obtained with the same strain supplied with leucine. This suggests that the amount of activity of the enzyme β-isopropylmalate dehydrogenase, encoded by *leuB*, is low but sufficient to support growth. The translation of *leuB*, the second gene of the leucine operon, suggests that the entire leucine operon is transcribed in yeast. The transcription of the leucine operon may be initiating from the tetracycline resistance gene promoter of pBR322 or from promoter sequences within the 2 μm circle portion. Experiments to determine the control region responsible for the expression of *leuB* as well as expression of *leuA*, *leuC* and *leuD*, in yeast are in progress.

ISBN 0-12-633360-2

Altered Patterns of Integrated SV40 DNA in Clonal Derivatives of tsA Mutant Transformed Mouse Cells.

John Hiscott, David Murphy and Vittorio Defendi
Department of Pathology, New York University School of Medicine
New York, NY 10016

The SV40 tsA58 transformed mouse clone A21 displays a temperature sensitive transformed phenotype, but colonies can be obtained in soft agar at 40°C with a frequency of less than 1%. Clonal derivatives of A21, isolated by growth in soft agar at 40°C, were analyzed by restriction endonuclease cleavage and Southern blot hybridization. In 10 of 12 clones examined, the conversion of A21 cells to anchorage independent growth at 40°C is accompanied by the acquisition of new sites of integrated SV40 DNA. By using the Bgl II endonuclease which lacks specificity for SV40, it was shown that the parental A21 clone contains SV40 sequences in 9 kb and 15 kb fragments of cellular DNA. The clonal derivatives generally retain the 9 kb and 15 kb integration sites but possess additional SV40 insertions within Bgl II fragments of 15 to 25 kb. The viral DNA monomer in both the 9 kb and 15 kb Bgl II inserts is a deletion fragment of 3.8 kb which is arranged in a tandem head to tail fashion in the cellular DNA. These results suggest that the change in biological expression of the transformed phenotype may be accompanied by the rearrangement of integrated SV40 specific sequences.

ISBN 0-12-633360-2

RNA SPLICING PATTERNS OF IMMUNOGLOBULIN GENES

J. Rogers, M. Gilmore-Hebert and R. Wall

Department of Microbiology and Immunology and Molecular Biology Institute, UCLA, Los Angeles, CA 90024.

In immunoglobulin genes, the homologous polypeptide domains are encoded by separate exons which become joined by RNA splicing. This splicing was first established for the κ light chain transcript in the MOPC21 mouse myeloma (1), where the final intermediate was a 4.8 kb species from which the V→C intron appeared to be removed in a single step. We have now carried out pulse chase experiments (2), and "Northern" blots with a V→C intron DNA fragment as probe, which confirm this splicing pathway.

Heavy chain RNA processing is more complex, since there are multiple constant region domains separately encoded (3). We have used restriction fragments from a cloned mouse γ1 gene (3) as probes for exons and introns in MOPC21 myeloma RNA on "Northern" blots. The species detected, all polyadenylated, suggest a possible splicing pathway in which the large V→C_H1 intron is excised in more than one step, and the smaller introns between the constant region domains are removed last.

Supported by NIH grants AI 13410 and CA 12800.

(1) Gilmore-Hebert M, Hercules K, Komaromy M and Wall R, PNAS 75, 6044-6048 (1978).
(2) Gilmore-Hebert M and Wall R, J.Mol.Biol., in press.
(3) Sakano H, Rogers J, Hüppi K, Brack C, Traunecker A, Maki R, Wall R and Tonegawa S, Nature 277, 627-633 (1979).

ISBN 0-12-633360-2

EXPRESSION OF THE HERPES SIMPLEX VIRUS THYMIDINE KINASE GENE AND THE rDNA GENE OF *Tetrahymena thermophila* IN *Saccharomyces cerevisiae*

G.B. Kiss, K.V. Cornish, J.D. Friesen and R.E. Pearlman. Department of Biology, York University, Downsview, Ontario, Canada.

A chimeric plasmid (pYF91) was used to clone and introduce into *Saccharomyces cerevisiae* the thymidine kinase(tdk) gene of Herpes Simplex Virus (HSV). The plasmid constructed (pGY14) contains a 2.4 mD *EcoR1* fragment of the yeast plasmid 2 µm circle, a 4.0 mD *BamH1 HindIII* fragment of the *leuB* gene of yeast, almost the entire *E. coli* plasmid pBR322 and a 2.2 mD *BamH1* fragment containing the HSV tdk gene that transforms tdk^- mouse cells to tdk^+. Plasmid pGY14 was used to transform Leu^- yeast to Leu^+. The transformants could not grow on thymidine as the sole source of dTMP precursor, suggesting the lack of functional tdk protein. Poly A^+ and poly A^- RNAs were isolated from transformants and analysed using "Northern" blot experiments. Three pBR322 specific bands were identified in poly A^+ RNA after hybridization to nick-translated pBR322 and autoradiography. Using isolated and nick-translated tdk gene as probe, hybridization was detected only to poly A^- RNA isolated from pGY14 transformed yeast. The above experiments showed that mRNA from procaryotic genes transcribed in eucaryotic yeast cells could be effectively polyadenylated. The absence of polyadenylation of HSV tdk specific RNA is at present not understood.

In vitro assays of thymidine kinase activity as well as attempts to identify tdk protein using antibody specific for HSV tdk protein will be described.

Plasmids were constructed containing the entire transcribed sequence of the rDNA gene of *Tetrahymena thermophila*. Experiments to determine the expression of rDNA genes in yeast using chimeric plasmids will be described.

Supported by grants from NSERC and National Cancer Institute of Canada.

ISBN 0-12-633360-2

CONSTRUCTION AND CHARACTERIZATION OF NEW CLONING VEHICLES. IV. DELETIONS OF pBR322 AND pBR325.

X. Soberon, L. Covarrobias and F. Bolivar
Departamento de Biologia Molecular, Instituto de Investigaciones Biomédicas, UNAM, México, D.F.

In vitro recombination DNA experiments involving restriction endonuclease fragments derived from the plasmids pBR322 and pBR325 resulted in the construction of two new cloning vehicles. One of these plasmids, designated pBR327, was obtained after an *Eco* RII partial digestion of pBR322. The plasmid pBR327 confers resistance to tetracycline and ampicillin, contains 3273 base pairs and therefore is 1089 base pairs smaller than pBR322. The other newly constructed vector, which has been designated pBR328, confers resistance to chloramphenicol as well as the two former antibiotics. This plasmid contains unique *Hind* III, *Bam* HI and *Sal* I sites in the tetracycline resistance gene, unique *Pvu* I and *Pst* I in the ampicillin resistance gene and unique *Eco* RI, *Pvu* II and *Bal* I sites in the chloramphenicol resistance gene. The pBR328 plasmid contains approximately 4900 base pairs.

Supported by Grant #1649 from PNCB, Consejo Nacional de Cioncia y Tecnologia, México.

*This article has been accepted by GENE, in press.

ISBN 0-12-633360-2

DIFFERENTIAL EXPRESSION AND INDUCTION OF VIRUS ANTIGENS IN CULTURED MOUSE MAMMARY TUMOR SUBPOPULATIONS.

J.C. Hager, J. Levy, L. Chou and G.H. Heppner, Department of Immunology, Michigan Cancer Foundation, Detroit, MI.

To understand the role of assembly of viral genetic information in tumor cells we compared the expression of mouse mammary tumor virus (MMTV) antigen and its induction with 5-iododeoxyuridine (IUDR) in a series of 7 spontaneously arising strain Balb/cfC_3H mouse mammary tumors. Subconfluent cell cultures of the 1st *in vitro* passage of spontaneous tumors were exposed for 24 hours to IUDR (20,40, 60 μg/ml) and evaluated over an 8 day course for cell surface antigen expression using rabbit anti-MMTV and fluoresceinated goat anti-rabbit IgG. Untreated cells from these spontaneous tumors varied in percentage of cells expressing MMTV antigen, and also in the peak day of that expression. Induction of viral antigen expression with IUDR varied in increase of the number of cells expressing antigen, the dose of IUDR which achieved maximal virus expression, and the day on which it was achieved.

In order to interpret this heterogeneity in virus expression in the spontaneous tumors, we examined antigen expression and induction in 4 distinct subpopulations isolated from a single spontaneous mouse mammary tumor, and a 5th subpopulation derived from a lung metastasis of the 10th transplant generation of the original heterogeneous tumor. Whereas cultures of the parent tumor consisted of 2-17% positive cells, one subpopulation was strongly positive (>90%), one variable and the other 3 were negative. IUDR (20-60 μg/ml) induced antigen expression in all 3 negative lines (50-85%) and enhanced it in the 2 others. To simulate a heterogeneous tumor, 2 of the negative cell lines were mixed together, and the strongly positive line was admixed with these in another combination. Under both conventional and induction culture conditions no interaction was detected. Thus, *in vitro* expression and induction by IUDR of MMTV surface antigens is independent in subpopulations of a heterogeneous tumor and the extent of viral antigen expression in such a tumor reflects the summation of the contribution of the component subpopulations.

ISBN 0-12-633360-2

NUCLEOTIDE SEQUENCE IN EARLY REGION E1 OF ADENOVIRUS DNAs

H. van Ormondt, J. Maat, C.P. van Beveren, B.M.M. Dekker and R. Dijkema.

Laboratory for Physiological Chemistry, State University of Leiden, 2333 AL Leiden, The Netherlands.

We have determined the nucleotide sequence of the leftmost early regions of adenovirus types 5 and 7 (Ad5 and Ad7). The genes in these regions have been shown to be responsible for the transformation of cells (1).

The sequences revealed a series of regulatory signals for transcription and translation which made it possible to predict the primary structures and molecular weights of the encoded polypeptides, as will be demonstrated for Ad5 DNA. The RNA mapping data of Chow et al. (2), and Berk and Sharp (3), and the mRNA sequence work of Perricaudet et al. (4) and Ziff (personal communication) are incorporated in our analysis.

The subregions E1a of Ad5 and Ad7 have been determined for homologies at the DNA and protein level.

Supported by grants from Z.W.O. and K.W.F.

(1) Van der Eb et al., Cold Spring Harbor Symp.Quant.Biol. 44 (1979) in press.
(2) Berk, A.J. and Sharp, P.A., Cell 14, 695-711 (1978).
(3) Chow, L.T., Broker, T.R. and Lewis, J.B., J.Mol.Biol. (1979) in press.
(4) Perricaudet, M., Akusjärvi, G., Virtanen, A. and Pettersson, U., Nature (1979) in press.

ISBN 0-12-633360-2

NUCLEOTIDE SEQUENCE ANALYSIS OF THE CHLORAMPHENICOL RESISTANCE TRANSPOSON Tn9.

N. Kirby Alton and Daniel Vapnek. Department of Microbiology, University of Georgia, Athens, Georgia 30602.

The chloramphenicol resistance transposon Tn9 is a 2800 bp segment of DNA composed of two directly repeated sequences of IS1 flanking a region of 1102 bp which carries the chloramphenicol transacetylase (CAT) gene (1). We have utilized the Sanger chain termination method of DNA sequencing to determine the nucleotide sequence of this region. Two methods were used to prepare the single-stranded templates needed for the primed-synthesis reaction. One of these involved the use of Exonuclease III digestion of double-stranded DNA and the other cloning in the single-stranded DNA bacteriophage M13.

By utilizing these templates the nucleotide sequence of the 1102 bp region bounded by the IS1 sequences was determined. Analysis of the sequence reveals an uninterrupted coding sequence for 219 amino acids beginning 224 base pairs from the left end of IS1. Amino acid sequencing data of CAT by W.V. Shaw confirms this as the coding sequence for the CAT gene. No other coding sequence for a protein of greater than 8,000 daltons is found within this region. Since IS1 itself is transposable, this result suggests that the 1102 bp region of Tn9 is not directly involved in the transposition. Transposition of Tn9 results in the duplication of a 9 bp sequence at the site of integration (2). The internal 9 bp sequences at the IS1 junction of Tn9 have no homology with each other nor with any of the 9 bp sequences found to be duplicated at the sites of Tn9 integration.

Supported by USPHS grant GM20160.

(1) MacHattie, L.A. and Jackowski, J. B. in DNA, Insertion Elements, Plasmids and Episomes (eds., Bukhari, A.I., Shapiro, J. A. and Adhya, S. L.) 219-228 (Cold Spring Harbor Laboratory, 1977).

(2) Johnsrud, L., Calos, M.P. and Miller, J. H. Cell 15, 1209-1219 (1978).

ISBN 0-12-633360-2

ISOLATION AND CHARACTERIZATION OF SV40 NONCONDITIONAL REPLICATION-DEFICIENT MUTANTS. Y. Gluzman, R. Frisque and J. Sambrook, Cold Spring Harbor Laboratory, New York.

A set of mutant SV40 DNAs, lacking the BgII restriction site (which is located at the origin of replication) were isolated and were shown by the Maxam-Gilbert procedure to have suffered deletions 4 to 58 nucleotides in length spanning the BgII site. S1 analysis of the RNA extracted from mutant transformed cells shows that irrespective of the size of the deletion the 5' end of the poly A^+ cytoplasmic early mRNA is apparently located at a fixed distance from the putative early promoter.

Mutant DNAs are unable to replicate in permissive simian cells although they code for a T-antigen that is able to complement V-antigen production by tsA mutants at $41^{\circ}C$. These origin-defective mutants transform rat cells with efficiency equal to that of wild-type DNA. All of the transformed lines produce T-antigen that is indistinguishable in size and comparable in quantity to that present in wild-type transformants.

Permissive CV-1 cells were transformed by one of the mutant DNAs (α6-1). These cells are able to complement the growth of T-antigen defective SV40 mutants and therefore can be used as hosts for the propagation of recombinant SV40 viruses with substituted early regions.

ISBN 0-12-633360-2